DECODING CHOMSKY

CHRIS KNIGHT is currently senior research fellow in the Department of Anthropology at University College London, exploring what it means to be human by focusing on the evolutionary emergence of language and symbolic culture. He is the author of *Blood Relations: Menstruation and the Origins of Culture* (Yale, 1991) among many other publications. His website is www.scienceandrevolution.org.

Further responses to *Decoding Chomsky*:

'The whole story is a wreck . . . complete nonsense throughout.'
Noam Chomsky

'One of the most exciting scholarly books I have read in years . . . *Decoding Chomsky* will be required reading for anyone at all interested in the history of intellectual and political thought since the 1950s.'
David Golumbia, author of *The Cultural Logic of Computation*

'Simply brilliant. Others have noted the systematic disjunct between Chomsky's Pentagon-funded linguistics and his political dissidence, but this is the first theoretically sophisticated analysis of a chasm between mind and body, theory and practice which has become profoundly symptomatic of postmodern culture as a whole.'
David Hawkes, author of *Ideology*

'History comes alive via compelling narrative . . . Knight is indeed an impressive historian when it comes to recounting the gripping personal histories behind Chomsky's groundbreaking contributions to science and philosophy.'
Sean O'Neill, *American Ethnologist*

'This extraordinary book will make uncomfortable reading for some because, while celebrating Chomsky's anti-racist and anti-imperialist politics, Knight reminds us of the other Chomsky, the world-famous linguist [who has been] . . . working in one of the Pentagon's most prestigious laboratories.'
Jackie Walker, author of *Pilgrim State*

'Few disagree that language has been a game-changer for the human species. But just how we came by language remains hotly contested. In *Decoding Chomsky*, Chris Knight strides into this minefield to bravely replace miraculous leaps and teleology with a proposal that actually makes evolutionary sense.'
Sarah Hrdy, author of *Mothers and Others: The Evolutionary Origins of Mutual Understanding*

DECODING CHOMSKY

SCIENCE AND REVOLUTIONARY POLITICS

CHRIS KNIGHT

YALE UNIVERSITY PRESS
NEW HAVEN AND LONDON

For information about this and other Yale University Press publications, please contact:
U.S. Office: sales.press@yale.edu   yalebooks.com
Europe Office: sales@yaleup.co.uk   yalebooks.co.uk

Typeset in Minion Pro by IDSUK (DataConnection) Ltd
Printed in Great Britain by Hobbs the Printers, Totton, Hampshire

Library of Congress Control Number: 2016945085

ISBN 978-0-300-22876-2 (pbk)

A catalogue record for this book is available from the British Library.

10 9 8 7 6 5 4 3 2 1

To all my grandchildren

'It is no wonder that "fraternity" has traditionally been inscribed on the revolutionary banner alongside "liberty" and "equality". Without bonds of solidarity, sympathy and concern for others, a socialist society is unthinkable. We may only hope that human nature is so constituted that these elements of our essential nature may flourish and enrich our lives, once the social conditions that suppress them are overcome. Socialists are committed to the belief that we are not condemned to live in a society based on greed, envy and hate. I know of no way to prove that they are right, but there are also no grounds for the common belief that they must be wrong.'

Noam Chomsky, *Class Warfare*, 1996

'Man is … an animal which can develop into an individual only in society. Production by isolated individuals outside society … is as great an absurdity as the idea of the development of language without individuals living together and talking to one another.'

Karl Marx, *Grundrisse*, 1857

'There was a time when language united people … "Us! One of us!" rings through the darkness with every word of their shared language. Language unites them like a familiar voice.'

Velimir Khlebnikov, 'Our fundamentals', 1919

CONTENTS

PREFACE TO THE PAPERBACK EDITION

These pages provoked the controversy I had expected. Responses have ranged from total condemnation to excitement and relief that a critic from the left has at last shed light on what has become known as the 'Chomsky problem'. Like any writer, I was buoyed by the enthusiastic support, and stung by the critical voices. Encouragement from so many scientists and activists whom I have always admired kept me going when Noam Chomsky publicly denounced my book as 'complete nonsense throughout', and called my whole account 'a wreck'.[1]

So what exactly is the 'Chomsky problem'? In essence, it's the difficulty of seeing the connection between Chomsky the scientist and Chomsky the political activist. Put more starkly, it's the difficulty of grasping just how Chomsky succeeded in reconciling his passionate hostility to the military with his lifelong employment in a military lab. I argue that it was Chomsky's commitment to political principle that impelled him to drive such a massive wedge between these two sides of his life.

As I delve into half-forgotten details, you may well ask, 'Why should we care?' My answer is this: Chomsky was and remains the towering figure in a philosophical movement – known today as the 'cognitive revolution' – the effects of which have come to dominate much of Western thought. As this post-war intellectual upheaval rippled through US academia and reverberated across the globe, the fissure between Chomsky's science and his activism widened steadily into a chasm which we all feel today. Under relentless sociological pressure, the notion of a wholly neutral, politically disinterested science is now popularly perceived as the preserve of an establishment elite set apart from the rest of us as we struggle to make sense of our lives. Meanwhile, the majority of activists have become more and more estranged from this entire way of perceiving and engaging in science. The outcome has been an institutionally enforced disconnect between science and activism whose damaging consequences, particularly in an era of potentially catastrophic climate change, have become all too clear.

Our current state of knowledge resembles a broken mirror, each fragment telling its own story. We need to put together the big picture, fighting for conceptual unification regardless of the political consequences. You can't get away from politics – from power differences and conflicts of interest. In principle, however, scientific research involves accountability and collaboration on a level transcending such things. The scientific community needs to defend itself against political interference, no matter how cleverly it is concealed. If science is to come first, we don't have a choice as to whether to become politically active. If you're inactive, you're colluding in someone else's politics.

To understand the crucial role played by Chomsky in all this, let me take you back to his very first job interview in 1955. It was for a post in a military lab in Boston, the Cold War by this stage well under way. The successful applicant would join a team at the Massachusetts Institute of Technology working on machine translation, a project heavily funded by the Pentagon with military applications in mind.[2] Noam Chomsky attended the interview but had other ideas. Detesting the whole concept of military research, he explained that he wasn't interested in machine translation and wouldn't help. On that understanding, he was given the job. From then on, according to Chomsky, he never touched machine translation at all.

His own portrayal of that paradoxical situation perfectly encapsulates my theme. Chomsky is in his surroundings yet outside them, bodily present yet mentally elsewhere, working for the military while determined to produce nothing they could use. The detached otherworldliness of his linguistics, if my argument is correct, served Chomsky well as he struggled with the moral dilemmas he faced. Only by redefining language as something utterly abstract and ideal – completely removed from social usage or any possible practical application – could he eliminate the danger that his work might assist the military in killing people. He would cooperate loyally with his laboratory colleagues, but confine himself to pure science.

Since Chomsky and his supporters deny my whole story, I can only respond by recalling details of documentary sources and memories on which all sides roughly agree.

It was the celebrated Russian linguist Roman Jakobson who had suggested to Chomsky that he secure that crucial interview. The interviewer that day was Jerome Wiesner, then the director of the Research Laboratory of Electronics (RLE) and one of the most influential military scientists in the country. Explaining his interest in the young linguist, Wiesner recalls: 'Professor Bill Locke suggested we use computers to do automatic translation, so we hired Noam Chomsky and Yehoshua Bar-Hillel to work on it.'[3] Wiesner had good reason to view Chomsky as an ideal catch. After all, the candidate already had experience with automatic rewrite rules and related procedures for arriving at logical operations accessible in principle to a machine. Illustrating his dry sense of humour, Chomsky recalls what he said to Wiesner in that interview:

> I told him, I don't think the project makes any sense. The only way to solve this problem is brute force. What's going to be understood about language is not really going to help and I'm just not interested, so I'm not going to do it. He thought that was a pretty good answer. So he hired me on the machine translation project. But mainly to do what I felt like.[4]

We may regret that Chomsky never got to write a beginner's manual for handling job interviews. 'Always turn the tables' might well have been his advice. From then on, nothing in his attitude to authority would change.

To savour the paradoxes of Chomsky's subsequent employment, it helps to know more about Wiesner. In terms of US military policy, Chomsky remembers him as a dove: 'He was on the extreme dovish side of the . . . Kennedy administration. But he never really accepted the fact that the students and the activists considered him a kind of a collaborator.'[5] This double-edged evaluation suggests to me that Chomsky knew from the outset about Wiesner's deep involvement with the Pentagon, but out of respect for his boss, left it to others to condemn him on those grounds.

A specialist in communication engineering, between 1952 and 1980 Wiesner rose from director of the RLE to provost and then president of MIT, in effect making him Chomsky's boss for over twenty years.[6] It is easy to see why, in Chomsky words, 'the students and the activists considered him a kind of collaborator'. After all, it was Wiesner who, in the 1950s, had brought nuclear missile research to MIT.[7] He was particularly proud of the fact that his Research Laboratory of Electronics – centrally situated on MIT's campus – had made 'major scientific and technical contributions to the continuing and growing military technology of the United States'.[8]

In 1971, the US Army's Office of the Chief of Research and Development published a list of what it called 'just a few examples' of the 'many RLE research contributions that have had military applications'. The list included 'beam-shaping antennas', 'helical antennas', 'microwave filters', 'ionospheric communication', 'missile guidance', 'atomic clocks', 'signal detection', 'communication theory', 'information and coding theory', 'human sensor augmentation' and 'neuroelectric signals'.[9] Given the military significance of all such projects, Chomsky would have his work cut out disentangling his own theories about language from any possible military use.[10]

But Wiesner's contribution to the US military was far greater than his involvement with MIT. One major achievement, he reminds us, was that he 'helped get the United States ballistic missile program established in the face of strong opposition from the civilian and military leaders of the Air Force and Department of Defense'. He adds that he 'was also a proponent of the Polaris missile system, the ballistic missile early warning system, and the satellite reconnaissance systems'.[11] By 1961, Wiesner had become President Kennedy's chief science adviser and it was from this influential position that he was able to insist that nuclear missile development and procurement 'must all be accelerated'.[12] To justify this military build-up, Kennedy invoked the myth of America's weakness compared with the Soviet Union – the total fiction of a 'missile gap' that, according to Wiesner's own account, 'I helped invent'.[13]

After Kennedy's assassination, Wiesner's power declined. However, he was still able to contribute to the US war effort by bringing together a team of leading scientists in a project to design and deploy a vast barrier of sensors, mines and cluster bombs along the border between North and South Vietnam.[14] Wiesner's long-standing involvement with nuclear decision-making, and his consequent awareness of just how flawed and dishonest the whole process was, did lead him to criticize the unrestrained stockpiling of nuclear missiles, particularly those equipped with multiple warheads. But this change of heart did not stop him from continuing to run MIT laboratories dedicated to research on just such developments.[15]

Wiesner stepped down as MIT's president in 1980, but what the university's representatives call its 'deep relationship' with the Pentagon continues to this day.[16] Since 1980, MIT's on-campus research has included work on missiles, space defence, warships, nuclear submarines, IEDs, robots, drones and 'battle suits'.[17] At one point in the 1980s, work on biological weapons was brought to the university by its provost, John Deutch, who, it is said, went as far as to pressure junior faculty into performing this research 'on campus'.[18]

MIT's military involvement is not in doubt. But then we come on to a quite different question. No matter what I say, neither Chomsky nor his supporters seem able to acknowledge that my argument throughout this book is not that Chomsky colluded with the military, but that he had to move mountains to avoid doing so – and resoundingly succeeded in that aim.

Responding as if I were condemning him, Chomsky, at times, comes close to denying that MIT had any involvement with war research at all. Early in 2017, I teamed up with Chomskyan linguists Neil Smith and Nicholas Allott to convene a conference in London to celebrate the fiftieth anniversary of Chomsky's legendary anti-war intervention, 'The Responsibility of Intellectuals'. Chomsky agreed to participate by live video link and, on the day, was asked to comment on a range of issues. One question concerned any conflicts he may have felt while employed within a Pentagon-funded institution famed for its military research. Chomsky replied that the question was based on a misunderstanding, since, with the exception of certain departments, 'MIT itself doesn't have war work'.[19]

'There was zero military work on campus', Chomsky again stated in direct response to my book, having adopted this line of argument as his standard reply.[20] As *Decoding Chomsky* was arriving in bookshops, a *New York Times* journalist asked him directly: 'How about Chris Knight? He connects your theory of language to Pentagon-funded work you did at MIT during the Cold War.' Chomsky agreed about the military funding but went on: 'Does this mean we were doing military work? There was a study in 1969, the Pounds Commission – I was a member of it – to investigate whether any military or classified work was being done on campus. Answer? None.'[21]

This frequently repeated claim is so strange – so starkly at odds with the truth – that it demands an explanation. To my mind, it simply confirms that when Chomsky says such things, it is his moral conscience that we hear speaking to us loud and clear.

Let me stress that I do believe Chomsky's moral claim, since I accept that none of his own research could possibly have aided the US military in any way. But this cannot apply to his institution as a whole. Chomsky apparently feels that in order to deny his personal involvement in war research, he has no choice but to deny MIT's involvement. He can make that wider claim only by invoking one particular study published in 1969.

As I show in Chapter 4, the Pounds Commission – named after its chair, William Pounds – was set up by Chomsky's employers in response to a virtual uprising by students against MIT's war research. In an attempt to pacify the students, MIT insisted that, although it did administer some labs working directly on weapons, these labs were not really part of the university because they were 'off campus'. As a member of the Pounds Commission, however, Chomsky signed an appendix to its final report which shows that as many as 500 students and academics worked at these 'off-campus' military labs.[22] In subsequent interviews, Chomsky is on record as stating that 'the labs were very closely integrated' with MIT, some being located only 'two inches off campus'.[23]

In another interview he was clearer still:

There was extensive [military] research on the MIT campus. There were
laboratories at MIT that were involved, for example, in the development
of the technology that's used for ballistic missiles, and so on. In fact, a
good deal of the missile guidance technology was developed right on
the MIT campus and in laboratories run by the university.

Now, for reasons of transparency, Chomsky wanted all such military work
kept on campus:

First, should this work be done at all? ... Assuming that the work
shouldn't be done, then the concrete and crucial question arises: Shall
we get it off campus? ... Now, my feeling is that if the work is going to be
done, I'd rather have it done on campus. That is, I'd rather have it be
visible, have it be the center of protest and activism, rather than moving
it somewhere else where it can be done silently, freely, the same people
doing it, often, in fact, by just changing the name of the connection ...

At this point, as I show in Chapter 4, Chomsky goes out of his way to reject
the whole idea of management smokescreens designed to hide what was
going on:

In fact, my proposal, and I meant this quite seriously, was that universities ought to establish *Departments of Death* that should be right in the
center of the campus, in which all the work in the university which is
committed to destruction and murder and oppression should be centralized. They should have an honest name for it. It shouldn't be called
Political Science or Electronics or something like that. It should be called
Death Technology or Theory of Oppression or something of that sort,
in the interests of truth-in-packaging.[24]

In the event, Chomsky's novel proposal was not accepted by either the
Pounds Commission or MIT's managers. Instead, the university decided to
divest itself of the labs working directly on nuclear missiles, while continuing to administer the huge Lincoln Laboratory which really was located
off campus – in an Air Force base well away from student protesters. But my
point here is that by taking a stand on the geographical location of MIT's
missile labs, urging their retention on campus, Chomsky was openly
acknowledging his university's deep involvement with them. Some years
before the Pounds Commission, in 1967, he had been worried enough to
consider quitting MIT altogether. In a letter to George Steiner, which I

quote in this book, Chomsky recounts how he had 'given a good bit of thought to . . . resigning from MIT' because of its 'tragic and indefensible' involvement in the Vietnam War.[25]

Let me now turn to Chomsky's second criticism of my argument in these pages. He says that the US military didn't care about the research they were paying for at MIT, adding that even if they did care, his own linguistic research would certainly have been of no interest to them. My error here shows that, in Chomsky's words, I am 'deeply confused about the work on linguistics that I and others are doing'. The military, Chomsky claims, were unconcerned with such things: 'There was no interest of the sort, and if there had been, it would have had nothing at all to do with our studies of universal grammar.'[26]

Chomsky is technically right when he says that the Pentagon did not involve itself in the day-to-day running of the university. But this is hardly surprising. As we have seen, Chomsky's boss Jerome Wiesner worked directly for the Pentagon and so was well able to represent its interests. Secondly, Wiesner had long known that what he called 'the anarchy of science' needed to be protected if the Pentagon's research institutes were to foster sufficient creativity for novel insights to emerge.[27]

Since publication, I have come across a 1946 directive by no less a figure than General Eisenhower, insisting that in order to enable military scientists across the US 'to make new and unsuspected contributions', they 'must be given the greatest possible freedom to carry out their research'.[28] Chomsky's recollection that he and his colleagues felt free from military interference does not mean that the military had 'no interest' in what was going on. To take just one example, in 1968 the deputy director of the Office of Naval Research in Boston wrote that his scientists 'have been close' to the RLE's research programmes 'from the beginning' and 'have provided much valuable advice and counsel to the directors and working scientists'.[29]

In yet another response to my argument, this time in the *London Review of Books*, Chomsky counters that his linguistic approach has remained consistent throughout his career, whether he had military funding or not. This, he says, constitutes sufficient proof that funding considerations had no influence on his work.[30] That is easy to answer. Even during Chomsky's early years at Harvard, when he had no military funding, both he and his wife, Carol, were already working in close association with philosophers and computer scientists at MIT's Research Laboratory of Electronics.[31] As I show in the book, well before he got his paid job at MIT, Chomsky's concept of language had been developed under the influence of an intellectual culture heavily shaped by military preoccupations. Even as Pentagon funding of linguistics declined, this intellectual culture lived on. And, of course, no one expects an academic who has committed his entire career to a particular paradigm to discard it and switch to some other idea just because the funding changes or stops.

Despite all this, Chomsky is insistent and continues to dismiss any suggestion of a connection between his own work and the militarily shaped intellectual culture then prevailing at MIT. In particular, he denies any link with Warren Weaver's ideas. In fact, Weaver was the Rockefeller Foundation director who had pioneered machine translation at MIT. It was he who first floated the idea that if an underlying logical structure common to the world's language could be found, automatic translation might proceed by encoding sentences first into this structure and then out again into whatever locally spoken language was required. I have argued that it was hardly surprising that when Chomsky offered his own apparently similar concepts – automatic 'rewrite rules', 'kernel sentences', 'deep structure', 'universal grammar' and so forth – his laboratory colleagues imagined these to be connected in some way with Weaver's visionary project and hence with their own technical concerns.

In reality, Chomsky had no interest in developing a universal language because he suspected that humans already possess just one underlying language. Neither did he go along with Weaver's enthusiasm for machine translation. But enough people in the 1950s shared Weaver's dream to make Chomsky's research programme seem a perfect way of turning it into reality. MIT's laboratory technicians and computer scientists would surely have been astonished had Chomsky informed them that there was no connection at all.

But Chomsky still holds his ground. When we exchanged emails, he told me that universal grammar was not remotely connected with anything Weaver may have dreamed of. While I accept the distinction, it was generally known that machine translation at MIT was Weaver's baby and that Chomsky had been recruited to work on it. Just about all those involved expected Chomsky's theoretical labours to be relevant in some way. In 1957, one MIT manager summed up the mood by reporting that Chomsky's new book, *Syntactic Structures*, 'may provide a new theoretical foundation for all work on the machine processing of verbally expressed information'.[32] In a widely read review of the same book, Robert Lees wrote that 'Chomsky's conception of grammar may prove to be of the utmost importance [in] the field of machine translation'.[33] In 1958, Yehoshua Bar-Hillel again voiced the assumption, shared by virtually all MIT insiders, that Chomsky's work on basic linguistic theory should, 'in due time, be turned into a new method of machine translation.'[34] So even if Chomsky didn't see a connection with Weaver's machine translation project, it's clear that others did.

When Weaver first glimpsed the possibilities of machine translation, a key part of his inspiration was Alan Turing's recent breakthrough in the mathematics of computing. Revealing his roots in the same intellectual culture, Chomsky stresses how his early formulations depended critically on 'the modern theory of computability, which was developed by Alan Turing and other great mathematicians of the 1930s and 1940s'.[35] To trace this common ancestry – to point out some relationship – is not to conflate Chomsky's approach with Turing's or Weaver's. Needless to say, Chomsky's concepts and theories were uniquely his own. Yet, to his admirers in the RLE in the early years, it seemed self-evident that Chomsky's symbolic strings and mechanical rewrite rules were designed to assist them in developing computer software for processing language.

My own explanation for all this confusion is that when Chomsky was specifically asked about the relevance of his work to software development or automatic processing, his impulse was to insist that he only ever worked on basic theory. I have no doubt that in Chomsky's case, his work genuinely did have no military applications. Despite this, I am reminded of Professor Jonathan King's description of how MIT graduate students described their weapons research during the 1980s:

There were hundreds and hundreds of physics and engineering graduate students working on these weapons, who never said a word, not a word . . . So you'd go and have a seminar on the issue they're just working on; you know, they're working on the hydrodynamics of an elongated object passing through a deloop fluid at high speed. 'Well, isn't that a missile?' 'No, I'm just working on the basic principle; nobody works on weapons.'[36]

So that's how it worked. Whatever you were doing for the military, there was always a 'basic principles' way of obscuring what was going on.

Not only does Chomsky deny that the military had any interest in his research, he goes an important step further when he claims that the myth of Pentagon funding was just an administrative device enabling state funding of basic research. An interview exchange between Chomsky and historian Howard Gardner runs as follows:

Gardner: 'But is it true that initially your work was funded because the military did want to take what you were doing and use it for translation and . . .'

> Chomsky: 'That's actually a widespread illusion . . . It's very widely believed but basically the military didn't care what you were doing . . . The military didn't care. What the military was doing was serving as a kind of a funnel by which taxpayer money was being used to create the high-tech economy of the future . . . This was just US industrial policy. The way you develop the economy of the future was by the government, meaning the taxpayer, funding research and ultimately handing it over to private corporations for profit . . . The Pentagon happens to be a natural way of funding electronics based research and development.'[37]

There may well be some truth to this argument. Yet the Pentagon itself has said that it makes 'a very thorough effort' to ensure that it funds 'only research projects directly relevant to the military's technological needs'. And, when exhaustively checked by a group of anti-militarist academics from Stanford University, this claim was found to be essentially accurate.[38]

Whatever we think of these conflicting arguments, we can be confident that when the US military invests resources in university research, it does so for reasons of its own. Contrary to his flat rebuttal of the entirety of my book, Chomsky often lets slip that, deep down, he has few illusions about any of this – as when he commented, 'I'm at MIT, so I'm always talking to the scientists who work on missiles for the Pentagon.' Or again, at the height of the student protests in 1969, when Chomsky criticised an MIT student for saying *'What I'm designing may one day be used to to kill millions of people. I don't care. That's not my responsibility. I'm given an interesting technological problem and I get enjoyment out of solving it'*, he commented that he could name twenty MIT faculty members 'who've said the same thing'.[39]

Like so many authors, I had no sooner committed to print than I discovered a wealth of resources I wish I had known about earlier. Let me single out an unpublished PhD thesis – 'Private knowledge, public tensions: Theory commitment in post-war American linguistics' – which was submitted by Janet Nielsen to the University of Toronto in 2010. From my standpoint, the value of this groundbreaking study is the way in which it describes, in unprecedented scholarly detail, the subtle mechanisms through which the Pentagon's funding priorities shaped the prevailing intellectual culture at MIT. Nielsen's insights have served to put the final pieces of the jigsaw puzzle into place.

Of particular interest were the chapters dealing with 'private knowledge'. Nielsen describes in vivid detail the extraordinary bubble of mutual reference, exclusivity and circularity inhabited by Chomsky and his colleagues during their early years at MIT. Partly, this reflected an earlier tradition of relying on internal reports. In a laboratory where technicians and engineers were experimenting with the design of military gadgets, there was neither the need nor the incentive to publish in international journals. On top of that, the need to classify projects as top secret severely restricted opportunities for publication conditional on peer review. For these and other reasons, the prevailing culture had been to rely heavily on private reports of conversations, mimeographed sheets of paper, unpublished notes and self-referencing, rather than peer-reviewed journal articles available for all to scrutinize and assess.

In fairness, it needs to be said that even for those working on basic theory, there were very few opportunities for journal publication at the time. Even where a suitable journal existed, Chomsky and his colleagues felt that developments were happening so fast that they could hardly be expected to wait while an out-of-touch editor sat on exciting new results for months on end. But Nielsen's point is that the combined effect of such a culture was to encourage in Chomsky and his followers a sense that they could legitimately escape all normal scholarly constraints. She shows how their reliance on private knowledge made them a closed inner circle, whose feelings of exclusivity and superiority led outsiders to perceive them as an intentionally secretive cult. Nielsen explains:

> The underground literature culture was a defining part of transformational grammar for over a decade, arising quickly in the late 1950s and persisting until the close of the 1960s. The existence of such a culture is, seemingly, at odds with the value placed on academic freedom and open scientific knowledge in the mid to late 20th century. In other disciplines from biology to physics, publication in mainstream journals and the open circulation of knowledge was the norm. Among other advantages, it was considered necessary in order to claim intellectual priority.[40]

Admittedly, post-war American physicists frequently exchanged pre-publication copies of papers submitted to or accepted by mainstream journals. But, unlike the transformational grammar literature, such preprints were not private knowledge but means of bringing exciting new results immediately to the physics community worldwide. 'In contrast,' writes Nielsen, 'transformational grammarians were riding on what they saw as the cutting edge of linguistics, and yet shunned mainstream journals and publication mechanisms.' Since external critics were thereby denied any way of evaluating developments at MIT, they tended to give up in disgust or despair. Their bad-tempered withdrawal from workshops and conferences served only to reinforce the conviction among believers that MIT outsiders were incapable of understanding the new results, hence unworthy of being granted access to this work.

Enlightened by Nielsen, I can better understand how it was that Chomsky and his followers were able to elbow aside all opposition, treating every other paradigm as deserving of scorn. They enjoyed substantial military funding, benefited from their location in a prestigious military lab, and felt no need to debate as equals with linguists from other backgrounds. It did not matter that each new incarnation of Chomsky's theory had a relatively short life. It did not matter that nothing, ultimately, made very much sense. Transformational grammar was so much the institutionally favoured paradigm that no rival approach could hope to compete. Add to that Chomsky's brilliance in negotiating the political landscape, condemning the US military from which his own institution derived funding and support, and we have the most influential intellectual of our age.

For anyone in my position as a lifelong activist, it feels risky to say things which can so easily be misunderstood. Certain of Chomsky's claims don't match the historical facts. Yet, viewed in context, they make sense as rational responses to genuine dilemmas. No part of my account can detract from Chomsky's record as a tireless anti-militarist campaigner, whether organizing draft resistance in the 1960s or giving inspirational lectures across the world in more recent decades. Neither can it detract from his persistence in withstanding the institutional pressures that he must have come under at MIT. Had he resigned in disgust in 1967, when he was thinking of doing so, he might never have gained the platform he needed to signal his dissidence across the world. There are times when all of us have to make compromises, some more costly than others. I have argued that in Chomsky's case it was his science rather than his activism which bore the brunt of those damaging pressures and costs. Despite everything, Chomsky the activist remains an inspiration to us all. If politicians were honest and governments told the truth, we would have no need of such a figure. But in an imperfect world, we do need him. If Noam Chomsky did not exist, we would have to invent him.

PREFACE AND ACKNOWLEDGEMENTS

When I first came across Chomsky's scientific work, my initial reactions resembled those of an anthropologist attempting to fathom the beliefs of a previously unknown tribe. For anyone in that position, the first rule is to put aside one's own cultural prejudices and assumptions in order to avoid dismissing every strange belief as incomprehensible nonsense. The doctrines encountered may *seem* absurd, but there are always compelling reasons why those particular doctrines are the ones people adhere to. The task of the anthropologist is to delve into the local context, history, culture and politics of the people under study – in the hope that this may shed light on the logic of those strange ideas.

The question of language and its origins is central to an understanding of what it means to be human. Although I immediately warmed to Chomsky's courageous politics, his assumptions about language just baffled me. I was ready to admit my own limitations here: I had no training in theoretical linguistics. But I suspected that the gap between us was also deeply philosophical and cultural. It soon became clear that the tribe whose culture I needed to study was the Pentagon-funded war science community clustered around Chomsky in the formative period of his career. I have no interest in conspiracy theories. Not for a moment did I believe that the Pentagon's initial funding of Chomsky's ground-breaking work cast doubt on that work's validity or implied some kind of master plan. It was also clear that nothing that Chomsky ever produced made the slightest practical contribution to American military power. Yet from the outset I suspected that Chomsky's 'revolution in linguistics' would make sense to me only if I could fathom the time and place in which it all occurred. That would mean reconstructing the intellectual climate prevailing in the United States immediately after the Second World War, when electronic computers, still in their infancy, were widely seen as the stuff of science-fiction fantasy.

When I began looking into all this, it emerged that the Pentagon's scientists at this time were in an almost euphoric state, fresh from victory in the recent war, conscious of the potential of nuclear weaponry and imagining that they held ultimate power in their hands. Among the most heady of their dreams was the vision of a universal language to which they held the key.

As early as 1946, Warren Weaver had the idea of reducing all the world's languages to 'basic elements' which a computer could handle. By 1955, Weaver was expressing the hope that a suitably designed machine might be able to use these elements to accurately translate from any one of the world's tongues into any other. The boldest route to the heart of things, in Weaver's view, was to delve right 'down to the common base of human communication' to discover whether a single computer code really did lie at the basis of the world's superficially different tongues. Although Chomsky vehemently denies any connection between Weaver's ambitious project and his own, the ferment around these ideas appears with hindsight to have anticipated the appeal of both 'deep structure' and 'Universal Grammar'.

In championing his project, Weaver reminded his audience of its Old Testament counterpart, the Tower of Babel. Humanity did originally speak with one voice, but so great was our consequent cooperative potential that God feared we might reach up to heaven, asserting ourselves as his equal. To keep us in our place, he confounded humanity by confusing our tongues, rendering us incomprehensible to one another.

Unbelievable as it may nowadays sound, American computer scientists in the late 1950s really were seized by the dream of restoring to humanity its lost common tongue. They would do this by designing and constructing a machine equipped with the underlying code of all the world's languages, instantly and automatically translating from one to the other. The Pentagon pumped vast sums into the proposed 'New Tower'.

The theme of mythical towers appears throughout this book. Prefiguring the Pentagon's new tower was a still more ambitious plan hatched by Russia's revolutionary artists and poets – who seriously aimed to build a vast tower reaching to the sky. Here, too, there was support from the state. Building in glass and steel, the young Soviet government planned to construct 'Tatlin's Tower' as its 'Monument to the Third International'. The genius behind this symbol of revolutionary internationalism and hopes for linguistic unity was the Russian poet Velimir Khlebnikov. Khlebnikov's name is not usually linked with Chomsky, but for me it was a sudden shaft of light that

Khlebhnikov was the primary inspiration behind Roman Jakobson, the renowned linguist whose insights about a universal phonetic alphabet led directly to Chomsky's ideas about a Universal Grammar.

The irony here is that while these Russians were anti-war anarchists and Bolshevik sympathizers, Chomsky found himself immersed in a political atmosphere of paranoid hostility to 'world communism' amid feverish attempts to master the theory of nuclear war. Against this background, Chomsky stood out as the leading figure in the Pentagon-funded cognitive revolution in linguistics, psychology and philosophy, while at the same time making himself heard politically as one of the few voices of sanity in a world gone mad. The more I researched this period, the more I was struck by the disconnect between Chomsky's politics – which seemed passionate and courageous – and his concept of science, which seemed the reverse on every count. It soon became clear to me that the scientist in Chomsky excluded social topics with the same scrupulous rigour that the activist in him excluded any reliance on science. This disastrous way of fragmenting human knowledge made no sense to me at all.

The following pages are the result of my efforts to take a step back and investigate what possible circumstances could have driven Chomsky to that damaging position. It was not long before I came up with a startlingly simple theory. Chomsky was working in a weapons research laboratory. As the Vietnam War intensified, his political conscience told him that criminal activities were under way. In order to speak out freely, he needed to preserve his complete autonomy with respect to anything he was doing in that lab. He could only do this by denying that his linguistic science had the slightest political or social relevance. On this level, too, his thinking – quite separate from his politics – was declared to be radically autonomous, being purely formal, purely abstract, purely neutral. In this way, Chomsky played a major role in strengthening the Western world's habit of detaching social issues from the remit of science. My aim here is to explain where this split came from so as to pick up the threads where Chomsky and his colleagues broke them off – restoring those essential connections between social action and science, practice and theory, body and mind.

Although most people know Chomsky either for his activism or for his linguistic theories, his output ranges far outside these subjects to include anarchist and Marxist politics, psychology, philosophy and recent developments in evolutionary theory. Consequently, this book has had to be just as wide-ranging, covering such topics as the history of computing, art movements during the Russian Revolution, McCarthyite US politics in the 1950s, student unrest in the 1960s, and gender relations among our evolutionary ancestors. I ask my readers to accompany me on this intellectual roller-coaster, hoping that by the end it will become clear that the journey was necessary in order to get a handle on the extraordinary influence of Chomsky's ideas.

So many people have helped me with this book. I sent Noam Chomsky the uncorrected proofs, mentioning that I was concerned lest my criticisms of his linguistic ideas might provide ammunition for the political right. Chomsky reassured me that having read through my book, he couldn't detect any criticisms of his linguistic ideas! Chomsky always situates himself to the left of his critics, and so is not used to criticism from that quarter. Following his usual political instincts, he described my misunderstanding of the relation of the Pentagon to MIT, and to advanced research in general, as a mistake common in mainstream ideology and in right-wing economics.

Among the supporters of Chomsky that I contacted for advice, Michael Albert – whose revolutionary activism I admire – was the most blunt in his criticism, stating that my effort wasn't even a book. George Katsiaficas gave me rich personal memories of student unrest at MIT during the Vietnam War years. Robert Barsky took pains to be helpful while objecting to my focus on what he called relatively inconsequential matters such as MIT's military funding, an emphasis which he considers misleading in the context of language studies of the era.

Neil Smith took immense trouble to work through the entire manuscript, making detailed comments. At his suggestion, we met up and spent an afternoon combing through every possible point of misunderstanding. I am grateful to Neil for his generosity in reaching out to someone who disagrees with his own authoritative assessment of Chomsky's work. In doing so, he introduced me to subtleties I had previously missed. I also owe much to Maggie Tallerman, Frederick Newmeyer, Norbert Hornstein and David Adger, all admirers of Chomsky, for pointing out mistakes and warning me that to criticize my subject is to tangle with a giant.

In a different camp are those intellectual historians who have put the Chomsky legend under a microscope. Randy Allen Harris helped me understand Chomsky's puzzling repudiation of senior colleagues – notably Zellig Harris and Roman Jakobson – who had previously given him unstinting support. Saussure's great biographer John Joseph responded with a series of engaging emails. I felt wonderfully understood when he mirrored back to me my subversive intention – to serve justice on Chomsky the scientist without doing an injustice to Chomsky the conscience of America.

Christina Behme read my entire manuscript and made numerous helpful suggestions. Where she disagreed with me, it was usually because she considered me too lenient with Chomsky. I also learned a great deal from David Golumbia, whose ultimate political verdict on the US post-war cognitive revolution closely parallels my own.

In opposition to my picture of two distinct Chomskys, George Lakoff responded that the two figures embody the same idealistic – Cartesian – philosophy. We seem likely to continue to disagree on this point, but I did learn from the exchange. Peter Jones reminded me that Chomsky has an impressive track record in demolishing critics from the Marxist left. In similar vein, Rudolf Botha warned me of the sophisticated techniques developed by Chomsky – whom he describes playfully as 'Lord of the Labyrinth' – to ensure that, in any public contest, the world's pre-eminent linguist must always be seen to win. I was happy to take heed of these warnings, resolving not to enter Chomsky's labyrinth at all, but instead to dig round and tunnel under from outside.

Among friends or colleagues who made valuable editorial suggestions, I would finally like to thank Ted Bayne, Iain Boal, Angelo Cangelosi, Jean-Louis Dessalles, Martin Edwardes, Ramon Ferrer i Cancho, Richard Field, Morna Finnegan, Robin Halpin, Keith Hart, Peter Hudis, Mark Jamieson, Dominic Mitchell, Ian Parker, Gregory Radick, Katrin Redfern, Luc Steels, Sławomir Wacewicz, Ian Watts and Przemysław Żywiczyński. Derek Bickerton gave me crucial encouragement at a time when I badly needed it. Michael Tomasello's appreciation of my decoding of Noam Chomsky came at a much later stage and greatly lifted my morale.

Outside the field of Chomskyan linguistics, it was my Sussex postgraduate tutor Robin Milner-Gulland who first sparked my interest in Khlebnikov. I am grateful to him for checking over the chapters dealing with Russia's revolutionary years. Ronald Vroon helped with some of the intricacies of the Russian language, in particular the word *sdvig* ('dislocation' or 'shift') as used by Khlebnikov.

Turning to evolutionary theory, I am indebted to Volker Sommer for helping me to understand modern sociobiology and its origins, making some needed corrections to my final chapter. Biological anthropologist Sarah Hrdy has been immensely supportive of the idea that language's evolutionary emergence should be attributed to profound social change. Sarah's authoritative insights into the evolution of distinctively human 'emotional modernity' and mutual understanding have recently revolutionized the whole field of human origins research; it was her interest and support which inspired me to make hunter-gatherer gender relations and cooperative childcare central to the conclusion of my book.

I try to make my writing uncluttered, straightforward and accessible to a wide readership. Where I have succeeded, I am heavily indebted to Hilary Alton, who worked with me closely on my previous book on humanity's evolutionary origins. If her influence is still discernible in this book, it is because, having written a complicated sentence or paragraph, I continue to ask myself: 'How might Hilary have expressed that idea?' It invariably works. Almost without exception, those who took the time to read my manuscript have told me how clearly the ideas are expressed and how much they enjoyed the adventure.

I have had the level of support from Yale University Press which most authors can only dream of. I really want to thank my editor Robert Baldock, who made it a pleasure to discuss with him all my political and intellectual anxieties. Once he had decided that Yale could take the risk of publishing me again, he gave me every encouragement. I would also like to thank Yale's Clive Liddiard and Rachael Lonsdale for their skilled editing, good humour and apparently endless patience.

My three brilliant offspring Rosie, Olivia and Jude have all offered encouraging comments and criticisms. Jude in particular has set up this website for anyone wanting to read further: www.scienceandrevolution.org.

Camilla Power has been a constant source of intellectual understanding and support over the years, her criticisms always searching and thought-provoking.

Needless to say, any mistakes that remain are entirely mine.

Chris Knight
London, 2016

CHAPTER 1

THE REVOLUTIONARY

Noam Chomsky began his career as a scientist employed ostensibly to research machine translation in an electronics laboratory built to replace an earlier one in which radar had been developed for the armed services during the Second World War.[1] Since childhood, he had been keenly aware of politics, identifying himself as a libertarian socialist. He hated the military in general and the Pentagon in particular. On the other hand, his income, once employed as a young scientist, came almost exclusively from the US Defense Department.

To align his scientific career with his political conscience, Chomsky resolved from the outset to collude neither politically nor practically with his employers' aims. He recalls that in the 1960s, during the US carpet bombing of Vietnam, there came a point when he felt so compromised 'that I couldn't look myself in the mirror anymore'.[2] Unless he took decisive action, he too would be complicit in that crime. The pressures he experienced had the effect of splitting him in two, prompting him to ensure that any work he conducted for the military was purely theoretical – of no practical use to anyone – while his activism was preserved free of any obvious connection with his science. To an unprecedented extent, mind in this way became separated from body, thought from action, and knowledge from its practical applications, establishing a paradigm which came to dominate much of intellectual life for half a century across the Western world.

If you know anything about Noam Chomsky, you will not be surprised that we begin with superlatives. Think Galileo, Descartes or Einstein. Chomsky is the foremost intellectual of modern times, who 'did for cognitive science what Galileo did for physical science', to quote a respected authority.[3] Chomsky has radically altered our perception of the human condition, overturning established thinking in what has been described as a Galilean revolution. Since launching his intellectual assault on the academic ortho-doxies of the 1950s, he has succeeded – almost single-handedly – in revolu-tionizing linguistics, elevating it to the status of a genuine natural science. 'If a Nobel Prize were offered for linguistics,' comments one intellectual histo-rian, 'he'd get the first one. Then they'd have to stop giving it out. No-one else comes close.'[4] Much has changed over the past six decades, but Chomsky remains to this day the most powerful force in contemporary linguistics.

But he is more than that. At one point, Chomsky was hailed as 'the most visited person on the internet' and 'the most quoted man alive today'.[5] In 2005, he was declared the world's top public intellectual after winning 4,827 out of 20,000 votes in a poll conducted jointly on both sides of the Atlantic.[6] In 1992, the Arts and Humanities Citation Index ranked him as the most cited person alive (the Index's top ten being Marx, Lenin, Shakespeare, Aristotle, the Bible, Plato, Freud, Chomsky, Hegel and Cicero). The Social Science Citation Index gave a similar picture, as did the Science Citation Index. 'What it means', according to the librarian who checked these statis-tics, 'is that he is very widely read across disciplines and that his work is used by researchers across disciplines . . .' She added that apparently 'you can't write a paper without citing Noam Chomsky'.[7]

Things may have changed since the 1990s, but to many people Chomsky's achievements remain unparalleled in modern times. Chomsky, we are told, is 'the scholar who is to the period initiated by the cognitive revolution of the mid-1950s what Descartes was to the first phase of the age of modern philosophy'.[8] It is widely held that 'nothing has had a greater impact on contemporary philosophy than Chomsky's theory of language'.[9] He has been described by the *New York Times* as 'arguably the most important intellectual alive'.[10] One of his biographers goes even further, describing him as someone 'who will be for future generations what Galileo, Descartes, Newton, Mozart, or Picasso have been for ours'.[11] In the words of a recent intellectual historian:

> More than any other figure, Noam Chomsky defined the intellectual climate in the English-speaking world in the second half of the 20th century . . . not only did Chomsky redefine the entire academic discipline of linguistics, but his work has been something close to definitive in psychology, philosophy, cognitive science, and even computer science.[12]

'He has shown', explains a senior figure in modern linguistics, 'that there is really only one human language: that the immense complexity of the innumerable languages we hear around us must be variations on a single theme . . .'[13]

Knowing Chomsky personally has been described as 'a bit like knowing Newton'.[14] Chomsky has also been called the 'Einstein of linguistics'.[15] 'Like Einstein's theory of relativity,' we are told, 'Chomsky's ideas about linguistics have spread in their influence, and their effects are gradually filtering down to the lives of ordinary people.'[16] 'Noam Chomsky is the closest thing in the English-speaking world to an intellectual superstar', writes *Guardian* journalist Seamus Milne. 'A philosopher of language and political campaigner of towering academic reputation, who as good as invented modern linguistics, he is entertained by presidents, addresses the UN general assembly and commands a mass international audience.'[17]

In 2014, Chomsky gave an invited talk to a Vatican foundation in Rome. There was a certain irony in this in view of the well-known atheist's perceived status as the Galileo of our age. According to *The Tablet*'s report, the Vatican's Science, Theology and the Ontological Quest Foundation – the body responsible for the event – goes back 'to the commission set up by Pope John Paul II to investigate the Galileo affair'.[18] Cardinal Gianfranco Ravasi, president of the Pontifical Council for Culture, introduced the speaker warmly as 'one of the princes of linguistics'.

It is for his work in linguistics that Chomsky is honoured and celebrated. Yet he seems to be not one person, but two – each as extraordinary as the other. Addressing a largely separate audience, Chomsky has over the years become far and away the Western world's best-known political dissident. His books and articles on political topics far outnumber his publications on linguistics. When we turn to his political writings, however, we find academics less enthusiastic. While popular with students, these publications, it has been pointed out, 'rarely appear on undergraduate reading lists nor do they, on the whole, enter the fray of mainstream debates about social and political organization'.[19] It would be hard to name a wealthy corporate funding agency or scholarly foundation that has honoured Chomsky explicitly for his politics. The Vatican, no matter how supportive in other ways, would hardly invite him to talk about revolutionary socialism or anarchism.

Again and again, Chomsky has stressed that the two great 'temptations' in his life – his politics and his science – pull him in opposite directions. He may try to connect the two, but it never works – his two interests 'just don't seem to merge'.[20] Chomsky views his hectic life, therefore, as a 'sort of schizophrenic existence',[21] made possible by a fortunate glitch in his brain which causes it to function 'like separate buffers in a computer'.[22]

Science, says Chomsky, is politically neutral – hence irrelevant to his activist concerns. The search for theoretical understanding, he explains, 'pursues its own paths, leading to a completely different picture of the world, which neither vindicates nor eliminates our ordinary ways of talking and thinking . . . Meanwhile, we live our lives, facing as best we can problems of radically different kinds.'[23]

Chomsky does not encourage his scientific colleagues to care too much about his politics. Neither does he need his activist supporters to worry about his science. If you haven't the necessary expertise, he advises, 'you're just not part of the discussion, and that's quite right.'[24] In this, he has been successful. Over the years, most of his activist supporters have accepted that his linguistics is simply none of their business.

Chomsky's colleagues and employers at the Massachusetts Institute of Technology (MIT) seem to have known little about his politics until the mid-1960s, when he first began taking to the streets in opposition to the Vietnam War. Already celebrated for his linguistics, he soon began commanding a mass audience, helping to organize draft card burning and other direct action against all aspects of the war. In October 1967, with many thousands of others, he attempted to form a human chain around the Pentagon – an event famously celebrated in Norman Mailer's book *Armies of the Night*.[25] 'The dominant memory', Chomsky later recalled, 'is of the scene itself, of tens of thousands of young people surrounding what they believe to be – I must add that I agree – the most hideous institution on this earth . . .'[26]

Walking up to a line of troops in front of the Pentagon building, Chomsky appealed directly to them through a loudhailer. As he was speaking, the soldiers advanced. Chomsky's activities that year shot him to prominence, propelling him toward his current status as the best-known academic dissident in the world. From the 1960s until the present day, it would be hard to think of a US military adventure that has not faced moral and intellectual opposition voiced passionately and effectively by Noam Chomsky.

More single-mindedly than any other Western academic, Chomsky has shone a spotlight on the high-tech terrorism inflicted on much of the planet since the United States displaced Britain as the world's leading superpower. He has a low opinion of most of his fellow intellectuals, especially the self-appointed experts who dominate the universities and media outlets. Accusing them of Orwellian double-speak, he describes how they replace the dictionary meanings of words with diametrically opposite doctrinal meanings, such that 'War is peace, freedom is slavery, ignorance is strength.'[27] Corrupting the meaning of words, Chomsky argues, is a good way of dumbing people down, preventing them from talking about shared problems and in this way keeping them under control.[28]

Chomsky illustrates the technique by showing how, following the 1962 American invasion of Vietnam, double-speak was used to cover up the crime. There had been no invasion. 'For the past 22 years,' he later explained, 'I have been searching in vain to find some reference in mainstream journalism or scholarship to an American invasion of South Vietnam in 1962 (or ever), or an American attack against South Vietnam, or American aggression in Indochina – without success. There is no such event in history.'[29] Any journalist mentioning the invasion would have met with incomprehension:

Such a person would not have been sent to a psychiatric hospital, but he would surely not have retained his professional position and standing. Even today, those who refer to the US invasion of South Vietnam in 1962 ... are regarded with disbelief: perhaps they are confused, or perhaps quite mad.[30]

This, for Chomsky, illustrates what he terms 'Orwell's problem' – the problem of explaining how people can know so little even when the evidence is before their very eyes. The explanation, he writes, lies in the extraordinary sophistication of the US media's steady stream of double-speak and propaganda. To solve Orwell's problem, Chomsky observes, 'we must discover the institutional and other factors that block insight and understanding in crucial areas of our lives and ask why they are effective.'[31] While noting the importance of this problem, however, Chomsky does not find it particularly intellectually exciting, because, in his view, it is not susceptible to the methods of science.

Having described Orwell's problem, Chomsky then tells us that he has another problem, which is the exact opposite. He calls it 'Plato's problem'. This time, it is not the ignorance of people which seems so baffling, but their extraordinary knowledge and understanding. Plato's problem belongs to the sciences and is 'deep and intellectually exciting' to Chomsky.[32] He cites the case of a child acquiring its first language, apparently knowing the essentials from birth without having had time to learn anything at all.

The problem is this: how does that child succeed in working out so complex a theoretical structure as the grammar of its native tongue when it receives no instruction, is not corrected for mistakes and hears only a fraction of the creative sentences that she/he will be able to express? The solution to Plato's problem is innate knowledge. Chomsky's 'argument from poverty of the stimulus', as he calls it, states that the child doesn't *need* to learn anything because she knows the basics already, thanks to her genetic nature.[33]

Chomsky is aware that talk of 'genetic nature' or 'human nature' upsets many intellectuals, especially followers of Michel Foucault and others on the political left. But he shows no patience with such people:

> Yes, I speak of human nature, but not for complicated reasons. I do so because I am not an imbecile, and do not believe that others should fall into culturally imposed imbecility. Thus, I do not want to cater to imbecility. Is my granddaughter different from a rock? From a bird? From a gorilla? If so, then there is such a thing as human nature. That's the end of the discussion: we then turn to asking what human nature is.[34]

Chomsky rejects the 'rather obscure' Marxist notion of the dialectic, dismissing it as 'a kind of ritual term which people use when they are talking about situations of conflict and so on'.[35] He shows no appetite for dwelling on contradictions: 'Plato's problem . . . is to explain how we know so much, given that the evidence available to us is so sparse. Orwell's problem is to explain why we know and understand so little, even though the evidence available to us is so rich.'[36]

How do we know so little? That's Orwell's problem. How do we know so much? That's Plato's. Chomsky makes no attempt to reconcile these two problems, leaving the contradiction between their flatly opposed assumptions unresolved. Which problem is chosen depends on who is speaking, whether activist or scientist. Chomsky's 'two problems' seem not only different but utterly unconnected with one another, as if to deliberately illustrate the gulf between the two compartments of his brain.

In his scientific role, Chomsky's commitment is to Plato, whose point of departure is the doctrine of the soul. Plato, Chomsky reminds us, asked 'how we can know so much, given that we have such limited evidence'.[37] 'Plato's answer', says Chomsky,

> was that the knowledge is 'remembered' from an earlier existence. The answer calls for a mechanism; perhaps the immortal soul. That may strike us as not very satisfactory, but it is worth bearing in mind that it is a more reasonable answer than those assumed as doctrine during the dark ages of Anglo-American empiricism and behavioral science – to put the matter tendentiously, but accurately.[38]

Chomsky acknowledges that talk of the soul does *sound* a bit medieval. To improve the way it sounds to modern ears, he rephrases it: 'Pursuing this course, and rephrasing Plato's answer in terms more congenial to us today, we will say that the basic properties of cognitive systems are innate to the mind, part of human biological endowment.'[39] Chomsky's purpose, then, is to bring Plato up to date. He terms his modernized version of the ancient philosophy, 'internalism' – the restriction of scientific attention to patterns inside the head.

It is interesting to ask how our subject's activist voice connects with that of the scientific linguist. A reviewer for the *New York Times* phrased the question this way:

On the one hand there is a large body of revolutionary and highly technical linguistic scholarship, much of it too difficult for anyone but the professional linguist or philosopher; on the other, an equally substantial body of political writings, accessible to any literate person but often maddeningly simple-minded. The 'Chomsky problem' is to explain how these two fit together.[40]

Leaving aside the 'simple-minded' jibe, it is true that reconciling the two Chomskys is no easy matter. In establishment circles, the linguist is celebrated, the activist ignored or even reviled. A good example is the conservative philosopher Roger Scruton, who praises the scientific Chomsky's 'important and original ideas' which 'nobody in his right mind' would dismiss – while writing off the political Chomsky as 'a spoilt brat'.[41]

Ask Chomsky to define his politics and he will typically describe himself as a libertarian socialist or 'some kind of anarchist'.[42] Anarchism, for him, means freeing people from authority, although he qualifies this by cautioning that no genuine anarchist will disregard the intellectual authority of Western science. He also clarifies that the freedom advocated by the libertarian right – the freedom of powerful people to do as they please – has no place in his political philosophy. We will always need rules: 'Any effort to create a more human existence is going to inhibit somebody's freedom. If a kid crosses the street in front of me when I have a red light, that inhibits my freedom to run him over and get to work faster.'[43] In many immediate contexts, Chomsky advocates not less regulation of individual or corporate behaviour, but *more*.

Chomsky's political positions 'haven't changed much since I was about 12 or 13'.[44] 'There is a remarkable consistency to Chomsky's political work', confirms his biographer, Robert Barsky, who adds: 'The same cannot, of course, be said of Chomsky's linguistic work.'[45] Indeed, to the outsider it seems that Chomsky discards and replaces his former scientific theories at almost breathtaking speed, making it difficult to pin down precisely what they are. As one critic complains:

> The history of Chomskyan theory is a study in cycles. He announces a new and exciting idea, which adherents to the faith then use and begin to make all kinds of headway. But this progress is invariably followed by complications, then by contradictions, then by a flurry of patchwork fixes, then by a slow unraveling, and finally by stagnation. Eventually the master announces a new approach and the cycle starts anew.[46]

One consequence is that if you browse through a textbook on modern linguistics – or perhaps a popular introduction to Chomsky's work – you are likely to find everything already out of date. Most authors begin, for example, with tree diagrams depicting noun phrases, verb phrases and their ordering in accordance with rules – notions abandoned long ago by Chomsky himself. Again, almost everyone still devotes page after page to Chomsky's insistence that core principles of grammar are part of our genetic endowment – despite the fact that, since turning to 'Minimalism' in the mid-1990s, he has been stressing just how *few* of these principles should be traced back to the genes.[47] A final example is the idea that Chomsky refuses to debate the evolutionary emergence of language in *Homo sapiens*. You are likely to hear much of this well-known self-denying ordinance – despite the fact that Chomsky has since changed his mind, in 2016 co-authoring a book devoted entirely to 'language and evolution'.[48]

Such apparent swings between extremes are characteristic of Chomsky's intellectual odyssey. Yet beneath all such fluctuations, one bedrock assumption underlies his work. If you want to be a scientist, Chomsky advises, restrict your efforts to *natural* science. Social science is mostly fraud. In fact, *there is no such thing as social science.*[49] As Chomsky asks: 'Is there anything in the social sciences that even merits the term "theory"? That is, some explanatory system involving hidden structures with non-trivial principles that provide understanding of phenomena? If so, I've missed it.'[50]

So how is it that Chomsky himself is able to break the mould? What special factor permits him to develop insights which *do* merit the term 'theory'? In his view, 'the area of human language . . . is one of the very few areas of complex human functioning' in which theoretical work is possible.[51] The explanation is simple: language as he defines it is neither social nor cultural, but purely individual and natural. Provided you acknowledge this, you *can* develop theories about hidden structures – proceeding as in any other natural science. Whatever else has changed over the years, this fundamental assumption has not.

Chomsky is famed, then, not only for his many changes of mind, but also for his lifelong commitment to that basic idea. His particular theories – 'auxiliary hypotheses' in the terminology of Lakatos[52] – keep changing from month to month, year to year; meanwhile, the protected core remains intact. So a further theme in this book will be to explore whether these seemingly incompatible characteristics – the variability on the one hand, the fixity on the other – are linked. If the auxiliary hypotheses keep getting abandoned and replaced, it may be *because* the theoretical core is preserved immune to change. After all, it is precisely when nothing works – when for deep reasons nothing can *possibly* work – that peripheral changes must continually be made.

While happy to keep changing his auxiliary hypotheses, Chomsky has at all times remained committed to the following:

- Insofar as linguistics is a truly scientific discipline, it is restricted to the study of 'I-language' – a system of knowledge internal to the individual. *('To summarize, we may think of a person's knowledge of a particular language as a state of the mind, realized in some arrangement of physical mechanisms. We abstract the I-language as "what is known" by a person in this state of knowledge. This finite system, the I-language, is what the linguist's generative grammar attempts to characterize.')*[53]
- At a deeper level, scientific linguistics is the study of Universal Grammar (UG), defined as the innate cognitive equipment enabling humans to acquire such an I-language. *(' . . . the study of generative grammar shifted the focus of attention . . . to the system of knowledge that underlies the use and understanding of language, and more deeply, to the innate endowment that makes it possible for humans to attain such knowledge . . . UG is a characterization of these innate, biologically determined principles, which constitute one component of the human mind – the language faculty.')*[54]
- Scientific linguistics is therefore a branch of natural science. Set apart from social anthropology or sociology, it excludes investigation of human social interaction, politics, communication or culture. In particular, it has no place for the popular notion of 'E-languages' ('external' languages) such as 'Chinese', 'Swahili' or 'English' conceived as culturally distinct traditions. *('Rather, all scientific approaches have simply abandoned these elements of what is called "language" in common usage . . .')*[55]

- Linguistic variation is superficial. *('The Martian scientist might reason-ably conclude that there is a single human language, with differences only at the margins.')*[56]
- Strictly speaking, a child does not need to 'learn' from others how to speak its native tongue, since it is equipped with the basics already. *('Learning language is something like undergoing puberty. You don't learn to do it; you don't do it because you see other people doing it; you are just designed to do it at a certain time.')*[57]

Extensions and elaborations include these:

- A child acquires its native tongue by discarding one language after another from the vast repertoire of tongues installed in its head from birth. *('It's pretty clear that a child approaches the problem of language acquisition by having all possible languages in its head but doesn't know which language it's being exposed to. And, as data comes along, that class of possible languages reduces. So certain data comes along and the mind automatically says "OK, it's not that language it's some other language."')*[58]
- Lexical concepts – including even industrial-age ones, such as *carbur-ettor* – are not variable products of history and culture, but are somehow natural givens. *('However surprising the conclusion may be that nature has provided us with an innate stock of concepts, and that the child's task is to discover their labels, the empirical facts appear to leave open few other possibilities.')*[59]
- Although language is a biological organ, it did not evolve by natural selec-tion. Language is simply too different from anything else in nature for Darwinian theory to be relevant here. *('There is no reason to suppose that the "gaps" are bridgeable. There is no more of a basis for assuming an evolutionary development of "higher" from "lower" stages, in this case, than there is for assuming an evolutionary development from breathing to walking; the stages . . . seem to involve entirely different processes and principles.')*[60]
- Unlike other biological adaptations, language is close to perfect in design, suggesting the work of 'a divine architect'. *('The language faculty interfaces with other components of the mind/brain . . . How perfectly . . .? If a divine architect were faced with the problem of designing something to satisfy these conditions, would actual human language be one of the candidates, or close to it? Recent work suggests that language is surpris-ingly "perfect" in this sense.')*[61]

Chomsky concedes that many people find such ideas utterly baffling. How can a biological organ exist if it did not evolve? How can a child acquire its first language without learning from others about the words and rules? By what conceivable mechanism can German, Guugu Yimithirr and all possible languages be deposited in a child's head? How can *carburettor* have been installed in the brain during the Stone Age, when people hadn't even invented the wheel? Why on earth would anyone expect a biological organ to be 'perfect'?

Chomsky retorts that any scientist must work not with the complexities of life, but with abstractions, whose advantage is their simplicity:

> It goes straight back to Galileo . . . for Galileo, it was a physical point – nature is simple, and it's the task of the scientist to, first of all, discover just what that means, and then to prove it. From the tides to the flights of birds, the goal of the scientist is to find that nature is simple: and if you fail, you're wrong.[62]

While philosophers of science might go along with this, not all would agree that Chomsky's own theories can be said to have passed this test. A widespread perception is that, far from simplifying things, Chomsky's interventions have immersed linguistics in tunnels of theoretical complexity, impenetrability and corresponding exasperation and interpersonal rancour without parallel in any other scientific field.[63]

It would be wonderful if language did reveal simple logical form, like the formula for a snowflake – Chomsky's current claim.[64] But demonstrating that an idea is simple is not the same thing as showing that it works. Some assumptions are just *too* simple. In much of what follows I will be exploring whether there are grounds for suspecting that, in Chomsky's case, what *looks like* oversimplified nonsense *really is* oversimplified nonsense.

I am certainly not the first to suggest the possibility that Chomsky's core assumptions are nonsensical – many others have done that before me.[65] But, to my knowledge, few have delved further to explore the sociological conundrums which arise. Why would the dominant military, corporate and academic institutions sponsoring cognitive science in post-war America value a contribution of baffling incomprehensibility which, on close examination, turns out to make no sense at all?

I agree that this is a difficult concept to digest. Why would such powerful institutions choose to value nonsensical doctrine at the expense of empirically based science? Various possibilities have been suggested, all worth exploring. The view that Chomsky is another Galileo or Einstein is far from universally held. Many critics view the aura surrounding him as essentially scientism – adherence to an idealized concept of science as eternal truth.[66] The philosopher John Searle remarks: 'But these guys think they're doing something called Science with a capital "S". And it's almost a religion.'[67]

When science becomes religion, Searle complains, no conceivable counter-evidence can possibly weaken the faith. Sacred postulates command loyalty not by providing evidence or logical argument, but by provoking endless wonderment.[68] The more striking and unlikely, the better. Religious beliefs have been defined as 'hard-to-fake commitments to counterintuitive worlds'.[69] Think of transubstantiation or virgin birth. Showing that you can believe in 'six impossible things before breakfast'[70] is a good way of demonstrating commitment. When converts express faith in far-fetched postulates and go on to proselytize, they have passed the all-important commitment test.[71] The more circular, meaningless or self-contradictory the beliefs, according to this view, the greater their value as tests of commitment, the more costly the work of maintaining them in people's heads – and, therefore, the more energetically and passionately they are defended and proclaimed.[72]

In their 1983 book, *Language, Sense and Nonsense*, two Oxford philosophers analysed Chomskyan linguistics and its offshoots not exactly as religion, but as 'the pseudo-sciences of the age, grounded in conceptual confusions and protected from ridicule only by a façade of scientific procedure and mathematical sophistication'.[73] In similar vein, the linguist Geoffrey Pullum describes the hold exercised by Chomsky's recent theorizing as 'the most influential confidence trick in the history of modern linguistics'.[74] Many view Chomsky's entire theoretical framework as, to quote Larry Trask, 'more a religious movement than an empirical science'.[75] Chomsky's former MIT collaborator Paul Postal likens the linguist and his followers to an end-of-the-world movement, noting how charismatic cult leaders so often feel obliged to appear undaunted as their predictions are repeatedly disconfirmed: 'People's ability to "save" their ideas from even the most devastating counterexamples is thus extraordinary. I suspect that that fact goes a long way toward helping us understand what goes on in a field like linguistics.'[76]

A related point is made by Rudolf Botha in his highly original book, *Challenging Chomsky*.[77] Botha pictures Chomsky as a skilled fighter at the centre of a vast intellectual labyrinth whose forks and hidden pitfalls are used aggressively to defeat anyone foolish enough to intrude. Nobody ever wins in a battle with 'the Lord of the Labyrinth', because the Master makes sure that each contest will take place on terrain which he himself has landscaped and designed. The cleverest feature of *Challenging Chomsky* is that Botha presents his book as a defence of the Master and his fighting skills. In the course of celebrating these skills, however, the author reveals them to be not conscientious scholarship, but devious, Machiavellian tricks designed to ensure victory by moving the goal-posts or tipping up the board – in other words, sheer foul play. Reading Botha's sobering account made me reflect that my own book is an attempt to avoid the fate of so many by declining to enter the maze.

But many problems remain. Why was that maze ever built? Who financed its construction and why? What secret concealed at the heart of that maze matters so much that none of us should ever be allowed to discover it? Botha is convincing in conceiving Chomskyan linguistics as an intentionally constructed maze whose awesomely complex features have been designed for the purpose of demoralizing and confusing all intruders. But no critic has satisfactorily explained why labyrinthine nonsense on such a scale – if nonsense it is – should be corporately funded and institutionally endorsed. This is the conundrum I will address in the chapters to follow.

CHAPTER 2

THE LANGUAGE MACHINE

Before Chomsky's entry onto the stage, linguists studied languages. In the United States, they focused in particular on the tongues of Native America – Navajo, Menomini, Kwakiutl and others. Although a small number of linguists had been getting interested in computers, very few imagined that this specialist interest indicated where the discipline was about to go.

Then, early in 1957, Chomsky published his first book. He might as well have thrown a bomb. 'The extraordinary and traumatic impact of the publication of *Syntactic Structures* by Noam Chomsky in 1957', recalls one witness, 'can hardly be appreciated by one who did not live through this upheaval.'[1] 'Noam Chomsky's *Syntactic Structures*', we are informed by an insider, 'was the snowball which began the avalanche of the modern "cognitive revolution".'[2] Two recent disciples put it more simply, echoing the opening words of Genesis: 'In the beginning was *Syntactic Structures* . . .'[3]

To the outsider at the time, it all seemed rather puzzling. How could a dry-as-dust technical book on syntax have produced such dramatic effects? Historians agree that Chomsky's approach 'was indeed revolutionary – cataclysmic in relation to earlier linguistics.'[4] Vicki Fromkin recalls:

> . . . the early years following the publication of *Syntactic Structures* were exciting ones; the 'revolution' had begun. The weekly linguistics seminars at the Rand Corporation in Santa Monica more resembled the storming of the Winter Palace than scholarly discussions. Passions rose . . . Any semblance of 'scientific objectivity' disappeared as the old guard took up arms against the views of the young upstart, Chomsky, and his followers . . .[5]

A young phonologist named Morris Halle quickly emerged as Chomsky's political organizer,[6] meticulously planning what the enemy camp perceived as a brazen seizure of power.[7] The revolution was formally consummated in 1962, on the final day of the Ninth International Congress of Linguists, located conveniently in Cambridge, Massachusetts. Everything fell into place. William Locke, well known for his championing of machine translation, was positioned as a key local organizer, working alongside Chomsky's close friend from his student days, Morris Halle. During the build-up to the event, a witness at MIT recalls Halle plotting the imminent coup 'as if he were Lenin in Zurich'.[8] Manoeuvrings behind the scenes can only be guessed at, but the upshot was that Chomsky unexpectedly replaced his former linguistics teacher Zellig Harris as the final plenary speaker. From the podium, Roman Jakobson introduced Chomsky as if he were already the principal spokesperson for linguistics in the United States.[9] If there was ever a coronation, that moment was it.

It was a curious situation. By his own admission, Chomsky knew little about the world's different languages. Indeed, he outraged traditionalists by claiming he didn't need to know. Chomsky was not interested in documenting linguistic diversity. Neither did he care about the relationship between language and other aspects of human thought or life. As far as his opponents could see, he was not interested in linguistics at all. Although he has always denied this,[10] he seemed to be more interested in computers.

No one disputes that a massively important factor fuelling Chomsky's meteoric rise was significant direct funding from the US military. The preface to *Syntactic Structures* reads:

This work was supported in part by the U.S.A. Army (Signal Corps), the Air Force (Office of Scientific Research, Air Research and Development Command), and the Navy (Office of Naval Research); and in part by the National Science Foundation and the Eastman Kodak Corporation.[11]

Two large defence grants subsequently went directly to generativist – that is, Chomskyan – research in university linguistics departments. One went to MIT in the mid-1960s, and the other, a few years later, to the University of California, Los Angeles. Chomsky's second book, *Aspects of the Theory of Syntax* (1965), contains this acknowledgement:

The research reported in this document was made possible in part by support extended the Massachusetts Institute of Technology, Research Laboratory of Electronics, by the Joint Services Electronics Programs

(U.S. Army, U.S. Navy, and U.S. Air Force) under Contract No. DA36–039-AMC-03200(E); additional support was received from the U.S. Air Force (Electronic Systems Division under Contract AF19(628)-2487), the National Science Foundation (Grant GP-2495), the National Institutes of Health (Grant MH-04737–04), and the National Aeronautics and Space Administration (Grant NsG-496).[12]

Several questions arise. Why did Chomsky – an outspoken anarchist and anti-militarist – take the money? And what did the military think they were buying? These questions are sharpened by the fact that MIT at this time had almost no tradition in linguistics. Why, then, was such military investment not directed to an institution with a proven record in linguistic research?

Explaining his choice of MIT, Chomsky recalls that he felt unqualified to serve in an established department of linguistics:

I didn't have real professional credentials in the field. I'm the first to admit that. And therefore I ended up in an electronics laboratory. I don't know how to handle anything more complicated than a tape recorder, and not even that, but I've been in an electronics laboratory for the last thirty years, largely because there were no vested interests there and the director, Jerome Wiesner, was willing to take a chance on some odd ideas that looked as if they might be intriguing. It was several years, in fact, before there was any public, any professional community with which I could have an interchange of ideas in what I thought of as my own field, apart from a few friends. The talks that I gave in the 1950s were usually at computer centers, psychology seminars, and other groups outside of what was supposed to be my field.[13]

But why were the military so supportive? In 1971, Frederick Newmeyer, then among the staunchest of Chomsky's supporters, took the trouble to find out what they imagined they were buying. Tracking down a key decision maker from the time, Newmeyer invited him to explain. Colonel Edmund P. Gaines was blunt:

The Air Force has an increasingly large investment in so called 'command and control' computer systems. Such systems contain information about the status of our forces and are used in planning and executing military operations. For example, defense of the continental United States against air and missile attack is possible in part because of the use of such computer systems. And of course, such systems support our forces in Vietnam. The data in such systems is processed in response to questions and requests by commanders. Since the computer cannot 'understand' English, the commanders' queries must be translated into a language that the computer can deal with; such languages resemble English very little, either in their form or in the ease with which they are learned and used. Command and control systems would be easier to use, and it would be easier to train people to use them, if this translation were not necessary. We sponsored linguistic research in order to learn how to build command and control systems that could understand English queries directly.[14]

Chomsky's followers were by then engaged in just such a project at the University of California, Los Angeles (UCLA), prompting the colonel to comment: 'Of course, studies like the UCLA study are but the first step toward achieving this goal. It does seem clear, however, that the successful operation of such systems will depend on insights gained from linguistic research.' He went on to express the air force's 'satisfaction' with UCLA's work.

A more recent historian of the period, David Golumbia, explains Chomsky's initial success in more general sociological and cultural terms. Rather than singling out machine translation, Golumbia discerns a link between Chomskyan linguistics in general and a certain *political* doctrine bound up with dreams of US technological supremacy:

Put most clearly: in the 1950s both the military and U.S. industry explicitly advocated a messianic understanding of computing, in which computation was the underlying matter of everything in the social world, and could therefore be brought under state-capitalist military control – centralized, hierarchical control.[15]

Against this background, Golumbia describes how – despite Chomsky's political misgivings – the new cognitive paradigm became embraced as part of 'a vibrant cultural current, an ideological pathway that had at its end something we have never seen: computers that really could speak, write, and think like human beings, and therefore would provide governmental-commercial-military access to these operations for surveillance and control'.[16]

The military, however, were not the only supporters of Chomsky's revolution. On the eve of the computer age, *Syntactic Structures* excited and inspired a new generation of linguists because it chimed with the spirit of the times. Younger scholars were becoming impatient with linguistics conceived as the accumulation of empirical facts about locally variable linguistic forms. Chomsky promised simplification by reducing language to a 'machine' or 'device' whose design could be precisely specified. Linguistics was no longer to be tarnished by association with 'unscientific' disciplines such as anthropology or sociology. Instead, it would be redefined as the study of a natural object – the specialized cognitive module which (according to Chomsky) was responsible for linguistic computation. Excluding social factors and thereby transcending mere politics and ideology, the reconstructed discipline would at last qualify as a mathematically based science akin to physics.

If a theory is sufficiently powerful and simple, reasons Chomsky, it should radically reduce the amount of knowledge needed to understand the field. As he explains:

> In fact, the amount that you have to know in a field is not at all corre-lated with the success of the field. Maybe it's even inversely related because the more success there is, in a sense, the less you have to know. You just have to understand; you have to understand more, but maybe know less.[17]

Syntactic Structures infuriated established linguists – and delighted as many iconoclasts – because its message was that much of the profession's work had been a waste of time. Why laboriously list and classify anthropo-logical observations on the world's variegated languages if detailed knowl-edge could be replaced by a short cut – a simple theoretical model? In an ice-cool, starkly logical argument that magisterially brushed aside most current linguistic theory, *Syntactic Structures* evaluated some possible ways of constructing the US military's longed-for 'language machine':

> Suppose we have a machine that can be in any one of a finite number of different internal states . . . the machine begins in the initial state, runs through a sequence of states (producing a word with each transition), and ends in the final state. Then we call the sequence of words that has been produced a 'sentence.' Each such machine thus defines a certain language; namely the set of sentences that can be produced in this way.[18]

As his argument unfolds, Chomsky rules out this first, crude design for the envisaged machine – it clearly wouldn't work. By a process of elimination, he then progressively narrows the range of designs that – on purely theoretical grounds – ought to work. Chomsky's argument seemed thrilling because it promised in effect the 'philosopher's stone' – a 'device' capable of generating consistently grammatical sentences not only in English, but in any language spoken or ever likely to be spoken on Earth.

Syntactic Structures, as it happened, proved unequal to this extraordinary task. In his next book, published in 1965, Chomsky proposed a different design for his machine, variously known as the Aspects model or – from 1971 onwards – as the Standard Theory.[19] While his earlier model had excluded appeals to 'meaning', Chomsky's Aspects approach placed meaning centre stage, transferring to semantics the approach developed by Russian formalist theoretician Roman Jakobson for the study of speech sounds.

Apart from numerous other problems, Chomsky's Standard Theory was shaken when two mathematical linguists – Stanley Peters and Robert Ritchie – demonstrated that the class of grammars described by Chomsky's new model was so all-encompassing as to be vacuous. A device built in such a way, they showed, would be quite extraordinarily stupid. In fact, it would be unable to distinguish between (a) any conceivable list of strings of symbols, arbitrarily selected and combined and (b) a list of actual strings used by humans for expressing themselves in, say, English.[20] As one critic put it, Chomsky's new model would be about as good as 'a biological theory which failed to characterize the difference between raccoons and light bulbs'.[21]

To remedy this, Chomsky offered what became known as the Extended Standard Theory, or EST. By the late 1970s, further changes seemed required, leading to the Revised Extended Standard Theory, or REST. Still dissatisfied, Chomsky in 1981 published his *Lectures on Government and Binding*, which swept away much of the apparatus of earlier transformational theories in favour of a much more complex design.[22] In its 'Principles and Parameters' incarnation, the envisaged machine becomes a box of switches linked to connecting wires:

We can think of the initial state of the faculty of language as a fixed network connected to a switch box; the network is constituted of the principles of language, while the switches are the options to be determined by experience. When the switches are set one way, we have Swahili; when they are set another way, we have Japanese. Each possible human language is identified as a particular setting of the switches – a setting of parameters, in technical terminology. If the research program succeeds, we should be able literally to deduce Swahili from one choice of settings, Japanese from another, and so on through the languages that humans can acquire.[23]

Without abandoning this extraordinary dream, Chomsky has since jettisoned most of the details (the machine got impossibly complicated) in favour of a radically simplified design – known as the 'Minimalist Program'.[24] It is hard not to suspect that should 'Minimalism' in turn be discarded, the patience of even Chomsky's most ardent supporters may begin to run out.[25]

To his academic colleagues in the humanities and social sciences, Chomsky's whole approach has caused predictable astonishment, exasperation and even outrage. How could such an approach possibly be used to deduce, say, English? Today's spoken English is a product of history – a complex intermingling of dialects and languages resulting from pacification under the Romans, intermarriage between Saxons, Vikings and other tribal populations, the Norman Conquest, the Puritan revolution, the industrial revolution, the rise of Empire and so forth. How could Chomsky imagine it possible, even in principle, to construct an electronic device from which to 'deduce' English by toggling its 'settings' this way or that?

In replying to such critics, Chomsky accuses them of misunderstanding the very nature of science. To do science, he explains, 'you must abstract some object of study, you must eliminate those factors which are not pertinent'.[26] How humans articulate their thoughts under concrete social or historical conditions is irrelevant. Instead of getting bogged down in such details, we must replace reality with an abstract model. 'Linguistic theory', Chomsky declares,

> is primarily concerned with an ideal speaker-listener, in a completely homogenous speech-community, who knows its language perfectly and is unaffected by such grammatically irrelevant conditions as memory limitations, distractions, shifts of attention and interest, and errors (random or characteristic) in applying his knowledge of the language in actual performance.[27]

In this deliberately simplified model, neither time nor history exists. Children acquire language in an instant.[28] The emergence of language in the species is also considered to have been instantaneous.[29] For Chomsky, lexical concepts (for example, *book*) are not historical products, but are possessed by each of us as part of a fixed repertoire from birth.[30] Humans speak not for social reasons, to communicate their thoughts and ideas, but simply because that is their biological nature.[31] Language is the natural, autonomous expression of a specialized computational mechanism – the 'language organ' – installed inside the brain of every human on Earth.

Picturing himself as an observer from Mars,[32] Chomsky sees people as 'natural objects', their language a 'part of nature'. Linguistics as a discipline therefore falls naturally within the scope of 'human biology'.[33] This is not, however, biology as normally understood. Discussing the evolution of language, Chomsky denies that Darwinism is the relevant mechanism here.[34] The properties characteristic of language, he suggests, may instead be 'simply emergent physical properties of a brain that reaches a certain level of complexity under the specific conditions of human evolution'.[35] Alternatively, it may be that 'a mutation took place in the genetic instructions for the brain, which was then reorganized in accord with the laws of physics and chemistry to install a faculty of language'.[36] As if willing to try anything, Chomsky later suggested that language's recursive structure may have emerged suddenly as a spandrel – an accidental by-product – of other, unspecified developments connected with, say, navigation.[37]

For Chomsky, linguistics can aspire to the precision of physics. This is because physics studies natural objects and language is just such an object.[38] Language approximates to a 'perfect system' – an optimal solution to the problem of relating abstract concepts to 'phonetic forms' or speech sounds. Scientists, according to Chomsky, do not normally expect to find perfection in the biological world – that is a hallmark of physics. He explains: 'In the study of the inorganic world, for mysterious reasons, it has been a valuable heuristic to assume that things are very elegant and beautiful.' He continues:

Recent work suggests that language is surprisingly 'perfect' in this sense . . . Insofar as that is true, language seems unlike other objects of the biological world, which are typically a rather messy solution to some class of problems, given the physical constraints and the materials that history and accident have made available.[39]

Language, according to Chomsky, lacks the messiness we would expect of an accumulation of accidents made good by evolutionary 'tinkering'. Characterized by beauty bordering on perfection, it is biology – yet cannot have evolved in the normal biological way.

What are we to make of all this? It is easy to understand why computer engineers might find it useful to treat language as a mechanical device. If, say, the aim were to construct an electronic command-and-control system for military use, then traditional linguistics would clearly be inadequate.

The requirement would be for a concept of language stripped free of 'meanings' in any human emotional or cultural sense, cleansed of politics and stripped also of poetry, humour or anything else not accessible to a machine.

But military figures such as Colonel Gaines were not the only people hoping to benefit from the new approach. We need to look at Chomsky's other institutional sources of support. Equally, we need to consider that very different constituency of socialists and anarchists who have always drawn inspiration from his fiercely anti-militarist politics. We might ask whether, for these radicals, the idea of anti-capitalist revolution connects up in some way with the revolution in linguistics. In fact, the issue is whether the two sides of Noam Chomsky can be reconciled at all. Was the young anarchist tailoring his theories to meet the requirements of his military sponsors – forcing us, perhaps, to question the sincerity of his radical political commitments? Or did he believe he was taking the money – refusing to let this influence his scientific results – in order to secure the best possible position from which to promote the anarchist cause?

Whatever the answer, one point is indisputable. Chomsky's 1957 *Syntactic Structures* marked the start of a massive intellectual wave which would quickly engulf psychology, linguistics, cognitive science and philosophy – a rupture now known as the 'cognitive revolution'.

CHAPTER 3

A MAN OF HIS TIME

U ntil Chomsky's arrival on the intellectual scene, language and speech had been conceptualized as 'culture' or 'learned behaviour'. During the 1940s and 1950s, the standard paradigm in 'scientific' psychology had been behaviourism – championed in the Soviet Union by Ivan Pavlov and in the United States most prominently by Burrhus F. Skinner. A psychologist describes the absolute dominance of behaviourism in America during those years:

> The chairmen of all the important departments would tell you that they were behaviorists. Membership in the elite Society of Experimental Psychology was limited to people of behavioristic persuasion; the election to the National Academy of Science was limited either to behaviorists or to physiological psychologists, who were respectable on other grounds. The power, the honors, the authority, the textbooks, the money, everything in psychology was owned by the behaviorist school.[1]

If you didn't subscribe to the cult, you most likely wouldn't get a job.

Skinner first came to prominence for his wartime experimental work harnessing pigeons to guide missiles to their target.[2] His magnum opus, *Verbal Behavior* – published in 1957 – claimed to explain human language as a set of habits built up over time.

Laboratory rats, Skinner showed, can be trained to perform extraordinarily complex tasks, provided two basic procedures are followed. First, the tasks must be broken down into graduated steps. Second, the animal under instruction must be rewarded or punished at each step. This type of learning was termed *operant conditioning*. Building on his work with rats, Skinner argued:

The basic processes and relations which give verbal behavior its special characteristics are now fairly well understood. Much of the experimental work responsible for this advance has been carried out on other species, but the results have proved to be surprisingly free of species restrictions. Recent work has shown that the methods can be extended to human behavior without serious modification.[3]

Skinner accordingly treated human language in stimulus–response terms, identifying 'meaning' with the habituated response of the listener to speech stimuli. Language was conceptualized as structured like a chain, learned by associating one link with the next, via appropriate approval or 'reinforcement'.

Planners and social engineers – among them Stalin in the Soviet Union – welcomed behaviourism because it seemed to promise enhanced techniques for mass education, pacification, political manipulation and control. Stimulus–response psychology, as one historian observes, 'encouraged industrial managers in the belief that securing compliance meant finding in the workforce which buttons to push and pushing them'.[4]

Or, as Chomsky succinctly puts it: 'Those who rule by violence tend to be "behaviorist" in their outlook. What people may think is not terribly important; what counts is what they do. They must obey, and this obedience is secured by force.'[5]

Although up to then Chomsky had defended behaviourism – ruling out mentalism for 'its obscurity and general uselessness in linguistic theory' – in 1958 he went into reverse.[6] Hostile now on deeply felt moral (as well as intellectual) grounds, Chomsky launched his campaign against behaviourism with a withering review of Skinner's *Verbal Behavior*. He was wise enough not to take issue with, say, the sophisticated school of child psychology pioneered in the Soviet Union by Lev Vygotsky or (for the time being, at least) the subtle and fruitful insights developed by the Swiss developmental psychologist Jean Piaget.[7] Despite major differences with psychoanalysis, these psychologists had echoed Freud in taking for granted that humans, like other animals, must have deep-rooted instincts of some relevance to a study of the mind. Chomsky, however, refrained from acknowledging the existence of such intellectual giants. By singling out behaviourism for attack and ignoring everything else, he succeeded in arranging the battleground to suit his own needs.

Chomsky's review of *Verbal Behavior* caused a sensation. Published in 1959 in the journal *Language*, the 'case against B.F. Skinner' set in motion a tidal wave of revolt against a school of thought increasingly perceived as Orwellian in its ultimate goal of moulding and manipulating all human life.

It was not difficult for Chomsky to associate behaviourist linguistics with Orwellian aims. Leonard Bloomfield, the dominant figure in American linguistics since the 1920s, spoke for his generation when he told the Linguistic Society of America in 1929: 'I believe that in the near future, in the next few generations, let us say, linguistics will be one of the main sectors of scientific advance, and that in this sector, science will win through to the understanding and control of human conduct.'[8]

Following the Second World War, the US government felt threatened by much political conduct which emphatically needed to be 'controlled'. The subversives deemed threatening to American interests included anti-fascist resistance partisans in Europe and guerrilla fighters in former colonies – all combining to convince linguists in the Bloomfield camp that they were living 'at a time when our national existence and possibly the existence of the human race may depend on the development of linguistics and its application to human problems'.[9] The wave of McCarthyite witch-hunting which swept the United States during the 1950s was in part premised on the belief that critics of 'the American way of life' must clearly have been brainwashed by communists. In this bitter Cold War context, linguistics became extraordinarily politicized, being viewed by US policy makers as a crucial weapon in the worldwide struggle for mastery and control.

Chomsky's view of behaviourism was straightforward: where B.F. Skinner was concerned, the whole thing was criminal fraud. This being so, he would expose the fraudulent edifice for what it was, reversing its premises point by point:

- Skinner taught that the entirety of a young child's linguistic knowledge is imparted to it by the environment in which it is placed. No, counters Chomsky, everything of theoretical interest or significance comes from inside – from the child's genetically constituted mind.
- Skinner taught that knowledge of language must be learned. Quite the opposite, responds Chomsky, none of the deep complexities of grammar can possibly be learned.[10]
- Skinner taught that 'mind' is a nebulous concept – we can know only the body and its observable behaviour. No, responds Chomsky, ever since Newton's discovery of gravity we have had no coherent concept of 'body' or 'matter'; where language is concerned, all that can be known scientifically is 'mind'.[11]
- Skinner taught that humans are essentially no different from rats. No, replies Chomsky, humans are so different from rats that Darwinian

natural selection is insufficient to explain how the gap between us and other animals could have been bridged.[12]

Chomsky is withering in his response to the notion that a child acquires language through social pressure, training and example. 'Attention to the facts', he wrote, 'quickly demonstrates that these ideas are not simply in error but entirely beyond any hope of repair.'[13] He appeals to what we might nowadays term 'critical theory' – ultimately traceable to Marx – to explain why such nonsense was ever touted as science:

> One has to turn to the domain of ideology to find comparable instances of a collection of ideas, accepted so widely and with so little question, and so utterly divorced from the real world. And, in fact, that is the direction in which we should turn if we are interested in finding out how and why these myths achieved the respectability accorded to them, how they came to dominate such a large part of intellectual life and discourse. That is an interesting topic, one well worth pursuing.[14]

The myth that language must be 'learned', Chomsky insists, became popular with the ruling elite because it encouraged the belief that humans can be manipulated at will – a convenient theory if you are a state commissar or manager hoping to control people's lives.

Chomsky asks how any sane person could imagine that language has to be learned. What kind of 'learning' is it when humans everywhere accomplish the task in basically the same way and to an equal extent? Languages, he points out, are not like other cultural patterns. They are not more or less complex, more or less sophisticated, according to the level of technological or other development. While superficially differing from one another, each language is an intricate, complex intellectual system; none can be described as more or less sophisticated or 'advanced'.

In all cultures, moreover, people speak fluently, regardless of social status, training or education. There is an innate biological schedule for language acquisition, specifying at what age a new language can be mastered and at what age the task becomes virtually impossible. While young children take quickly and easily to the task, adults encounter immense difficulties, often making recurrent basic errors and revealing a permanent tell-tale accent, despite years of trying. Young children not only learn easily: in linguistically impoverished environments, they may creatively invent improvements, developing a language more systematic than any they have heard. It is as if they knew by instinct how a proper language should be structured, anticipating regularities and establishing them inventively where necessary.[15]

The syntactic skills of children mastering a language, Chomsky points out, are acquired with extraordinary rapidity and in unmistakably creative ways. The child is not just assimilating knowledge or learning by rote: on the contrary, what comes out seems to exceed what goes in. Children hear relatively few examples of most sentence types, are rarely corrected, and encounter a bewildering array of half-formed sentences, lapses and errors in the language input to which they are exposed. Yet, despite all this, they are soon fluently producing sentences never heard before, knowing intuitively which sequences are grammatical and which are not. In Chomsky's words:

> The fact that all normal children acquire essentially comparable grammars of great complexity with remarkable rapidity suggests that human beings are somehow specially designed to do this, with data-handling or 'hypothesis-formulating' ability of unknown character and complexity.[16]

Chomsky concluded that knowledge of language is conferred by genetics, with the corollary that linguistics must be reconstructed as a *natural* science.

CHAPTER 4

THE MOST HIDEOUS INSTITUTION ON THIS EARTH

In the last chapter, I tried to present Chomsky's arguments in as positive a light as I could manage. But to understand how it came about that Chomsky so accurately hit the mood of the times, we need to recall the way in which, during the Second World War, the United States military brought together computer scientists, psychologists, linguists and manufacturers of armaments and other equipment, pressing them to integrate human operators into their projects and designs. We need to understand why the challenges they faced prompted them to reject – or at least to question – behaviourist stimulus–response psychology long before anyone had heard of Noam Chomsky.

Too often, Chomsky is depicted by his supporters as the genius who overthrew behaviourism single-handedly. That narrative overlooks how the ground had already been prepared by influential sections of the US military and industrial establishment which, as I will now attempt to show, had been bridling at the narrowness of behaviourist dogma for their own reasons since the early years of the war. Well before Chomsky intervened, glimpses of a new, mind-focused cognitive science were appearing within the Pentagon-funded think tanks and laboratories, although in most cases these were still bound up with lingering behaviourist assumptions.

Behaviourism's fundamental claim had been that mental states could not be studied; only behaviour can be known. But the development of computers changed all that. The revolutionary new 'computational theory of mind' – keystone of the 'cognitive revolution' – was based on the idea that the human brain is a digital computer or 'information-processing device'. Mental states, correspondingly, could be viewed as the variable discrete states of such a machine. But as 'mind' was being reinstated during and immediately following the Second World War, the secrecy surrounding developments in the think tanks of the US military left the world at large mostly unaware.

The Pentagon's scientists had little incentive to air their military agendas in public and had been keeping most of their important discoveries under wraps. While Chomsky appeared to be slaying the behaviourist monster single-handedly, the truth is that wartime requirements of command and control had long been tipping the scales his way. It was Corporate America's urgent need for a mind-centred psychology – not Chomsky's later eloquence in championing it – that at the deepest level spelt doom for behaviourism and guaranteed the cognitive revolution's rapid and stunning success.

Chomsky's actual role can best be appreciated by taking a step back from immediate events. How might the situation have looked without him? At a formal ceremony on 15 February 1946, just before pressing a button that set the latest digital computer (ENIAC) to work on a new set of hydrogen-bomb equations, Major General Gladeon Barnes spoke of 'man's endless search for scientific truth'.[1] When people heard a beribboned general mouthing such phrases, there was a danger that they would respond with cynicism. For well over a decade, news of intriguing developments in cognitive science had been seeping into the public realm, but only in the vaguest terms and mostly in the words of uninspiring military figures such as Major General Barnes. How much better to find a genuinely idealistic spokesperson who could step in and do the necessary public relations job, speaking with real passion and conviction. Clearly, any such standard-bearer would need to be a civilian, with absolutely no military connections.

In all these respects, Noam Chomsky turned out to be the ideal candidate. Far from sharing his colleagues' excitement about the prospects of nuclear weaponry, his instincts had always recoiled from the very idea:

> I remember on the day of the Hiroshima bombing . . . I remember that I literally couldn't talk to anybody. There was nobody. I just walked off by myself. I was at a summer camp at the time, and I walked off into the woods and stayed alone for a couple of hours when I heard about it. I could never talk to anyone about it and never understood anyone's reaction. I felt completely isolated.[2]

So it was in many ways a cruel irony that this genuinely idealistic teenager should find himself, years later, in an electronics laboratory whose publicly stated mission was to develop command-and-control systems for nuclear and other military purposes.

In the early 1950s, Chomsky still had no expectation of an academic career and, as he says, 'was not particularly interested in one'.[3] Perhaps his strongest motive for getting academically qualified in the first place was to avoid being drafted to fight in the Korean War.[4] He did manage to get a fellowship at Harvard, but had no expectation of a job there. As he later explained, 'Jewish intellectuals couldn't get jobs' in many universities and, at Harvard, 'you could cut the anti-Semitism with a knife'.[5] One reason he ended up at MIT was because, being Jewish, he could not get a job anywhere else. 'I had no thought of going into the academic world because there was nowhere to go', he recalls. 'But Jakobson suggested I talk to Jerry Wiesner, and so I did.'[6] With engaging modesty he elaborates: 'When I got my degree, PhD, which was total fraud incidentally, I hadn't done any work or anything else . . . I had no professional qualifications at all and that's why I went to MIT because they didn't care . . . because they were getting a ton of Pentagon money.'[7]

Even after he began working at MIT, Chomsky was not necessarily committed to the job and still had in mind that he might leave America to live on a left-wing kibbutz with his partner, Carol. But the difficulties of life in Israel after only a few weeks seem to have persuaded the young couple to remain in the United States.[8]

Having decided to stay on at MIT, it wasn't long before Chomsky found himself being honoured, promoted, financed and hoisted aloft by the very establishment whose foreign policy crimes he so passionately opposed. It was a favourable wind that would prevail for the next 60 years. Given the paradox that an anti-militarist activist was working in the world's foremost military research centre, there were going to be challenges in keeping both parties on board. But Chomsky's relaxed and informal demeanour and his appealing stance as a military outsider undoubtedly helped. Watching him deliver a lecture or seeing him on television, it was often difficult for an onlooker to keep the other side of his life in mind. His political passion was directed against the US military. But, as he said about MIT as it was in 1969: '[The university] was about 90 percent Pentagon funded at that time. And I personally was right in the middle of it. I was in a military lab.'[9] Chomsky may have been an anarchist, but he was also in the belly of the beast.

The largest of America's wartime university research programmes, MIT's Radiation Laboratory was the renowned institution where radar (invented in Britain, but for security reasons hastily shipped over to the United States) had been successfully developed. Armies of academics from across America had been coordinated via this laboratory to work for industrial and military research groups. Industrial scientists collaborated with their university colleagues, and generals were frequent visitors. In this way, the war had brought about the most radical centralization and reorganization of science and engineering ever accomplished in the United States.[10] It would be hard to exaggerate the long-term ideological impact of MIT's Radiation Laboratory as a vast interdisciplinary effort restructuring American scientific research, solidifying the trend to science-based industry – and, in the process, introducing unprecedented levels of government funding and military direction of intellectual life.

Chomsky was the right man in the right place at exactly the right time. MIT emerged from the Second World War with a staff twice as large as the pre-war figure, a budget four times as large, and a research budget magnified tenfold – 85 per cent of it from the military services and the Atomic Energy Commission.[11] Eisenhower famously named this sprawling corporate monster the 'military-industrial complex', but – as historian Paul Edwards notes – the nexus of institutions is perhaps better captured by the concept of an 'iron triangle' of reciprocally self-perpetuating academic, industrial and military collaboration.[12] If you had military funding, as Chomsky notably did, that gave you not only plentiful money, but also enviable academic status, from which you could hope to attract still more financial and other institutional support.

But why did this military institutional and funding context lead Chomsky to what is most characteristic about his philosophical outlook – the extreme way in which he splits what is pure and ideal from what is messy and real? Later, it will become clear how such bifurcation was intrinsic to the computational theory of mind, which pictured information as independent of the material which carried it. But my suggestion is that, in Chomsky's case, there was more to it than that. He found in the mind/body split a means of escape – escape from the potentially painful moral implications of the work he was doing.

Chomsky himself has written of his need to be able to look himself in the mirror each morning 'without too much shame'.[13] Employed to conduct research on a Pentagon-funded device – a 'language machine' – he needed to keep his conscience clear. There was no problem here, he would sometimes argue, since his own institution embodied 'libertarian values':

> It's true that MIT is a major institution of war research. But it's also true that it embodies very important libertarian values . . . Now these things coexist. It's not that simple, it's not just all bad or all good. And it's the particular balance in which they coexist that makes an institute that produces

weapons of war willing to tolerate, in fact, in many ways even encourage, a person who is involved in civil disobedience against the war.[14]

But Chomsky may not have been entirely happy with this line of thought. True, MIT did espouse 'libertarian values' – most universities do. But in the United States, 'libertarian' can mean almost anything, being frequently associated with the far right. Beyond that, Chomsky could hardly describe MIT's primary sponsor, the Pentagon – 'the most hideous institution on this earth'[15] – as libertarian. The weapons research conducted on campuses like his own was explicitly designed, according to him, 'to harm people, to destroy and murder and control'. 'As far as I can see,' Chomsky continued, 'it's elementary that that kind of work simply should not be done.'[16]

All this shows just how strong were Chomsky's misgivings, even while he was doing the intellectual work which he loved. In mitigation, he could argue that it was scarcely possible to conduct scientific research in an American university *without* military funding:

Ever since the Second World War, the Defense Department has been a main channel for the support of the universities, because Congress and society as a whole have been unwilling to provide adequate public funds. Luckily, Congress doesn't look too closely at the Defense Department budget, and the Defense Department, which is a vast and complex organization, doesn't look too closely at the projects it supports – its right hand doesn't know what its left hand is doing. Until 1969, more than half the MIT budget came from the Defense Department, but this funding at MIT is a bookkeeping trick. Although I'm a full-time teacher, MIT paid only thirty to fifty percent of my salary. The rest came from other sources – most of it from the Defense Department. But I got the money through MIT.[17]

Despite all this, Chomsky's conscience remained active. He once urged the renaming of military laboratories, perhaps including his own:

In fact, my proposal, and I meant this quite seriously, was that universities ought to establish Departments of Death that should be right in the center of the campus in which all the work in the university which is committed to destruction and murder and oppression should be centralized. They should have an honest name for it. It shouldn't be called Political Science or Electronics or something like that. It should be called Death Technology or Theory of Oppression or something of that sort, in the interests of truth-in-packaging. Then people would know what it is; it would be impossible to hide. In fact, every effort should be made to make it difficult to hide the political and moral character of the work that's done.[18]

Changing the name in this way, continued Chomsky, might provoke sufficient outrage to disrupt or even halt the work:

I would think in those circumstances it would tend to arouse the strongest possible opposition and the maximal disruptive effect. And if we don't want the work to be done, what we want is disruption: maybe the disruption will be the contempt of one's fellows.[19]

As we shall see, Chomsky had no appetite for the contempt of his fellows, whether these were university administrators or radical students.

Chomsky is right in saying that his own work at MIT was in one sense nothing special: military sponsorship was (and remains) essential to many areas of Western scientific research. But, whether in linguistics or any other field, the problems accorded priority do tend to follow the funding. As a laser scientist at Stanford University explained: 'Nobody likes [military] support less than we do . . . The problem is, when the military supports everything, they're the people who come around with the problems and so you think about those problems.'[20] At MIT in the 1960s, well-funded research laboratories were working on helicopter stabilization, on radar for tracking bomb targets and on various counterinsurgency techniques – all useful technologies for the ongoing war in Vietnam.[21] At the same time, as Chomsky explained in an interview in 1977, 'a good deal of [nuclear] missile guidance technology was developed right on the MIT campus'.[22]

By late 1968, the student unrest of the period had spread to MIT, largely in opposition to the Vietnam War, but specifically in opposition to MIT's role in the whole US war machine. Chomsky's university soon became, in his own words, 'one of the most militant campuses in the country'.[23] A confrontation erupted between the administration and a radicalized student body demanding the removal of all defence-related research from the campus. Late in 1969, at the height of the protests, activists in the 'November Action Committee' explained:

MIT isn't a center for scientific and social research to serve humanity. It's a part of the US war machine. Into MIT flow over $100 million a year in Pentagon research and development funds, making it the tenth largest Defense Department R&D contractor in the country. MIT's purpose is to provide research, consulting services and trained personnel for the

US government and the major corporations – research, services, and personnel which enable them to maintain their control over the people of the world.[24]

These activists went on to explain that their recent action was 'directed against MIT as an institution, against its central purpose'.[25]

Clearly, these students were aiming at an all-out confrontation. But despite this, Chomsky's recollection was different. According to him, 'Nobody wanted the confrontation on either side. The dynamics of that kind of thing are pretty obvious. So how do you get out of a confrontation? You set up a committee. So the committee was set up to investigate it and I was kind of implored by the Dean and the students to be on it'.[26]

The dean, or more accurately the president of MIT, was Howard Johnson, who made clear that he needed Chomsky on his committee 'to satisfy the radicals'.[27] This had the effect, it seems, of persuading at least one of these radicals – Jon Kabat – to drop his former resistance and agree to join as well.[28] The subsequent lengthy deliberations, of course, helped Howard Johnson take 'the steam out of the lab situation'.[29] As one student told a journalist at the time, '[Johnson] is smart. He'll co-opt you before you know what's happening'.[30] Between Chomsky and Johnson, astute politicking won the day.

Five months later, the committee produced a final report, to which Chomsky added a dissenting appendix. Chomsky's delicate compromise position was that 'we should not sever the connection with the special laboratories but should, rather, attempt to assist them in directing their efforts to "socially useful technology"'. He continued:

In my opinion, the special laboratories should not be involved in any work that contributes to offensive military action. They should not be involved in any form of counterinsurgency operations, whether in the hard or soft sciences. They should not contribute to unilateral escalation of the arms race. They should not be involved in the actual development of weapons systems. They should be restricted to research on systems of a purely defensive and deterrent character.[31]

Years later, when Chomsky recalled these events, he put a slight gloss on them:

The students and I submitted a dissident report disagreeing with the majority. The way it broke down was that the right-wing faculty wanted to keep the labs, the liberal faculty wanted to break the relations (at least formally), and the radical students and I wanted to keep the labs on campus, on the principle that what is going to be going on anyway ought to be open and above board, so that people would know what is happening and act accordingly.[32]

But Chomsky's suggestion that he and the radical students formed a unified bloc does less than justice to the complexities of the situation. The key student activists of the time have different memories.

One of those activists was Stephen Shalom, who complains that Chomsky's recollection (as recorded by Robert Barsky) 'obscures the fact that most radical students, as well as many liberal students, wanted first and foremost to stop the war research.'[33] A second activist, George Katsiaficas, who actually sat on the MIT committee along with Chomsky, simply recommended 'the immediate curtailment of weapons research.'[34] A third dissenting activist was MIT's student president at the time, Michael Albert, for whom Chomsky's compromise, however well intentioned, threatened to take the steam out of the whole campaign: 'Given attitudes at MIT, we could successfully organize around closing down research . . . We could not develop support for preserving war research with modest amendments.'[35] The three most prominent student activists – Albert, Katsiaficas and Kabat – all described the final report as a 'smokescreen'.[36] Considering all this, it is not surprising that Chomsky himself says that he had 'considerable conflict' with radical students in this period.[37]

Adding to this picture of Chomsky on the horns of a dilemma, some MIT faculty members remember him as not particularly radical in this period. We are told that Chomsky was in fact one of the signatories when the initial, May 1969, version of the MIT committee report recommended keeping both the military labs functioning – although he did later change his mind.[38]

The complexities of these events become still more apparent when other perspectives are recalled. Asked about the atmosphere at MIT in the late 1960s, one professor recalls: 'Things were brewing; the radical student vibrations were going on. They were in a siege mentality. All of the people in the upper administration carried walkie-talkies with them.' He then offers his own explanation as to how and why the students were defeated:

All radical revolutionary movements of any kind, if you read history, make the same mistake . . . They have to decide whether to move closer to the establishment – or to become more radical. And they always make the same mistake. They become more radical, and the establishment destroys them.

See, when we were standing in the president's office trying to protect the president's office from the radical students, I'd look up and I'd be standing right beside Noam Chomsky.

The interviewer intervenes at this point, remarking that Chomsky was 'perceived as quite a radical himself'. The professor replies: 'Yes. But not if you're going to mess around with our institution.' He continues: 'And the time that the students interrupted a class? Oh, my God. The faculty came down with a giant iron fist. It was a wild time. But we got through it; we got through it.'[39]

By early 1970, Michael Albert had been expelled from MIT, provoking other students to occupy and damage the president's office. Having begun the process of expelling more students and securing the imprisonment of two of them for the alleged crime of 'disruption of classes',[40] the grateful professors then organized a surprise party for Howard Johnson, during which they cheered their long-suffering president and presented him with an engraved clock.[41] Somehow, Chomsky managed to sign a letter calling for an amnesty for the protestors, while also turning up at this celebratory party, at least according to Johnson's recollection.[42]

Whatever the accuracy of these scattered reminiscences, Chomsky has stated candidly that he considered the student rebels of the period 'largely misguided' and their actions, in some cases, 'indefensible'.[43] Rather than urging the students forward at this time, he advised postponement and delay on the basis that 'The search for confrontations is a suicidal policy'.[44]

But while Chomsky was in two minds about disrupting university campuses, he showed rare courage, intransigence and extraordinary confrontational spirit on the wider political stage. In the spring of 1970, at the height of the Vietnam War, he accepted an invitation from the North Vietnamese government to visit Hanoi, giving a seven-hour lecture on linguistics to an enthusiastic audience at the University of Hanoi. During the trip he also spoke to Premier Pham Van Dong, to the editor of the Communist Party newspaper, and to various intellectuals and peasants, receiving a cordial welcome everywhere as a 'progressive American'.[45]

Chomsky's was merely an extreme case of a more general predicament for liberal academics in the period. But the situation at MIT was particularly strained. President Howard Johnson ended up imposing an injunction and welcoming in riot police to protect what he termed a 'free and open university'. In fact, Johnson was protecting the right of workers at the university to work on guidance systems for nuclear missiles.[46] Johnson's subsequent replacement, Jerome Wiesner, then ended up doing his best to get more anti-war students sentenced to prison while, at the same time, insisting that he, like them, was opposed to the war in Vietnam. In fact, Wiesner continued to oversee a huge military research programme at MIT, naturally justifying this on grounds of 'academic freedom'.[47]

The strain on Chomsky was clearly showing as early as March 1967, when he admitted in a private letter (immediately published in the *New York Review of Books*) that 'I have given a good bit of thought to . . . resigning from MIT, which is, more than any other university, associated with activities of the department of "defense". . . As to MIT, I think that its involvement in the war effort is tragic and indefensible. One should, I feel, resist this subversion of the university in every possible way'.[48] Then, only a few weeks later, he had second thoughts about such blatant criticism of his employer. In a new letter, he chose to 'reformulate' his earlier statement against MIT: 'This statement is unfair, and needs clarification. As far as I know, MIT as an institution has no involvement in the war effort. Individuals at MIT, as elsewhere, have a direct involvement, and that is what I had in mind'.[49] Did something happen in between these two letters – some pressure from above which made him fear for his job?

Chomsky was genuinely grateful to the student activists of the 1960s for transforming the atmosphere at MIT and he acknowledged that they suffered things 'that should not have happened'.[50] But he always defended MIT as an institution, describing Howard Johnson as 'an honest, honorable man'.[51] In one interview in 1996 Chomsky claimed that:

> MIT had quite a good record on protecting academic freedom. I'm sure that they were under pressure, maybe not from the government, but certainly from alumni, I would imagine. I was very visible at the time in organizing protests and resistance. You know the record. It was very visible and pretty outspoken and far out. But we had no problems from them, nor did anyone, as far as I know, draft resisters, etc.[52]

In another interview, he said:

> I have been at universities around the world and this is the freest and the most honest and has the best relations between faculty and students than at any other university I have seen. That may be because MIT is a science-based university. Scientists are different. They tend to work together, and there is much less hierarchy.

He then went on to claim that MIT has 'quite a good record on civil liberties. That was shown to be particularly true during the sixties.'[53]

These remarkable statements make more sense when we appreciate that Chomsky's position on academic freedom uncannily resembled the MIT management line on these issues. You can research what you like – provided you don't actually do anything about it. That line was succinctly expressed by Howard Johnson in October 1969 when he was desperately trying to prevent students from disrupting MIT's military labs: '[The university] is a refuge from the censor, where any individual can pursue truth as he sees it, without any interference.'[54]

It is only against this background that we can properly understand what happened when Chomsky later took up a position which laid him open to charges of anti-Semitism. Over the years, Chomsky's right-wing detractors have made much of his defence of holocaust denier Robert Faurisson. In 1979, Chomsky famously signed a petition stating: 'We strongly support Professor Faurisson's just right of academic freedom and we demand that university and government officials do everything possible to ensure his safety and the free exercise of his legal rights.'[55] Whatever we may think of the wisdom of Chomsky's intervention, his detractors ought to acknowledge that his position did not imply any sympathy towards holocaust denial. It was simply a logical extension of a principle common to all Western universities – one which his management at MIT felt obliged to uphold with special tenacity in view of what its own researchers were doing.

To take an example, one of MIT's researchers was the economist Walt Rostow who had produced influential theories on how to counter the 'disease' of communism in the Third World. Rostow had since left MIT to work as an adviser to the US government, where he had become one of the major architects of the Vietnam War. In fact, Rostow had advocated escalating that war more vigorously than almost anyone else in the US government, even urging it to risk a nuclear stand-off with Russia. With Nixon's election victory in 1968, however, he lost his government job and was looking to return to MIT.[56] In the event, MIT decided not to offer Rostow a job, probably because it already faced enough campus unrest over the Vietnam War.[57] It was at this point that Chomsky took it upon himself to demand that his university stick to its own liberal principles by letting Rostow return. Let Chomsky himself explain:

> In fact, as a spokesman for the Rosa Luxemburg collective, I went to see the President of MIT in 1969 to inform him that we intended to protest publicly if there turned out to be any truth to the rumours then circu-

lating that Walt Rostow (who we regarded as a war criminal) was being denied a position at MIT on political grounds.[58]

In short, at a time of mounting anti-war unrest, Chomsky seriously proposed that he could lead MIT's most radical students in a campaign to defend the right of someone he regarded as a 'war criminal' to rejoin the university community.

When Rostow visited the campus for just one day that year, the improbability of this scenario became crystal clear. Although some anti-war students were reluctant to prevent Rostow from speaking, others angrily disrupted his talk. One complained, 'He has a lot of blood on his hands', while Michael Albert argued that 'nobody has the right to listen to him'.[59] George Katsiaficas did support Rostow's right to free speech,[60] but remains to this day uncompromisingly critical of MIT's abuse of this ideal:

Academic freedom when it was originally developed in Europe was to protect the rights of dissident people who disagreed with the church, who disagreed with the government, to speak up without sanctions. It was never intended to protect the rights of war makers to make weapons of mass destruction or to harm other people. It was always intended for professors and the university community to be a place of free speech in order to have free debate ... I think Harvard today and MIT and large universities hide behind the veil of academic freedom to mask the fact that they are prostituting universities to big government and to the military.[61]

Michael Albert's words on MIT are even harsher: 'War blood ran through MIT's veins. It flooded the research facilities and seeped even into the classrooms.'[62] While a student, he described the place as another 'Dachau', explaining years later that 'MIT's victims burned in the fields of Vietnam'. 'I'd certainly have lit a match,' he added, 'if I thought it would have done any good.'[63] Asked about his expulsion, Albert replied: 'Well, for me, it didn't mean much, in the sense that it's a little like being expelled from a cesspool.'[64] Despite their contrasting overall verdicts on MIT, Chomsky never fell out with any of these former students, Albert in particular remaining a close colleague to this day.

How do you collude with the military-industrial complex, work for it – and express moral opposition to its objectives at the very same time? I hope I have shown that Chomsky was not alone in experiencing such contradictory pressures. But the truth is that while many liberal academics at MIT felt under similar pressures, Chomsky was torn more than most – and for that reason reflected the contradictions of his institutional situation more than most. On the one hand, he needed to retain the confidence of MIT's president, keep his job and reassure his colleagues in the Research Laboratory of Electronics that he was a loyal member of their team. On the other, he needed to eliminate the slightest suspicion that his research in that laboratory could possibly be aiding the US military in any practical way.

Many a lesser mind might have given up, concluding that it was impossible to reconcile such flatly opposing demands. Chomsky's genius was to fathom a way. His solution, fitting nicely with a core tenet of the so-called 'cognitive revolution', was to separate knowledge in the head from its social or political use. In his particular case, that meant separating theory from practice in a ruthlessly consistent and far-reaching way.

In his role as a linguist, perhaps the most familiar fact about Chomsky is that he isolates knowledge of language as something separate from its use. What lies quietly in the head differs absolutely and categorically from anything going on in the social or political world. This well-known Chomskyan dichotomy – conventionally the distinction between *competence* and *performance* – has traditionally been viewed as a reflection of the linguist's rationalist philosophy. But my aim here is not merely to give that philosophy a label, but to offer an explanation for it. My conclusion is straightforward. In Chomsky's hands, a careful reformulation of Descartes's celebrated distinction between body and soul would free him, as one interviewer put it, 'to have it both ways'[65] – to get up each morning at ease with his own conscience, while continuing to work in that Pentagon-funded laboratory at MIT.

There is no need to picture Chomsky as double-dealing or politically insincere. I prefer to think that his elimination from linguistics of all things social was a necessary move, given his situation. As a thought-experiment, imagine a social anthropologist recruited to work alongside Chomsky in his MIT laboratory. Chomsky would have suspected that person of collusion with unsavoury pursuits, such as psychological warfare, intelligence gathering and counterinsurgency – rightly, as it turns out.[66] Suspicious of their politics, Chomsky has never shown much sympathy for sociologists or social anthropologists. Only if he could extricate linguistics from its former position as a discipline within anthropology – cutting it off from *any possible* social meaning or content – would he feel on safe ground. Pursuing this logic to the very end, Chomsky eliminated from linguistics not only real human beings, but – just to be safe – all social life and interaction. Since linguistic communication connects a speaker with a listener, not even communication could be allowed.

If language could be reduced to pure mathematical form – devoid of human significance – its study could be pursued dispassionately, as a physicist might study a snowflake or an astronomer some distant star. As Chomsky himself puts it, the various aspects of the world might then all 'be studied in the same way, whether we are considering the motion of the planets, fields of force, structural formulas for complex molecules, or computational properties of the language faculty'.[67] In 1955, when Chomsky first got his job at MIT, he set out with that goal in mind; in subsequent years, he has refused to abandon the project. Linguistics, he asserts, should be as formal and free of political contamination as Galilean astronomy, Newtonian physics or Einstein's theory of relativity.

Chomsky's ideal of linguistics as pure natural science has always seemed inspirational and liberating. But his activist and academic admirers were not the only people to be pleased. Paradoxically, as I will argue here, the project for a wholly non-political linguistics – cut off in particular from Marxist-inspired social science – suited not only its inventor, but, more significantly, the US military-industrial establishment responsible for the funding and development of science and technology in the immediate post-war years. Critics from the left have always complained that hidden agendas can be far more damaging than those which are public and explicit. It is 'our belief', write John Joseph and Talbot Taylor in their book *Ideologies of Language*, 'that any enterprise which claims to be non-ideological and value-neutral, but which in fact remains covertly ideological and value-laden, is the more dangerous for this deceptive subtlety'.[68]

Evidently, Chomsky does not agree: he believes that his linguistic work really is neither ideological nor political at all. When Einstein was working on his theory of relativity, he did not consider himself to be colluding with militaristic ideology or encouraging the production of an atomic bomb. Provided he restricted himself to pure science, he hoped that his conscience in these respects would not burn. Chomsky clearly saw things the same way. If he could separate his science from any possible social or practical application, he could keep his conscience clear – even if it meant bifurcating his own mind into utterly separate spheres.

CHAPTER 5

THE COGNITIVE REVOLUTION

Chomsky's overthrow of behaviourism was just the beginning of an intellectual revolution. This momentous event has today reached into all areas of popular culture, changing profoundly our view of the world without our being aware of its origins or even its existence. I am often surprised at how subtly this paradigm change has worked its effects on us all, given that in my experience, few socialist or anarchist activists have even heard of 'the cognitive revolution'. Nearly everybody knows quite a lot about computers, virtual realities and concepts of digital encoding and transmission. The majority of younger people, wherever they are on the planet, know about hard disks or memory sticks and the possibility of sending data from one source to another over any distance, instantaneously and without damage or loss. In culture and the arts, the idea that information follows its own laws, apparently independently of matter and energy, has been central to science fiction in ways that have reached virtually everyone. 'Beam me up, Scotty', from the 1968 *Star Trek* television series, captures perhaps the most seductive idea – that a person's individuality consists not of matter but of information, all of which might in principle be transported between universes at the speed of light.

Chomsky did not invent computers, design software or write science fiction. More than anyone else, however, it was Chomsky who convinced the computational science community that their ideas about computers related directly to human beings. It was Chomsky, after all, who claimed that the defining feature of humanity, our capacity for language, was itself a digital machine. There was not really much evidence for this (Chomsky says it is so obvious you don't need evidence), but that hardly mattered. The astounding and exciting claim, never before seriously imagined, *just had to be true* – or at least needed to be asserted and generally believed – for the simple reason that, without it, none of the other claims would carry weight. Without Chomsky's digital device – the one alleged to be inside each one of us – those computer nerds preoccupied with their laboratory devices would have been just that, technicians designing inanimate machines. They welcomed Chomsky's revelation because they needed it to elevate their work to a level where it mattered – that of people. This idea subsequently licensed a cascade of logical implications, not all of which Chomsky himself would endorse.

One logical implication is the idea of beaming people up. Behind this is the deeper assumption that we are just software, in which case we can discard our physical body while preserving who we are. Take cognitive scientist Marvin Minsky, brilliant co-founder in 1958 of MIT's Artificial Intelligence laboratory. Described as the 'father of artificial intelligence', Minsky's main interest lay in building computer models capable of replicating the activities of human beings. Among other things, he was the scientist who advised Stanley Kubrick on the capabilities of the HAL computer in his 1968 film *2001: A Space Odyssey*. Like Chomsky, Minsky entertained beliefs that were unshakeable because they stemmed logically from the assumptions taken for granted by everyone at the time. If the mind really is a digital computer, then our bodies no longer matter. Our arms, legs and brain cells are just perishable and imperfect hardware, essentially irrelevant to the weightless and immortal software – the information – which constitutes who we really are.

At a public lecture delivered by Minsky in 1996 on the eve of the Fifth Conference on Artificial Life in Nara, Japan, he argued that only since the advent of computer languages have we been able to properly describe human beings. 'A person is not a head and arms and legs,' he remarked. 'That's trivial. A person is a very large multiprocessor with a million times a million small parts, and these are arranged as a thousand computers.' It is not surprising, then, that Minsky dreams of banishing death by downloading consciousness into a computer:

> The most important thing about each person is the data, and the programs in the data that are in the brain. And some day you will be able to take all that data, and put it on a little disk, and store it for a thousand years, and then turn it on again and you will be alive in the fourth millennium or the fifth millennium.[1]

A cultural critic who has offered an insightful overview of these science-fiction implications is Katherine Hayles. The take-home message of the cognitive revolution, she explains, is that 'information is in some sense more essential, more important, and more fundamental than materiality'.[2] By 1948, the underlying idea had coalesced sufficiently for Norbert Wiener – MIT mathematician and inventor of cybernetics – to hold it up as a criterion that any adequate theory of materiality would have to meet: 'Information is information, not matter or energy. No materialism which does not admit this can survive at the present day.'[3]

Whereas scientists in the past had respected the material world, deploying their abstract models to make sense of its complexities, the new cognitivists treated their models as actually the more fundamental reality, messed up all too often by the material world's unfortunate complexities. In an insightful passage, Hayles notes that the classic move of the scientist had been to infer from the world's noisy multiplicity a simplified abstraction:

> So far so good: that is what theorizing should do. The problem comes when the move circles around to constitute the abstraction as the originary form from which the world's multiplicity derives. Then complexity appears as a 'fuzzing up' of an essential reality rather than as a manifestation of the world's holistic nature.[4]

Whereas the first move goes back to the Greeks, the circling round is more recent. To reach fully developed form, it required the assistance of powerful computers. This move starts from simplified abstractions and, using simulation techniques, generates a multiplicity sufficiently complex that it can be mistaken for the real world.[5] When the world doesn't fit, it's the world that's wrong.

Wiener's concept of information was the one invented by Bell telephone engineer Claude Shannon, who defined it as something with no dimensions, no materiality and no necessary connection with meaning. A message could exist without being sent or encoded in any material or energetic form. Information is information whether transmitted or not.[6] Even in Shannon's day, malcontents were grumbling that to divorce information from meaning or context made the theory so narrowly formalized as to be useless as a theory of communication in general. Shannon himself cautioned that it was meant to apply only to certain technical situations. But in the United States in the aftermath of the Second World War, it was the right idea arriving at exactly the right moment. 'The time was ripe', as Hayles explains, 'for theories that reified information into a free-floating decontextualized, quantitative entity that could serve as the master key unlocking secrets of life and death.'[7] The new concept of information undermined not just Marxism but the entire materialist view of the world. 'The central event of the 20th century is the overthrow of matter' is how one influential document later summarized the paradigm change.[8]

While expressing reservations, Chomsky likens this post-war shift in perspective – effectively the overthrow of scientific materialism – to 'the first cognitive revolution' of the seventeenth and eighteenth centuries, when the achievements of Copernicus, Galileo, Descartes and Newton inaugurated the modern age of science.[9] Exploding onto the scene in the late 1950s, this second wave of revolution brought together a noisy, mostly youthful coalition of radicalized intellectuals whose opposition to behaviourist dogma often amounted to visceral loathing. The pluralism of the new movement in its early years, before Chomsky's paradigm took over, is recalled by the psychologist Jerome Bruner:

> Now let me tell you first what I and my friends thought the revolution was about back there in the late 1950s. It was, we thought, an all-out effort to establish meaning as the central concept in psychology – not stimuli and responses, not overtly observable behavior, not biological drives and their transformation, but meaning ... We were not out to 'reform' behaviorism, but to replace it.

Bruner continues: 'The cognitive revolution as originally conceived virtually required that psychology join forces with anthropology and linguistics, philosophy and history, even with the discipline of law.'[10] So this was no narrow movement sponsored by the US military merely to serve its push for global control: it was a genuinely liberating alliance between the humanities and natural sciences – its goal almost a 'theory of everything'. Its unifying rallying cry was an idea so simple that it seemed extraordinary that it even needed saying: the human mind is by nature active and intricately structured. Contrary to Skinner's 'blank slate' doctrine, likening humans to rats who can be controlled by punishment and reward, there is such a thing as distinctively human – often rebellious – nature.

Yet, despite the pluralism and intrinsic appeal of the new movement, the role of the military was central from the beginning. During the war, Alan Turing's use of early computer technology to crack the ENIGMA code had been wrapped in secrecy. As news of his methods filtered out during the early post-war years, it provoked ripples of interest and excitement. Before long, primarily in the United States, the notion of thinking machines had gripped the imagination of an influential circle of psychologists and philosophers.

Turing himself never argued that the human brain really is a digital computer – only that a man-made computer might one day pass itself off as a human being.[11] But, in 1945, the Hungarian mathematician John von Neumann went further.

In a report on the EDVAC – the Electronic Discrete Variable Computer – von Neumann depicted the machine as built not from vacuum tubes, but instead from idealized neurons. He was convinced that the mathematical rules governing the input and output of signals in EDVAC might help explain neuronal activity in the human brain. Because of the binary character of the nervous system – each neuron at any given moment either firing or not firing an electrical charge – the brain (he believed) really did work on digital principles. Neurons were the vacuum tubes of nature's own digital computer, the human brain.[12]

Although von Neumann was realistic enough to note important differences between vacuum tubes and neurons, this did not stop the philosophers from going wild. 'The proper way to think of the brain is as a digital computer', proclaimed Chomsky's high-school classmate and undergraduate colleague Hilary Putnam in 1960,[13] unleashing upon the world a new and highly influential trend in philosophy – the so-called 'computational theory of mind'.[14] In a seminal paper entitled 'Minds and Machines', Putnam in fact went the whole way – to the point of claiming, on Chomsky's authority, that 'the whole of linguistic theory' now applied not to the linguistic creativity of conscious human beings, but to the output of one of the new digital machines:

> It is important to recognize that machine performances may be wholly *analogous* to language, so much so that the whole of linguistic theory can be applied to them. If the reader wishes to check this, he may go through a work like Chomsky's *Syntactic Structures* carefully, and note that at no place is the assumption employed that the corpus of utterances studied by the linguist was produced by a conscious organism.[15]

In other words, the philosopher most centrally responsible for the '*mind equals digital computer*' idea considered himself to be building on the precedent set by Chomsky in his *Syntactic Structures*.

If Putnam and Chomsky were right, the new cognitive community began to realize, it might explain a mystery that had baffled René Descartes – and indeed philosophy in general since the ancient Greeks. That mystery is the so-called 'mind–body' problem. If mind is immaterial, how can it produce material effects? How can the mind influence the body at all?

Turing's theoretical breakthrough suggested a solution: mind may be weightless and intangible, yet it still needs hardware on which to run. In the human case, of course, the hardware is the physical brain. Whether or not brain stuff is itself intrinsically digital, the fact is that it transmits *information* – 'bits' of which are digital by technical definition. Telecommunications engineers such as Claude Shannon had long noted that when transmission is digital rather than analogue, all messages conveyed are independent of whether the physical medium is a length of string, a copper wire, an optical cable or whatever. It makes no difference what kind of thing is switched 'off' or 'on', so long as you can tell the difference. 'Mind', translated as 'information', is now apparently independent of matter. Writing of the 'computational theory of mind', Steven Pinker evaluates its significance as follows: 'It is one of the great ideas in intellectual history, for it solves one of the puzzles that make up the "mind–body problem": how to connect the ethereal world of meaning and intention, the stuff of our mental lives, with a physical hunk of matter like the brain.'[16]

From the idea that mind is independent of matter, it is only a small step to a further idea, namely that you can study mental states without worrying too much about the material brain. Indeed, from the new theoretical standpoint, the details of mind appear *more* relevant – and certainly easier to understand – than details of its material embodiment. After all, in a well-designed computer, the clean and crisp software determines what the messy hardware will do. Chomsky himself is forever discriminating between those aspects of the human mind/brain which are crisp, clean and perfect – and those which are 'messy' and, consequently, beyond the scope of mathematically based science.[17] Whenever he draws this distinction, it is always the clean – that is, the machine-like or digital – components which he celebrates, treating these as the focus of 'biolinguistic' investigation. Bearing all this in mind, it is easy to see how mentalism – the idea of mind over matter – became a defining feature of the new cognitive science. With the mind–brain distinction now conceptualized as the difference between hardware and software, it was declared that the age-old philosophical mind–body problem had been solved.

That, in brief, is the standard account of the cognitive revolution – the one you will find in the textbooks. It serves its purpose, but tends to hide the critical fact that the entire intellectual upheaval was driven by *industrial* and *military* imperatives bound up with the Cold War.

A good example is the wartime development of cybernetics, devised initially as part of a wider attempt to treat human beings as fully functional components of weapons guidance systems.[18] As historian Paul Edwards explains:

The airplane, the tank, and the submarine were primitive examples of what would eventually be labelled 'cyborgs': biomechanical organisms made up of humans and machinery. Their internal and external linkages took the form of electronic feedback circuits. These included not only interphones and radiophones for communicating with other humans but the dials, controls, and bombsights through which humans communicated with the machines. Should any of these information linkages fail, a cybernetic weapon could be totally disabled.[19]

The overall mantra was C3: Communication, Command and Control. This meant what it said: top-down instruction requiring obedience and no answering back. In the past, neither psychologists nor linguists would have chosen such despotism as an environment in which linguistic creativity was likely to flower. In the aftermath of the Second World War, however, the military needed a version of linguistic science radically detached from previous scholarship – far removed from, say, anthropological studies of Native Americans conversing in their diverse traditional tongues.[20]

The post-war use of computers for military purposes led naturally to the assimilation of humans as conveniently low-cost and available thinking machines. 'Man', in the words of John Stroud, research psychologist at the San Diego Naval Electronics Laboratory, 'is the most generally available general purpose computing device.'[21] A favourite image was that of the human operator sandwiched between a radar-tracking device on one side and an anti-aircraft gun on the other. The human being, Stroud observed, is 'surrounded on both sides by very precisely known mechanisms and the question comes up, "What kind of a machine have we put in the middle?"'[22] 'The human being is the most marvelous set of instruments', Stroud observed, 'but like all portable instrument sets the human observer is noisy and erratic in operation. However, if these are all the instruments you have, you have to work with them until something better comes along.'[23] Good or bad, the human operator is now an input–output device: information comes in from the radar, travels through the man and goes out through the gun.[24] Fully automated aircraft, submarines and other weapons were as yet only a distant prospect; in the meantime, human operators would have to be relied on for crucial functions.

But humans, it was feared, might pose problems of their own. Introducing the crew of the USS *Missouri* to a group of scientists, a Second World War admiral spoke bluntly of his worries: 'Twenty-five hundred officers and men: gentlemen, twenty-five hundred sources of error.'[25] The navy's increasingly complex weapons systems would work only if their human components acted in drastically simplified ways – not as complex emotional beings, but as rapid-fire cyborgs, integrated efficiently with their electronic surroundings. Man needed to be conceptualized as one more of those digital machines, stripped of unwanted emotional complications. Chomsky's admirers and supporters would soon be doing just this, arguing, for example, that it is legitimate to 'view the speaker as being essentially a machine'.[26]

The military's navigators, gunners, pilots, bombers and other human components needed to communicate fluently and easily with the dials, controls and computational devices surrounding them. During the 1950s, the dream of separating out information from noise morphed into the ultimate command-and-control fantasy – the dream of a universal digital code comprehensible alike to man and machine, an idea popularized by science-fiction writers and current to this day. It was assumed that the underlying code would be logical and mathematical. If English and all other natural languages, deep down, were written in that universal code, then a number of problems might be solved. Computers could be programmed to think in that code. Instead of needing special training, human operators might then simply chat in their vernacular to their electronic gadgetry and be understood. It was to an MIT project of precisely this kind – sponsored as usual by the Pentagon – that Noam Chomsky was assigned when he got his first job there in 1955.[27]

A key task faced by US laboratory technicians was to find new ways of making sure that engine sounds, explosions and other noises were not confused with spoken exchanges between military personnel. The Electro-Acoustic Laboratory at Harvard specialized in the physical and electronic elements of this problem; the Psycho-Acoustic Laboratory focused its efforts on 'those problems arising from the fact that a human being is part of the total circuit'.[28] Receiving almost $2 million from the National Defense Research Council – a vast amount by the standards of the time – Harvard's Psycho-Acoustic Laboratory became America's largest university-based programme of psychological research.[29] It got such privileged treatment because communication was so evidently essential for victory in war. Against this background, psychologists directed their efforts towards 'specifying the design parameters of the human organism, in order to insert that organism into electromechanical military systems'.[30] 'Cognitive psychology' was born.

Although Chomsky was to become the most prominent public face of this intellectual revolution, he was flanked by two towering figures with whom he collaborated closely in the early years. One was Jerome Bruner, a polymath whose interests ranged from law through primatology and anthropology to child language acquisition and social play.[31] Like everyone of his generation, Bruner's early experiences were massively influenced by the war. Having received his PhD at Harvard, he spent much time with the US army in France, directing military analysis of Axis radio broadcasts from Italy and designing programmes of psychological warfare.

The other major figure to back Chomsky in the early years was the experimental psychologist George Armitage Miller, who with Bruner co-founded the Center for Cognitive Sciences at Harvard. His specialized field was speech and communication. His most widely cited paper, 'The Magical Number Seven, Plus or Minus Two', cited experimental evidence of an average limit of seven items for human short-term memory capacity. Miller received his doctorate from Harvard's Psycho-Acoustic Laboratory, where, during the Second World War, he had studied voice communications for the Army Signal Corps. His 1946 PhD thesis, 'The Optimal Design of Jamming Signals', was classified as top secret. From psycho-acoustics and the wartime problem of turning meaning into noise, Miller went on to address the problem of integrating humans with machines in the computer age. During the war, he had used rats to perform learning experiments, publishing his findings in the standard behaviourist language of stimulus and response. Yet, soon after the war, he and other scientists involved in command and control began groping their way towards a subtly different paradigm. Soon after meeting Chomsky in 1958, Miller began a period of close collaboration with him, persuaded – at least momentarily – that the mind was some kind of digital computer. Years later, in a famous retraction, he denounced the computer metaphor as disastrously misleading.[32]

In the new age of computers and information technology, mental states could no longer be ignored. Behaviourism had perfectly suited administrators, content to manipulate people through punishment and reward, targeting not their minds, but their emotions and bodies. From the standpoint of these people, it had hardly mattered what their powerless victims thought. But now that communication had become a central concern, mental states *did* matter. Experiments had established, for example, that the effects of subjecting people to loud engine noise all day, for weeks on end, were militarily significant only at the level of *information*:

> *Environmental* noise did not affect the motor skills of soldiers; it did not alter their bodies or behavior. But it did interfere with the transmission of messages . . . Noise was not a physical but an abstract threat: a threat to the mind, not the body – a threat to 'information' itself.[33]

Such findings posed an existential threat to behaviourism, which, as we have seen, denied the very existence of mind. Against this background, the US military had little choice but to nourish a new psychological paradigm – one which would place mind centre stage.

Jamming enemy speech signals was one military priority; accurately transmitting friendly signals was another. To improve transmission, the Psycho-Acoustic Laboratory compiled lists of especially intelligible words for use in battlefield communication. 'These lists,' notes Edwards, 'still employed today, are familiar from war movies in which radiomen shout hoarsely into their microphones such phrases as "Charlie Baker Zebra uh-One, do you read me?"'[34]

This, then, was the heavily militarized intellectual scene encountered by Chomsky on his arrival in Cambridge, Massachusetts, in 1951. Here lay the roots of the impending cognitive revolution.

CHAPTER 6

THE TOWER OF BABEL

In 1951, the 22-year-old Chomsky arrived in Cambridge, Massachusetts, to take up a fellowship at Harvard University. He now found himself at the heart of intellectual life in the world's pre-eminent centre of technological innovation and research, mingling with colleagues destined to become giants of post-war Western philosophy, psychology and cognitive science. Chomsky recalls how cybernetics, sound spectrography, psycholinguistics, experimental psychology and 'the mathematical theory of communication' were all in a state of rapid development and exuberance, lending 'an aura of science and mathematics' to the study of language.[1] He would attract that aura to his own work on language in a way that no one had ever done before.

Electronic computers were just beginning to make their impact. Alan Turing's top-secret work (as yet scarcely known to the public at large) had raised deep philosophical questions that no one had previously asked. Can a machine really be said to think? If so, does it need language? If language is needed, what kind of language? Are artificial computer codes remotely like ordinary natural languages such as English? Everyone knew that, superficially at least, human languages are extraordinarily complex – far more so than the formal languages invented by computer engineers. But, if the mind really is a digital computer, it should operate on similar principles. It was hoped that if linguists got to work on, say, English, its underlying digital code would prove accessible to a suitably designed machine. Most exciting of all was the possibility that this underlying code might turn out to be universal, enabling every language in the world to be handled by a machine of just one design.

From this it was only a small step to a breathtaking idea: a Universal Language Machine that could read a portion of text in any language and automatically translate it into any other. Of course, twenty-first-century readers familiar with the fumbling approximations of 'Google Translate' might wonder why this seemed such an ambitious project; but we must remember that in the 1950s the whole idea seemed Utopian. One reason for the excitement was that if a universal code existed, there would be no need for approximations. If translation could be based on an underlying grammar of pan-human thought, it might potentially prove astonishingly accurate. Enthusiasts claimed that if a machine could be invented with access to the universal code, it would be an extraordinary achievement for humanity, bringing within range previously undreamed-of possibilities for international cooperation, mutual understanding and peace.

The extraordinary project, at first sight utterly improbable, might have been dismissed as far too Utopian but for the corporate influence and connections of its most enthusiastic early champion, the information theorist, scientific bureaucrat and top-level fixer Warren Weaver. In 1955, building on progress so far, Weaver penned the foreword to the first-ever collection of papers on machine translation. Remarkably, for so earthbound a bureaucrat, he chose to uplift his readers by reminding them of the story of the Tower of Babel. Recognizing the ambitious nature of the project, he conceded that God might once again be annoyed.

With obvious reference to the Holy Bible, Weaver entitled his essay 'The New Tower'.[2] Shortly after the Flood, he reminded his readers, King Nimrod of Shinar – Noah's great-grandson – began erecting a tower to reach heaven:

> And the Lord said, Behold, the people is one, and they have all one language; and this they begin to do; and now nothing will be restrained from them, which they have imagined to do. Go to, let us go down, and there confound their language, that they may not understand one another's speech. So the Lord scattered them abroad from thence upon the face of all the earth: and they left off to build the city. Therefore is the name of it called Babel; because the Lord did there confound the language of all the earth: and from thence did the Lord scatter them abroad upon the face of all the earth.[3]

A crucial motif here is the Tower, a vastly ambitious scientific and technological project on a par with what was happening in Cold War America. The language of these city-dwellers is inseparable from their science and technology, prompting them to aspire to reach heaven – to conjoin Earth and sky.

Weaver comments that whatever God's true involvement, the languages of humanity really have become mutually unintelligible, just as the story depicts, causing problems which have accumulated and intensified up to the present day. 'Our own generation', he explains, 'has experienced the second stage of this process, whereby modern methods of travel and communication have made the whole world so small that the confusion of tongues, a second time in man's history, becomes a vital issue.' It was against this background that Weaver sought to justify work so far in erecting the New Tower:

> Students of languages and of the structure of languages, the logicians who design computers, the electronic engineers who build and run them – and especially the rare individuals who share all of these talents and insights – are now engaged in erecting a new Tower of Anti-Babel. This new tower is not intended to reach to Heaven. But it is hoped that it will build part of the way back to that mythical situation of simplicity and power when men could communicate freely together, and when this contributed so notably to their effectiveness.[4]

Weaver's poetic appeals to human solidarity, mutual understanding and brotherhood need to be viewed in context. Some years earlier, he had been influential in establishing the 'Research ANd Development' (RAND) Corporation – a well-funded, massively influential military think tank. Set up in 1946, its mission was to explore long-range planning of future weapons for use in intercontinental warfare. One of its contributions became known, appropriately enough, as MAD – the doctrine of nuclear deterrence by Mutually Assured Destruction. In his 1960 book *On Thermonuclear War*, RAND chief strategist Herman Kahn championed the idea of a 'winnable' nuclear exchange,[5] which led to his becoming a model for the sinister Dr Strangelove in Stanley Kubrick's film of that name.

The idea that all the world's languages might share the same digital code seems originally to have been Weaver's. As early as 1947, he was discussing possible applications of the new digital computers with a British electrical engineer named A.D. Booth. Suspecting that mechanical translation might be feasible, Booth recalls that, from the outset, 'Weaver suggested that all language might contain basic elements which could be detected by means of the techniques developed during World War II for the breaking of enemy codes'.[6] The reference here, of course, is to Alan Turing's celebrated cracking of the code of the ENIGMA machine used by the German High Command to encrypt U-boat and other military communications. Three years later, Weaver returned to this idea in a famous memorandum which he circulated among a select group of scientists. Having surveyed possible solutions to the challenges of machine translation, Weaver outlined what he considered the most promising idea:

> Think, by analogy, of individuals living in a series of tall closed towers, all erected over a common foundation. When they try to communicate with one another, they shout back and forth, each from his own closed tower. It is difficult to make the sound penetrate even the nearest towers, and communication proceeds very poorly indeed. But, when an individual goes down his tower, he finds himself in a great open basement, common to all the towers. Here he establishes easy and useful communication with the persons who have also descended from their towers.
>
> Thus may it be true that the way to translate from Chinese to Arabic, or from Russian to Portuguese, is not to attempt the direct route, shouting from tower to tower. Perhaps the way is to descend, from each language, down to the common base of human communication – the real but as yet undiscovered universal language – and then re-emerge by whatever particular route is convenient.[7]

Weaver did not coin the terms 'deep structure' or 'Universal Grammar', but his allegory of a vast common basement hosting in its depths a previously undiscovered 'universal language' unmistakably anticipated such ideas.

CHAPTER 7

THE PENTAGON'S 'NEW TOWER'

Among the recipients of Weaver's memorandum was the man who first inspired Chomsky to study linguistics, Zellig Harris. Harris was the star of the Linguistics Department at the University of Pennsylvania; his life-long ambition is accurately summarized by Chomsky: 'For as long as I have known Harris, he has been trying to work out formal techniques that, in principle, could be applied by a computer to a corpus of linguistic data.'[1] Harris, then, was hoping to redesign linguistics for its new and exciting mission – rendering sentences accessible to an electronic machine.

In 1954, he wrote to Weaver reminding him of his memorandum and declaring an interest in the 'New Tower of Babel' project. The following year, MIT published that ground-breaking edited volume, *Machine Translation of Languages*, for which Weaver had written his 'New Tower' introduction. Although the authors floated a range of different approaches, the most widely favoured remained Weaver's 'New Tower of Babel' idea – the search for a universal basic code. In the words of one contributor:

> . . . if a *coded general syntax* or common syntax programming were developed, which we could call syntax *X*, with characteristics embodying all the syntactic operations of each of the natural languages dealt with, the problem of syntax programming would be greatly simplified.

Using such a procedure, this author continued, the syntax of each natural language would be 'programmed into the common syntax, or *syntax X*, and also, in reverse, we would program *syntax X* in terms of each natural language system. Translation could then be handled from or into any language so programmed.'[2]

This was exactly what Weaver had proposed. 'If you could reduce a language to a structure,' one former student recalls of Harris's similar idea, 'then you could find a way of translating that structure into another structure.'[3] In 1955, having benefited enormously from Harris's generous help and encouragement, Chomsky was assigned to work on Weaver's project in MIT's Research Laboratory of Electronics, explaining his latest thinking to undergraduate science and engineering students.

For reasons already explored, Chomsky felt deeply ambivalent about the world in which he was now immersed. He distanced himself from the practicalities of machine translation, aware that any advances were likely to be shaped by military priorities which he detested. He was determined not to assist the military in any practical way. He was also right to suspect that realistically, developments in this area would probably be based on trial-and-error statistical methods, reliant on the development of vast memory banks. The rise of so-called 'connectionism' in the past 30 years can be seen as corporate America's eventually lucrative alternative to generative grammar in any of its versions. The big corporations wanted something which worked. But in the early years, when Chomsky first realized that such statistical approaches might prove popular, he dismissed them as piecemeal tinkering, quite unlike his immeasurably more exciting (even if currently impractical) 'Universal Grammar' idea. His betwixt-and-between position – simultaneously inside and outside the machine translation project – set a pattern which would characterize his subsequent career.

Being formally and physically on the inside undoubtedly had its advantages. Chomsky's cramped office in MIT's Building 20 may not have looked impressive, but no rival institutional setting could have conferred more prestige on his work. 'The affiliation', comments Frederick Newmeyer, 'guaranteed that vast sums of money (largely military in origin) would trickle down into the department, enabling the kind of support for a linguistics program that no other university could hope to match.'[4]

The benefits went beyond finance. Pentagon endorsement on such a scale gave the impression that, even if much of Chomsky's work seemed incomprehensible, it must surely be good science. The military are hard-headed folk; they would not be wasting their money on some left-wing activist's nonsense. The most awe-inspiring and incomprehensible of all Chomsky's writings at the time was his full-length work *The Logical Structure of Linguistic Theory*, written during his years as a junior fellow at Harvard. In 1955, the University of Pennsylvania awarded Chomsky his PhD for submitting a thesis consisting of just one chapter, which he had been allowed to excerpt from that fabled work. Rumours of the complete manuscript's vast erudition and significance depended in part on the potent factor of mystery. First, it was written in dense mathematical formulae. Secondly, it was almost impossible to get hold of a copy. Even the copies which did exist seemed to have been changed on each reprinting, so that no one knew quite which version was supposed to be the authentic one. The magic evidently worked:

> As dense as the book is, as restricted as its availability was, and as partially read as it was, even by those few who had access to copies, the very existence of the manuscript stood as guarantor to any gaps of coverage or generality of explication that might be perceived in *Syntactic Structures*.[5]

Those who loathe Chomsky and all his works include, predictably enough, a number of prominent linguists on the political right. Two of these, however – Paul Postal and Geoffrey Sampson – were formerly ardent champions of the 'Chomskyan revolution', which complicates any simplistic attempt to dismiss their intellectual scepticism as political ideology. In recent years, Postal has taken to denouncing Chomsky for his un-American activities, condemning his political output as all of a piece with his 'junk linguistics'.[6] Such diatribes do little to foster rational debate; but not all Postal's criticisms of Chomsky and his 'biolinguistics' followers are so easily dismissed.[7]

Sampson's disillusionment first became evident in a 1979 book review. Scandalously, he described Chomsky's 1955 dissertation – now at long last published for all to see – as:

> a long book full of algebraic notation which may look impressive to the mathematically naïve, but which when carefully examined turns out to be mathematically semi-literate, containing various expressions which are meaningless, or say something other than what the author evidently wants to say, or at best choose a gratuitously obscure way of saying something which a competent mathematician would express straight-forwardly.[8]

Whatever we think of Sampson's verdict, it is difficult to dismiss it on grounds of the author's politics, if only because, just four years earlier, he had introduced his own book – *The Form of Language* – by announcing that it 'will strongly support Chomsky's novel way of looking at language'.[9]

Chomsky's institutional setting helped convey the impression that his bafflingly unfamiliar methods must somehow be on a par with radar and other inventions incubated by scientists in the same shabby building – technological breakthroughs celebrated for having made a substantial contribution to victory in the war. All this enhanced Chomsky's prestige, encouraging students to enrol on the basis that careers were to be made. In his 1972 review article, 'Chomsky's Revolution in Linguistics', the philosopher John Searle observed: 'Chomsky did not convince the established leaders of the field but he did something more important, he convinced their graduate students.'[10] On graduating in the early 1960s, none of Chomsky's students found any difficulty in finding a job.[11]

Although Chomsky's *Syntactic Structures* propelled him to prominence in 1957, many historians attribute greater significance to an earlier date. On 11 September 1956, an almost unknown Noam Chomsky delivered a paper at an MIT symposium on information theory – and triggered the cognitive revolution. Chomsky's key paper – entitled 'Three Models for the Description of Language' – thrilled his peers by treating language as a formal system that is as logical and precise as mathematics. It was not the first time a theorist had mused in public along such lines, but, as George Miller later recalled, Chomsky was perceived as 'the first to make good on this claim'.[12]

Alan Turing's wartime achievement had been to reconstruct the internal architecture of the ENIGMA machine by analysing its output. Although today he claims not to have been interested in computers, Chomsky in those early years advocated precisely this method of reconstructing the internal architecture of the mind:

Think of the computer as a model for the human mind. A scientist who didn't know what the program, or input, of the computer was would assume that the only way to find out would be to analyze the output, language. He would construct an idealization, a model, and develop an explanatory theory in the hope that in time he would be able to deduce the input ... In fact, his way would be the only way to discover the input.[13]

At the time, the eminent psychologist George Miller – like most of Chomsky's colleagues – held essentially the same idea, although in later years he confessed that, on reflection, 'how computers work seems to have no real relevance to how the mind works, any more than a wheel shows how people walk.'[14]

But we must return to the early history of machine translation. In 1951, Chomsky had formed a close intellectual friendship with Yehoshua

Bar-Hillel, an Israeli logician and cybernetics enthusiast who had met up with Chomsky's future teacher Zellig Harris in Palestine four years earlier and become excited about machine translation. Trained by Harris, Chomsky's first forays into generative grammar treated 'transformations' as unconscious, automatic procedures such as a machine might be able to replicate. But then he began moving in a novel direction. When, in 1951, Chomsky completed his MA thesis – 'Morphophonemics of Modern Hebrew' – hardly anyone understood a word of it. While Nelson Goodman, his former philosophy teacher, thought the whole approach 'completely mad' and his thesis supervisor Harris considered it 'crazy', Bar-Hillel was intrigued. He not only read the thesis, but suggested amendments to an early draft – improvements which Chomsky accepted.[15] The puzzling question is what Bar-Hillel found so exciting about Chomsky's 'mad' or 'crazy' thesis.

Part of the explanation must surely lie in the fact that Bar-Hillel at the time was in charge of all machine translation research at MIT. Intellectual historians sympathetic to Chomsky vehemently insist that machine translation was never a project of any interest to him, his encounter with this project being just a coincidence. This narrative is not easy to believe. Bar-Hillel – the man who first persuaded Chomsky to break out on his own – had recently made history by becoming the first scientist in the world to be employed full time on mechanical translation techniques. In the spring of 1952, thanks to a grant from Warren Weaver's Rockefeller Foundation, he organized at MIT the first-ever international symposium on machine translation (MT), proclaiming at regular intervals that 'if a human being can do it, a suitably programmed computer can do it too'.[16] In the following year, Bar-Hillel published a paper concerning 'a quasi-arithmetical notation for syntactic description', suggesting 'a method whereby a machine could analyze the syntactic structure of a sentence'.[17] Other Bar-Hillel publications included 'Machine Translation', 'The Present State of Research on Mechanical Translation' and 'The Treatment of "Idioms" by a Translating Machine'. Although he later lost confidence in the whole project, machine translation in those early years was Bar-Hillel's central preoccupation, and his position at MIT quickly established him as the world's most effective early propagandist for the craze.[18]

At first sight, it might seem difficult to understand quite why Bar-Hillel was so intrigued by Chomsky's thesis. 'Morphophonemics of Modern Hebrew' suggested a logical basis for switching between alternative pronunciations of selected Hebrew phonemes without altering their contribution to word or sentence meaning. That combination of invariance and change is what morphophonemics is all about: you hold constant the combinatorial role of a speech sound while altering how it is produced in the mouth and perceived by the ear. In English, for example, the past-tense ending [d] (as in 'hugged') becomes pronounced [t] (as in 'kissed') when it immediately follows an unvoiced consonant. Whether it is pronounced [d] or [t] makes no difference to the meaning, but it follows a consistent rule. Assuming there is some logical basis to this, then a computing machine might in principle be designed to internalize that logic. In that case, quite mechanically, it ought to be capable of replacing each instance of pronunciation (a) with pronunciation (b) while preserving the original meaning.

As yet, this may seem a far cry from machine translation, but note that the principle is already there. To fully realize the potential, all that is needed is to assume that languages differ from one another only in *how they sound* – their inner logic remaining invariant. If that belief turned out to be valid, translating a sentence from one language into another would be merely a matter of progressing from the initial way of pronouncing it to the appropriate alternative way. If the trick worked, needless to say, it would be the wildest dream of any machine translation engineer.

As it happens, Chomsky really does seem to have believed that a language could be built up by performing transformations on just 'a small, possibly finite kernel of basic sentences'.[19] And he has continued to claim that the apparent differences between languages mostly concern pronunciation – that is, superficial differences in how they sound.[20]

In his early years, Chomsky treated language as characterized by a hierarchy of entities operating on different levels, the system's unconscious mechanisms providing means of mapping or translating between them – from phones, to phonemes, to morphophonemes . . . all the way up to the very highest levels of syntax and semantics. From a logical standpoint, everyone realized that it would be ideal if it turned out that the 'deepest' levels – those furthest removed from mere sound – were genetically fixed and universal. If Chomsky's ideas seemed thrilling at the time, it was because he was by far the most persuasive proponent of this very idea, returning to it again and again.[21] Bar-Hillel was excited, then, for good reason. Although no one could be sure, there was always the possibility that Chomsky's strange approach would enable machine translation to be placed on an entirely new – and vastly more promising – theoretical footing.

Writing long after these events, when the machine translation craze was no more than an embarrassing memory, Chomsky claimed never to have been interested, attributing confusion on this score to others.[22] But if people were confused, it was hardly surprising. Throughout the 1950s, virtually all

Chomsky's scientific colleagues and friends were optimistic about machine translation and confident that Harris-style 'transformations' would help. Chomsky's notion of 'transformational grammar', as one historian observes, was 'rampantly' taken up by the entire cognitive community 'as a program implemented on a computer to generate, understand, translate, and otherwise deploy natural languages. This usage falls directly into line with *grammar* as "a kind of machine" in [Robert] Lees' terms, and as a "device", in Chomsky's regular phrasing.'[23]

The machine translation people – including notably those working with Chomsky in the Research Laboratory of Electronics – certainly understood *grammar* in this way. Chomsky's game-changing *Syntactic Structures* was widely perceived as relevant to machine translation, owing precisely to its terminology, filled as the text was with references to Markov sources, information theory, automata and so forth. Chomsky claims that he was trying to lead his MIT students away from all this and toward real language itself, but the misunderstandings fostered in virtually all his colleagues were understandable enough. Whatever his personal doubts about what they were doing, Chomsky at the time had little choice but to cooperate with these colleagues and live with any consequent ambiguities – otherwise there would have been no job for him in that electronics laboratory at MIT.

In the end, he managed not only to tolerate the ambiguities, but over the years to discover more and more ingenious ways to make a virtue of them. While stating that scientific proposals should normally be couched in precise and unambiguous language, when it came to his own work, ambiguity inexplicably moved centre stage. In the opening pages of his *Aspects of the Theory of Syntax*, to take just one example, Chomsky explained that he would be using the term 'grammar' at all times with 'systematic ambiguity':

> Clearly, a child who has learned a language has developed an internal representation of a system of rules that determine how sentences are to be formed, used and understood. Using the term 'grammar' with a systematic ambiguity (to refer, first, to the native speaker's 'theory of his language' and, second, to the linguist's account of this), we can say that the child has developed and internally represented a generative grammar, in the sense described.[24]

It would be hard to imagine any other linguist with sufficient authority to state in advance that his theory of grammar must *by definition* be true: true because he was now deliberately using the very term 'grammar' to mean both his own *theory* of the topic under discussion and *the topic itself*! But Chomsky had now found a way.

Returning to his involvement with machine translation, Chomsky says that, during his interview for the MIT post, he protested from the outset that the project had 'no intellectual interest and was also pointless'. It was a low-level engineering job: 'It may have some utility; it could be on a par with building a bigger bulldozer, which is a useful thing. It's nice to have big bulldozers if you have to dig holes.'[25]

In the end, if we accept this account, Chomsky agreed to cooperate on the understanding that his contribution would be limited to pure theory – in his case, to the theoretical study of the human language 'device', not a man-made copy or electronic counterpart. If he could assist at all with the laboratory's work, in other words, it would not be by tinkering with existing procedures, but by starting afresh, taking a step back and exploring an exciting idea. On being invited to join the project, then, Chomsky's response was to decline and yet simultaneously agree, the ambiguity of his decision reflecting the curiously betwixt-and-between status of his language 'device' or 'machine'. Was it something really present in the human brain – or was it rather an ideal to be developed at some point by computer engineers in his laboratory? It seemed at the time that either or both interpretations might be made. The quite extraordinary ambiguity and other-worldliness of Chomsky's subsequent intellectual work – detached as it became not only from real human beings or their brains, but also from any known or even humanly constructible machine – accurately reflected that contradiction.

This was the context in which Chomsky published his *Syntactic Structures*. Convinced that the complexities of syntax far outstripped the capabilities of any currently envisaged machine, he demolished the over-simplified theories which had prevailed among computer engineers until that time. But although the short book's most impressive and enduring arguments were negative, their paradoxical effect was to promote extraordinary confidence and optimism within the machine language community. Whatever his personal doubts, Chomsky's colleagues naturally assumed that his work would be useful in clarifying just what lay within the bounds of possibility. In 1958, Bar-Hillel himself expressed confidence that Chomsky's novel approach to the study of grammar 'may well turn out to facilitate the mechanization of translation from new angles'.[26]

It is well known that Chomsky treats language not as social communication, but as internal computation – in effect, as the speaker talking to himself. If we trace his thinking to its earliest stages, the reasons for this become clear. The assumptions of his closest colleagues – Bar-Hillel foremost among them – profoundly shaped and limited the way in which he would define language as a technical term. 'Language' in Chomsky's new sense had nothing to do with two or more people engaged in social communication. Not even a hint of social usage or communication was permitted or implied. Reflecting the limitations of computers in those days, Chomsky's envisaged machine – his 'Language Acquisition Device' – was not connected by wires to similar machines in a wider network. Like a Turing machine – or like ENIGMA itself – it was envisaged as a standalone device. The corresponding philosophical assumption – namely that linguistic computation occurs under conditions of complete social isolation – had long been internalized by all those working on natural language processing at MIT. No member of the team would have envisaged two or more digital computers talking to one another. Machine translation 'was seen as a two-step process: a decoding of the input text into a representation of the "message" (a "transition language") and its encoding into an output text in another language'.[27]

In this new picture, 'speaker' and 'listener' are abolished in favour of 'input' and 'output'. Any notion of communication has completely disappeared.

Throughout most of the 1950s, the quest for the right kind of machine became an extraordinary craze, likened by more than one bemused critic to the search for the Holy Grail. In 1958, a prominent devotee trumpeted that the prize had been won, the solution at last found. 'While a great deal remains to be done,' he announced, 'it can be stated without hesitation that the essential has already been accomplished.' Judging by the sophistication of its design, wrote this author, the new machine looked powerful enough to achieve anything: 'Will the machine translate poetry? To this, there is only one possible reply – why not?'[28]

Whatever Chomsky's personal feelings about all this, the fact remains that his acclaimed 'revolution in linguistics' broke out *within* this hopelessly misguided specialist community and, thanks to the stature of its leader, this revolution 'cast a mystique over the whole field of MT'.[29] As vast sums of CIA, Pentagon and other state funding poured in, intellectuals with ambition competed to be part of the action. Despite his misgivings, Chomsky's ideas had soon come to underpin the theoretical foundations of the machine translation project at MIT.[30] Chomsky and his supporters have always denied that he had any interest in or involvement with machine translation. But to say, as so many of his followers do, that his encounter with the project was purely coincidental is misleading. There was nothing random or accidental about Chomsky's involvement with this project. Any claim that this turning point in Chomsky's career was a product of chance should be treated with scepticism. It has the effect of discouraging us from exploring the underlying economic, political, sociological and cultural factors responsible for Chomsky's quite extraordinary dominance over post-war linguistics. Only by reference to the events of his time can we appreciate how and why Chomsky's peculiarly dehumanized, bizarrely mechanistic conception of language wielded such authority as it began emanating from that Research Laboratory of Electronics at MIT.

CHAPTER 8

MACHINE TRANSLATION: THE GREAT FOLLY

The deeper we delve into this early period, the more clearly we see the connection between machine translation and Transformational Grammar. Chomsky wants us to believe that his involvement with MIT's machine translation project was based on a misunderstanding. His biographer Robert Barsky comments on how ironic it was that Chomsky got assigned to precisely the kind of project he had always denounced.[1] A stronger version of the chance involvement theory comes from Carlos Otero, who, in a short biographical essay, interprets not only this episode, but Chomsky's entire career, as a sequence of improbable accidents:

> It would be hard to deny that the most extraordinary accident in Chomsky's case is that someone with his brain appeared on the surface of the earth precisely in 1928. The gene lottery did not have to come up with such a prize in 1928 rather than, say, 1988 or 2198, if ever . . .
>
> But the accident of his genetic endowment is not the only unlikely accident. His early years constitute an almost unbelievable series of lucky accidents.[2]

Having listed accidents one to five – date of birth, place of birth, earliest school and so forth – Otero turns to the sixth alleged accident. This is that Morris Halle, a student of Roman Jakobson, finds a way of getting Chomsky to work (officially on machine translation) at MIT – gaining entry 'through the back door, so to speak'. Otero continues: 'He was thus able to become a full professor in a very distinguished institution at 32, to receive an endowed chair at 37, and to be appointed Institute Professor (a rank reserved for scholars of special distinction, mostly Nobel Prize winners) at 47.'[3]

According to this account, had it not been for this culminating lucky accident, Chomsky would never have found himself working in a Pentagon-funded electronics laboratory, would never have received such massive institutional funding – and, therefore, might never have risen so high.

But there is another way of looking at this. We have seen how Bar-Hillel was, first, the one person in the world equipped to appreciate Chomsky's 1951 thesis and, secondly, the one person in the world whose full-time job was machine translation. Was there really no meaningful connection between these two facts? When he first got to know Chomsky, Bar-Hillel was the founding organizer in overall charge of MIT's machine translation project, located in the Research Laboratory of Electronics – precisely the place where his young friend would be working four years later. Just an accident? These institutional details, too often overlooked, surely tell us that Chomsky's brush with machine translation was no random coincidence, but was in fact a logical extension of all that had gone before.

In his effort to avoid practical applications which his conscience would not tolerate, Chomsky steered clear of all work on man-made computers or other machines, choosing instead to redesign *linguistic theory* in order to make it fit for purpose in its new technological role. The newly appointed head of machine translation when Chomsky arrived, Victor Yngve, justified Chomsky's presence in the laboratory on precisely these grounds. He pointed out that he and his team as yet lacked adequate grammars of Russian, German and other important languages – adequate from the standpoint of the 'mechanical recognition routines' being developed. 'It is these problems,' reported Yngve, 'that have been occupying N. Chomsky. He has been working on a theory of grammar that gives many and powerful insights into the structure of language.'[4]

In the past, grammar had all too often been treated as something produced and comprehended by human beings. The task was to make it accessible to a machine. Chomsky duly redefined the term 'grammar', conceptualizing it now as a sentence-generating 'device'.[5] Whatever his subjective intentions, his offering of Transformational Grammar was taken by the entire cognitive community as a program to be implemented on a computer, the aim being to mechanically generate, interpret, translate and otherwise deploy English, Russian and other sample sentences, the 'immense complexity' of language having been reduced to 'manageable proportions'.[6]

Software is one thing, hardware another – as different from one another as 'mind' from 'body', 'theory' from 'practice'. In order to keep his conscience clear, Chomsky was adamant that he would keep strictly to theory. Unlike

Alan Turing, who had always been eager to explore how his theoretical machine might one day be transformed into a functioning electronic device, Chomsky would steer an other-worldly course – a policy which inevitably attracted criticism. As one commentator notes:

> Chomsky resembles an inventor who has designed a new 'device', which he equates with both a 'grammar' and a 'theory', but, instead of actually building it, puts only the design on the market with confident or technical reassurances that the device will function and save labour . . . It will 'produce' or 'generate' sentences . . . It will take 'data' (or 'signals') as 'input' and give 'grammar' as 'output' . . . It will help things get 'automatically' done . . .

The inventor is so busy telling the prospective purchaser about the 'power' and 'superiority' of his 'device' that crucial steps in its construction get left out.[7]

This critic seems to think that Chomsky was simply negligent, forgetting to include crucial steps. But there was nothing forgetful about Chomsky. He intentionally declined to help produce hardware, because that would have blurred the boundary between abstract theoretical linguistics and hands-on computer engineering. To have tackled the problematic interface between computational states and hardware development would have meant helping to produce gadgets of direct military use. True to his principles, Chomsky had no intention of assisting the military in that way.

Chomsky often claims to have been misunderstood. A prime example, he says, is the way in which virtually all commentators have described him as a linguist obsessed with computers. No, he says, he has never been interested in computers: 'I have been surprised since to read repeated and confident accounts of how work in generative grammar developed out of an interest in computers, machine translation, and related matters. At least as far as my own work in concerned, this is quite false.'[8]

Statements of this kind seem hardly credible – until we recall where the lines drawn by Chomsky were always drawn. Chomsky is right when he says he was not interested in computers. He genuinely wasn't. The distinction becomes clear when we recall how he and his MIT colleagues saw the *human mind* as itself a digital computer. Note how this was the mirror-image reverse of Alan Turing's most celebrated notion. Turing contemplated getting machines to 'learn' clever tricks by using behaviourist principles, similar to those used to teach human children. MIT's psychologists turned this idea on its head when they tried to teach human children using the very methods that had worked when 'teaching' electronic computers.[9]

Chomsky was interested only in the natural (in Descartes's terms, the 'God-given') computer responsible for language. He was not interested in the man-made, inevitably low-quality electronic counterparts which so fascinated his colleagues in MIT's Research Laboratory of Electronics. By the late 1950s, the computer metaphor of mind had become so deeply internalized that Chomsky genuinely believed that a human child comes into the world with a digital computer already in its head. So when he protests – despite all appearances – that he has never been interested in computers, he sees no contradiction. Chomsky is telling us, quite truthfully, that his focus of interest has always been that mysterious digital computer which nature has installed inside the skull.

It is easy to see how, among its other advantages, his insistence on this distinction may have helped him square his professional work with his conscience. For admirable reasons, Chomsky refused to build or even help work on any material object which the US military might use. Needless to say, a hard-line socialist or anarchist might still have objected that this was not enough. Would not Chomsky have better served the cause by refusing to work under Pentagon sponsorship at all?

Chomsky has, on various occasions, replied that any such course would have meant turning his back on science. Everywhere in the industrialized West, he points out, scientific research is dependent on large-scale corporate – including military – funding. Yes, he might have chosen to prioritize abstract purity of principle. Other dissidents might have decided to set up intellectually on their own, even at the risk of receiving no paid employment at all. But Chomsky saw no point in taking that risk. Happy in his MIT office – constructed quite deliberately as something of an ivory tower – he decided that the best he could do was accept military sponsorship, while declining to roll up his sleeves. As he would later clarify, designing and building a working model of the human language faculty would seem to require a 'divine architect' anyway.[10]

In the event, CIA and Pentagon enthusiasm for machine translation turned out to be a passing phase. Ironically, it was Bar-Hillel more than anyone who convinced the authorities that their euphoria had been embarrassingly misplaced. In 1960, he pointed out that for a machine to genuinely understand even a simple sentence – 'The box is in the pen', for example – it would require virtually infinite knowledge about human life and experience in the world. How can a box – which is a large object – be contained *inside* a small one, such as a pen? The machine would have to know all about play-pens and their size relative to writing pens, in order to grasp what such a sentence could possibly mean.[11]

In 1961, a best-selling popular book entitled *Computers and Common Sense* concluded that, instead of trying to build a machine capable of understanding sentences, the nation's monetary resources would be better devoted to research 'looking toward the Second Coming'.[12] The following year, Bar-Hillel was in Venice for a NATO Summer School on machine translation. While emphasizing the usefulness and relevance of Chomsky's new approach, he confirmed that no computer could ever be equipped with full common-sense knowledge and understanding of the world, adding: 'So long as we are unable to wire or program computers so that their initial state will be similar to that of a newborn infant, physically or at least functionally, let's forget about teaching computers to construct grammars.'[13]

In other words, to elucidate the correct wiring, you would have to be God. Bar-Hillel's devastating sudden scepticism prompted a profound collective mood swing, culminating in a groundswell of bitter disillusionment that included suspicions of corruption – and even fraud. People were asking how, exactly, all those millions of dollars had been spent? In 1963, the CIA withdrew its funding from a highly publicized experimental project at Georgetown University.

In 1964, Victor Yngve – now in charge of MIT's machine translation programme – resolved to press ahead regardless of the faint-hearts and doubters:

> We have come face to face with the realization that we will only have adequate mechanical translations when the machine can 'understand' what it is translating and this will be a very difficult task indeed . . . Many of the former workers in mechanical translation are giving up in the face of the tremendous difficulties . . . But some of us are pressing forward undaunted.[14]

By now, however, Chomsky had absolutely no appetite for 'pressing forward undaunted'. In principle, he realized, one might equip a machine with sufficient memory to learn certain superficial features of, say, English, through trial and error. But it would be unreasonable to expect it to discover the underlying grammar of a natural language in that way. He therefore suggested a more realistic compromise – install from the outset the full range of *possible* grammars, leaving the machine merely to evaluate which of these worked best.[15] As Chomsky explained: 'There would be no difficulty, in principle, in designing an automaton which incorporates the principles of universal grammar and puts them to use to determine which of the possible languages is the one to which it is exposed.'[16]

If no machine could possibly 'learn' grammar, what did this tell us about the learning capacities of a human child? Other scholars – those not dependent on Pentagon funding – might have concluded that, since the resources of the child must far surpass those of any machine, it might be best to forget about machines entirely and focus instead on developmental processes among real children. Unable to escape from his chosen institutional framework, however, Chomsky did the opposite. He would restrict the scope of linguistics to such tasks as a *sufficiently sophisticated* machine might in principle be able to manage. As if the child's language organ really was a machine, he concluded that, when presented with raw data, it could not be expected to *discover* the grammar of its ambient language. Instead, the full range of theoretically possible grammars would need to be installed inside that child's brain in advance. Chomsky's core notion of Universal Grammar was invented to explain how such an idea might work.

Chomsky's envisaged machine – his language organ – recalls 'Donovan's Brain' in the classic science-fiction book of that name. Cut surgically from its deceased owner's still warm body, the glistening hunk of flesh is restricted to disembodied computation, attached to wires and tubes and kept alive at blood temperature in a special glass vat.[17] While life in an ivory tower, cut off from the world, might seem lonely and dispiriting enough, the 'brain in a vat' metaphor takes things a horrific step further.

Given the assumption of a machine modelled on an amputated brain, the otherwise curious features of Chomskyan theoretical linguistics begin to make sense:

- *Naturalism.* The theoretical framework acknowledges no socially constructed persons, no communities, no traditions – only a biological object whose properties need to be elucidated.
- *Internalism.* The framework acknowledges no outside world, no environment or context in which speaking takes place. As in pure mathematics, there is no reference relation.
- *Individualism.* Apart from minimal input during language acquisition, there is no relationship between speaker and hearer. Everything of interest is internal to the mind of the individual considered in isolation.

These strange doctrines were unlikely to have prevailed, had not Warren Weaver's machine translation fantasies captivated Chomsky's mentors, colleagues and sponsors during the formative years of his scientific career. The extravagant craze was to prove short lived. In 1966, the National Science Foundation published a report on all machine translation projects being conducted across the United States. It recommended that all state support for such projects be terminated forthwith.[18] As machine translation became condemned by almost everyone as an embarrassing failure, researchers raced to distance themselves from the stigma of their former association with it.[19] But by that stage, the episode had left its mark. It went so deep as to fix the initial trajectory of the cognitive revolution, progressively divorcing the scientific Chomsky from his own political activism, while on other levels wrenching competence from performance, mind from body and theory from practice. In these critical respects, these events determined the course of post-war US linguistics, psychology and philosophy.

CHAPTER 9

A UNIVERSAL ALPHABET OF SOUNDS

Chomsky needed to construct a firewall between his science and his politics, keeping each compartment of his life autonomous with respect to the other. Being free to say what he liked meant ridding his linguistics of any evident social content or meaning, and by the same token purging his politics of any obvious connection with his science. While this neat arrangement may have seemed appealing to him even in the 1950s, it is easy to see why it became an absolute imperative once the wider world had become aware of his political dissidence.

The need to keep his politics at arm's length from his science fitted with, and in every way reinforced, Chomsky's most extraordinary assumption – the initial one from which all else stemmed. This was his uncompromising insistence that language is a *natural organ*, hence something to be studied purely within the framework of *natural* science. He was certainly not the first linguist to insist on the need to be 'scientific': in 1950s America, that was a constant mantra. But why just 'natural' science? Why not allow some space, at least, for social and cultural dimensions? Chomsky was unusual in insisting that for linguistics to be scientific, socio-cultural factors had to be excluded completely.

From one point of view, language is a system consisting in its entirety of abstract units of information. An intriguing question was this: can a vast range of abstractions be attributed to a biological mechanism? During any earlier period in the history of biological science, the idea might have seemed far-fetched. But, in 1953, an article was published in the journal *Nature* which revolutionized biology almost beyond recognition, for the first time placing digital combinatoriality centre stage. 'A Structure for Deoxyribose Nucleic Acid', by James Watson and Francis Crick,[1] explained for the first time the molecular structure of the genetic code. The article's dense text was accompanied by a sketchy pen-drawing of the double-helix structure, an image which quickly took the world by storm.

Suddenly, genetics had become exciting, quasi-mathematical, aesthetically beautiful and utterly revolutionary. When, three years later, Chomsky presented his legendary 'Three Models for the Description of Language' paper to an audience of artificial intelligence engineers, the Watson–Crick precedent was still fresh in everyone's mind.[2] Intellectual historian Randy Allen Harris indeed argues that Chomsky carefully phrased his concluding words – epoch-making in their own way – so as to resonate with and recall that inspiring precedent. '[W]e picture a language', declared Chomsky,

> as having a small, possibly finite kernel of basic sentences . . ., along with a set of transformations which can be applied to kernel sentences . . . to produce new and more complicated sentences from elementary components. We have seen certain indications that this approach may enable us to reduce the immense complexity of actual languages to manageable proportions and, in addition, that it may provide considerable insight into the actual use and understanding of language.[3]

While 'certain indications' may not sound like incendiary talk, Harris detects here an echo of the concluding line in Watson and Crick's revolutionary DNA paper – 'It has not escaped our notice that the specific pairing we have postulated [i.e. the double-helix structure of DNA] immediately suggests a possible copying mechanism for the genetic material.' Harris comments:

> Chomsky does not reach quite the level of smug faux-humility that Watson and Crick achieve in this most famous of litotes in the history of science. His paper is nowhere near as earth-shattering as theirs, his understatement not so elaborate. But his tone, like theirs, is of conspicuously muted triumph: the 'certain indications' are for reducing 'immense complexity' into 'manageable proportions', thereby yielding 'considerable insight' into a notoriously difficult problem, 'the actual use and understanding of language'.[4]

The impact of Watson–Crick on Chomsky's later thinking is undeniable. The fact is illustrated by a frequently cited article published in 2002, in which Chomsky and colleagues state that 'the human faculty of language appears to be organized like the genetic code – hierarchical, generative, recursive, and virtually limitless with respect to its scope of expression'.[5]

But a molecule is not an organ. So we need to explain how Chomsky arrived at the idea of studying language in much the same way that a biologist might study the digestive system or the heart. By way of explanation, Chomsky himself describes how, during the early 1950s, he became aware of the vibrant new discipline of ethology (the science of animal behaviour under natural conditions), a field whose most prominent exponent was the right-wing naturalist Konrad Lorenz.

Captured by the Russians and kept as a prisoner of war during the 1940s, Lorenz's unusually respectful prison guards had permitted him extraordinary freedom to continue his researches – and indeed to write and circulate an extensive manuscript on the subject of human nature. Lorenz had always ferociously opposed behaviourism, pouring scorn on the idea that nestling birds, for example, are condemned to laboriously try out all possible limb movements before figuring out that their wings are useful for flying, their legs best for running on the ground: 'Just a few minutes' observation of a chick or duckling freshly removed from the incubator is sufficient to demonstrate conclusively the utter untenability of this hypothesis. The behaviourist school nevertheless doggedly clings to it.'[6]

Lorenz argued that genetically determined patterns of behaviour in humans and other animals – walking, flying, nest-building and so forth – are just part of their nature. No one needs to *learn* how to grow legs or arms – they just develop of their own accord. Environmental factors may help trigger developments, but, apart from that, the environment is essentially irrelevant. The requisite motor patterns, being genetically fixed, exist quite independently of trial-and-error learning: 'They exist, so to speak, as "emancipated" sequences of impulses generated by the central nervous system that are independent of all *external* stimuli and contingencies.'[7] Excited by these ideas, Chomsky pondered whether the same might apply to the human child as it acquires its first language.

Chomsky and his circle of friends were wondering how to detach linguistics from its former association with social and cultural anthropology. He recalls: 'We were starting to read ethology, Lorenz, Tinbergen, comparative psychology; that stuff was just becoming known in the United States . . . We were interested, and it looked like this was where linguistics ought to go.'[8]

Lorenz – celebrated as 'the father of ethology' – had long been arguing that animal behaviour patterns should be studied as if they were organs.[9] Together with his colleague Niko Tinbergen,[10] he developed the idea of an 'innate releasing mechanism' to explain how each such organ might work. Chomsky was excited: could language be treated in the same way – as a biological organ? If so, linguistics could be reconstructed as a discipline much like ethology – a purely natural, purely biological science.

Chomsky was among the first to view the human mind – or at least the component of it responsible for language – as a digital computer. Recalling that it was Hilary Putnam who in 1960 announced that the 'proper way to think of the brain is as a digital computer',[11] we can see how difficult it must have been to claim that this computer was somehow made of flesh and blood, like any other organ.

The digital aspect of a computer is its operating system and software. This consists entirely of abstract units of information. Abstract units of information are weightless, take up no space, and cannot be chemically analysed. As Paul Postal points out, it is quite different with an organ such as the heart, which a surgeon might cut from the body and weigh.[12] A biological organ, unlike a sentence, can be investigated by physically handling it or placing it under a microscope. So it really is philosophically incoherent to conflate the two types of entity. For this reason, describing the digital language computational mechanism as a biological organ – a component of the body – was an idea which no one had previously imagined. No anatomist or neurophysiologist had ever claimed to discern a *real* digital computer hidden somewhere in the human brain.

Yet one recent discovery did suggest to Chomsky the possibility of conceiving language as an organ of the body. Roman Jakobson – then teaching and researching in phonology and structural linguistics at MIT – had recently become widely known for an extraordinary finding. A key component of the human body indeed seemed to function like a digital computer. Still better, that computer was the secret of speech.

Known to his friends as 'The Man Who Could Do Everything',[13] Jakobson was a linguist in the old tradition of scholarship,[14] interested in phonology and morphology, metric patterns in Indo-European languages, Slavic epic tales, the biblical Tower of Babel myth, Old Church Slavonic – and, above all, modern Russian and other Slavic experimental poetry. Unfamiliar with the technicalities of computing, Jakobson hated the whole idea of a 'scientific' linguistics carved out as a narrow specialism: 'I don't think one can have a theory of language without studying other sign systems – painting, sculpture, cinema, theatre, music, pantomime. All must be taken into account.'[15]

Despite these cultural interests, however, Jakobson had always considered linguistics to be no less scientific than any other branch of science:

I decided when I was at school that I would become a linguist. My father was a chemist, and he asked me, 'Roman, why have you decided to become a linguist?' And I said, 'Linguistics is no different from chemistry. I want to find the finite elementary units in the structure of linguistic elements.'[16]

Becoming acquainted with computers only during the First World War,[17] Jakobson had independent reasons for viewing the human speech apparatus as a system of digital switches.[18] He realized that, in purely acoustic terms, nothing prevents a consonant from falling somewhere between, say, 'voiced' and 'unvoiced', the term 'voicing' here indicating 'vocal cord vibration'. The vocal cords, in principle, might vibrate – but only very slightly. While this was true in acoustic terms, however, Jakobson's point was that, for the combinatorial system to work, the output had to be psychologically *perceived* as categorically one or the other, as if the voicing feature could only be fully 'off' or else fully 'on'. In English, for example, the whole system relies on the fact that no vocal output can be recognized as occupying a position intermediate between, say, 'pat', which starts with an unvoiced consonant, and 'bat', which starts with a voiced one.

Jakobson's soberly worked-out version of his celebrated theory concerned essentially the *sound patterns* of speech. But the idea intrigued Chomsky for larger reasons. If the human organism contains one digital component – the one responsible for sounds – might not additional components exist? Might not the language organ in its entirety be constructed along similar lines, its digital principles extending to a 'syntactic component', a 'semantic component' and so forth?[19]

Chomsky recalls that in 1951, when he first arrived in Cambridge to take up his Harvard fellowship, Jakobson had established through his lectures at MIT an intellectual milieu 'that had very powerful attraction and that I found myself quickly drawn into, as were many others of that academic generation'. Above all, Chomsky admired Jakobson's 'seriousness of purpose': 'For Roman, linguistics was a science that sought to discover something fundamental, something real and invariant, in the real world – something analogous, let's say, to the laws of physics.'

This thrilled Chomsky, whose teacher Zellig Harris had set out from quite different assumptions. 'In my own training', recalls Chomsky,

linguistics was a system of ingenious analytic techniques that could be used to yield a systematic organization of data, which in principle could be done in many different ways. For others, language was an adventitious habit structure mirroring the environment, and so on. As far as Roman was concerned, all of this was deeply wrong from the start.[20]

Chomsky welcomed Jakobson's discovery of 'a fixed inventory of atomic elements, and universal laws governing their combinations – a view that was often rejected at the time as "absolutist".[21] He comments:

The significance of structuralist phonology, as developed by Troubetzkoy, Jakobson, and others, lies not in the formal properties of phonemic systems but in the fact that a fairly small number of features that can be specified in absolute, language-independent terms appear to provide the basis for the organization of all phonological systems. The achievement of structuralist phonology was to show that the phonological rules of a great variety of languages apply to classes of elements that can be simply characterized in terms of these features; that historical change affects such classes in a uniform way; and that the organization of features plays a basic role in the use and acquisition of language. This was a discovery of the greatest importance, and it provides the groundwork for much of contemporary linguistics.[22]

Chomsky, then, readily acknowledges his intellectual debt to Jakobson. And as Frederick Newmeyer confirms, 'Roman Jakobson probably exerted a greater influence on transformational grammar than any other linguist.'[23]

To put this in context, Jakobson was also the linguist who gave Claude Lévi-Strauss the idea of Structural Anthropology.[24] Jakobson first met Lévi-Strauss in 1942, shortly after arriving as a refugee in New York's École Libre. Lévi-Strauss remembers his first meeting with Jakobson:

It was enormously important. At the time I was a kind of naïve structuralist, a structuralist without knowing it. Jakobson revealed to me a body of doctrine that had already been formed within a discipline, linguistics, with which I was unacquainted. For me it was a revelation.[25]

For Lévi-Strauss, linguistics was unquestionably the master discipline – the only part of social science which (thanks largely to Jakobson himself) had proved itself truly scientific. In his 1956 essay 'Structure and Dialectics', Lévi-Strauss observed that 'the method I am employing is simply an extension to another field of structural linguistics, which is associated with the name of Jakobson'.[26] Assuming the mind to be some kind of digital computer, he began looking for evidence of its internal architecture in Native American kinship systems, totemic beliefs, magico-religious myths and much else.

A few years later, Roman Jakobson, Claude Lévi-Strauss and Noam Chomsky met together in Harvard, Chomsky sitting in on lectures given by

Jakobson and sometimes also by Lévi-Strauss.[27] Inspired by Jakobson, Chomsky was hoping to discover the system of Universal Grammar underlying the world's languages. Meanwhile – equally inspired by the same revered Russian genius – Lévi-Strauss was hoping to discover some kind of universal grammar behind the world's superficially variable kinship systems and myths.

In 1958, just a year after Chomsky published his *Syntactic Structures*, the founder of structural anthropology claimed to have discovered the formula. It seems, Lévi-Strauss declared,

> that every myth (considered as the aggregate of all its variants) corresponds to a formula of the following type:
>
> $F_x(a):F_y(b) \simeq F_x(b):F_{a-1}(y)$

Explaining that *a* and *b* are 'functions' while *x* and *y* are 'terms', he announced that:

> a relation of equivalence exists between two situations defined respectively by an inversion of *terms* and *relations*, under two conditions: (1) that one term be replaced by its opposite (in the above formula, *a* and a-1); (2) that an inversion be made between the *function value* and the *term value* of two elements (above, *y* and *a*).[28]

Years later, when writing the second volume of his *Science of Mythology*, Lévi-Strauss reminded his readers that he had realized all this back in 1958, invoking the same formula 'as proof of the fact that I have never ceased to be guided by it since that time'.[29]

For Lévi-Strauss, evidently his mathematical formula was anthropology's answer to Einstein's $e = mc^2$. Some years later – once the euphoria around structural anthropology had died down – Dan Sperber mocked all this as laughable pseudo-science, making Lévi-Strauss look 'not like a scientist but rather like a transcendental meditator claiming to be guided by his mantra'.[30] Similar comments were made in the 1960s in response to the pages of extraordinarily complex quasi-mathematical axioms, definitions and formulae invented by Chomsky in his mimeographed treatise, 'The Logical Structure of Linguistic Theory'. The linguist Charles Hockett responded from the outset that Chomsky's claims to mathematical precision were 'as worthless as horoscopes'.[31] The 'transformationalists', Hockett complained, had 'retreated into mysticism, into a sort of medieval scholasticism'.[32] But such was Jakobson's prestige that anyone who invoked his celebrated 'distinctive features' theory for their own purposes, no matter how divergent from his own, was likely to be heard. This would prove to be the case even when Jakobson himself recoiled in horror from the consequences. 'The more Chomsky's epigones talk about deep structure,' he later complained, 'the more remote and incomprehensible it gets.'[33]

Picking up the threads of our narrative, Jakobson's uncompromising hostility to the Nazis meant that, during the final years of the war, he was happy to work in close technical support of the US war effort. In his case, this meant assisting laboratory technicians to develop systems of speech recognition and transmission for military use. Volunteers, including conscientious objectors, were monitored with sensitive recording instruments as they carefully pronounced vowels and consonants. It soon became clear that, when produced against a noisy background, the consonants mattered most. Jakobson recalls: 'It is noteworthy that the authorities of acoustic laboratories in the United States were ready to disclose the images of vowels in the "visible speech experiment", whereas those of consonants were concealed until the end of World War II in order to hinder the deciphering of secret messages.'[34]

Partly because so much of it was top secret, much of this work remained unknown to the public until the publication in 1951 of Jakobson's 'Preliminaries to Speech Analysis', co-authored with Gunnar Fant and Jakobson's student, Morris Halle.

Conducting their research in MIT's Acoustics Laboratory, the authors had set out in search of phonology's Holy Grail. They dreamed of discovering a fixed repertoire of irreducible building blocks – in effect, a universal alphabet of sounds – behind the phonological systems of the entire world. In 1951, the authors triumphantly announced that they had discovered nothing less:

The inherent distinctive features which we detect in the languages of the world and which underlie their entire lexical and morphological stock amount to twelve binary oppositions: 1) vocalic/non-vocalic, 2) consonantal/non-consonantal, 3) interrupted/continuant, 4) checked/unchecked, 5) strident/mellow, 6) voiced/unvoiced, 7) compact/diffuse, 8) grave/acute, 9) flat/plain, 10) sharp/plain, 11) tense/lax, 12) nasal/oral.[35]

These twelve features were depicted as the hidden digital code underlying the languages of the world. As yet, the results were restricted to the level of sound. But who knew where all this might lead? Given time, might not syntax and semantics be incorporated as well?

Jakobson, Chomsky and Halle all agreed that vowels and consonants, although superficially important, are not the ultimate building blocks from which speech sounds are made. Instead, as Jakobson had been the first to teach, the human organism is equipped from birth with something resembling a digital device – an arrangement of biologically fixed 'on/off' switches. These are the various speech articulators – lips, tongue, soft palate and so forth – each of which can be set in one or another binary-digital state. Take, for example, the English consonant <d>. Switch 'voicing' from 'on' to 'off' and the consonant becomes <t>. Restore voicing to the 'on' position, and once again we have <d>.

Previously, linguists had viewed speech sounds as locally distinctive, culturally learned and unpredictably variable from one language to the next. If that were so, any search for an underlying digital code would be doomed from the start. Jakobson's breakthrough was momentous because it suggested the existence of a universal code, adding plausibility to Warren Weaver's dreams about a 'New Tower'. 'All of this', recalls Chomsky, 'presented an exciting vista, which challenged the imagination in a way that made other approaches to the study of language seem pallid in comparison.'[36] He recalls:

> I arrived at Harvard in 1951, after a couple of years of graduate work in linguistics, feeling quite confident that I knew my way about the field. One of the first things I did, naturally, was to go to see Roman Jakobson, who was of course a legendary figure. Our first meeting was rather curious – we disagreed about everything imaginable, and became very good friends.[37]

In these words, Chomsky clarifies what must have happened. At the time, Chomsky still shared many of the basic assumptions of Zellig Harris, for whom – as we have seen – linguistics involved preparing texts in the light of non-linguistic criteria, the aim being to isolate patterns detectable to a machine. That had been the whole point of Harris's 'transformations' – they were tricks for 'normalizing' real sentences, standardizing imaginatively created verbal expressions to yield fixed stereotypes capable of being handled by a machine. As Chomsky later observed, Harris 'thought of linguistics as a set of procedures for organizing texts, and was strongly opposed to the idea that there might be anything real to discover'. Jakobson, from that perspective, could not have been more different. For him, language was *already* logical and mathematical, *before* being fed into any machine. In 1951, Chomsky was apparently new to that inspiring idea and not quite ready to believe it. Soon after those early meetings with Jakobson, however, it became – and to this day has remained – Chomsky's central article of faith.

A few years later, in 1955, Jakobson – by this stage a visiting lecturer at MIT – arranged with his student Morris Halle to secure for Chomsky his first job. Since 1953, as we have seen, Chomsky had been collaborating closely with Halle. Both were equally excited and inspired by Jakobson's theory of 'distinctive features'. But, whereas Halle, over the years, would restrict his focus to sound patterns, Chomsky's aims were much more ambitious from the very start. If Jakobson's 'distinctive features' are universal, then – in Chomsky's own words – 'perhaps much more is universal. If language were biologically based, it would have to be . . .'[38]

These words explain everything. Having discovered Jakobson's work, Chomsky glimpsed the possibility of extending the approach from the study of sounds to the rest of linguistics, including ultimately syntax and semantics. It should also be emphasized that Jakobson himself had no objection in principle to extending 'distinctive features' theory to semantics. Having discussed how sounds may be perceived as varying in their colour – a prominent Russian futurist theme – he commented:

> Even more instructive have been the results of experiments by linguists on sound symbolism, that is, the evident and direct association between the oppositions of speech sounds and the fundamental semantic oppositions of basic meaning, such as high/low, light/dark, sharp/blunt, joyous/sad, etc. It has turned out that phonic oppositions . . . possess their own direct yet latent significance, which exerts considerable influence on the organization of the lexicon.[39]

But Jakobson was here thinking of the mythology of language, especially as it inspired Russian futurist and other poetic imagery. No language machine could be expected to cope with poetry, and Chomsky was not interested. He would place Jakobson's 'fundamental semantic oppositions' elsewhere in his envisaged machine, far removed from sound, sealed up in a separate compartment somewhere deep inside syntax.

There were clearly difficulties here. In *Syntactic Structures*, Chomsky had insisted that 'grammar is autonomous and independent of meaning'.[40] He had insisted on radical autonomy precisely to justify the policy of setting 'meaning' to one side. Any back-tracking on this – any return to the topic of 'meaning' as a focus of scientific interest – risked embroiling Chomsky in the murky complexities of actual language *use*:

> There is no aspect of linguistic study more subject to confusion and more in need of clear and careful formulation than that which deals

with the points of connection between syntax and semantics. The real question that should be asked is: 'How are the syntactic devices available in a given language put to work in the actual use of this language?'[41]

Chomsky had good reason to avoid venturing too far down that road. Language's social uses might fascinate some scholars, but no mechanical device can possibly have social uses in mind. Chomsky saw no way of addressing this topic within the framework of natural science.

But he soon found a neat solution. Meanings, he decided, are not a matter of usage after all. They are hard-wired already somewhere *inside* the computational device. If the linguist were to probe to a sufficiently 'deep' level, semantics would be discovered hiding somewhere within language's 'syntactic component'. As he explained in 1962: 'In general, as syntactic description becomes deeper, what appear to be semantic questions fall increasingly within its scope.'[42]

Chomsky's most celebrated idea – 'deep structure' – took some years to develop, but can be discerned already in these words. If the plan succeeded, Chomsky could claim that 'meanings' were, after all, independent of social purposes, arising instead out of something natural and universal – language's genetically determined *syntactic* structure. As we have seen, Jakobson himself had long suspected a 'direct association between the oppositions of speech sounds and the fundamental . . . oppositions of basic meaning', so in a sense the idea was not new. But Chomsky had to eliminate all trace of Jakobsonian sound symbolism in favour of something he could convincingly present as a separate 'component' somehow 'installed' inside his 'machine' or 'device'.

With these requirements in mind, and to a significant extent inspired by two of his best students – Paul Postal and Jerrold Katz – Chomsky envisaged a universal semantics made up of 'fixed properties (semantic features) of some sort, for example, animate-inanimate, relational-absolute, agent-instrument, etc.'[43] He imagined each lexical item as a bundle of such features, some phonological, some semantic, some grammatical. Jakobson had shown that the phonological features are drawn from a universal alphabet. Presumably, surmised Chomsky, the semantic ones must equally be drawn from such an alphabet, this time made up of atomic units of meaning. While he conceded that 'little is known' about the semantic alphabet, Chomsky expected it to consist of irreducible digital choices – human/non-human, abstract/concrete, animate/inanimate and so on – each switched this way or that to produce whatever concept was required.

Seeking an authoritative classical precedent for all this, Chomsky invoked John Wilkins, bishop of Chester and a founder member – and first secretary –

of the Royal Society. Wilkins' 1668 'Essay Towards a Real Character and a Philosophical Language' attempted, in Chomsky's words, 'to develop a universal phonetic alphabet and a universal catalogue of concepts in terms of which, respectively, the signals and semantic interpretations for any language can be represented'.[44] The bishop reasoned that any new idea must really have been constructed by combining elements from a finite catalogue – an atomic table of all possible thoughts or components of thought. Chomsky agreed: 'Although the defects in execution in such pioneering studies as that of Wilkins are obvious, the general approach is sound'.[45]

By the mid-1970s, this entire framework had begun to collapse, triggering alarm and despondency among Chomsky's followers.[46] When it became clear to everyone that there was no real analogy between semantics and phonetics – no universal system of semantic features corresponding to Jakobson's well documented system of phonetic ones – Chomsky disowned the whole idea of deep structure, attributing the confusion to Paul Postal and Jerrold Katz, the two students who had originally persuaded him to adopt the idea as his own.[47]

It was a damaging U-turn, casting doubt on Chomsky's long-standing promise[48] to explain semantics internally, by delving sufficiently deeply into syntax. He now conceded that it was impossible to ignore external matters – 'beliefs and knowledge about the world' – when figuring out what a sentence means.[49] The consequences were serious. Beliefs are held by flesh-and-blood humans: machines don't hold beliefs. But if meanings depend on beliefs, on what basis could Chomsky continue to claim linguistics as a branch of natural science? To readmit human agency into the picture would be to jettison this entire approach. Since such a move was unthinkable, deep structure would have to go. In some ways, it was back to square one – to Jakobson's theory in its original form, shorn of Chomsky's attempt to apply it where it did not belong. The project to extend 'distinctive features' theory directly to syntax and semantics – vital to Chomsky's *Aspects of the Theory of Syntax* theoretical edifice – had sadly come to naught.

Years later, two of Chomsky's closest colleagues – Sylvain Bromberger and Morris Halle – made the damaging admission that the whole attempt to extend the impressive achievements of Jakobson-inspired phonology to the study of syntax and semantics had been a serious mistake. Not sparing him, they concluded in 1989 that Chomsky's central project during his early years at MIT had been seriously misconceived, remarking that it took 'two decades of intensive research' for him to realize his mistake.[50]

CHAPTER 10

RUSSIAN FORMALIST ROOTS

We have seen how, in his early years, Chomsky stood on Roman Jakobson's shoulders. But Jakobson in turn had been standing on the shoulders of earlier intellectual giants, many of them Russian, and it seems worth investigating why this earlier ancestry has so often been ignored. It seems likely that, in America during the 1950s, there were political reasons for the apparent amnesia – reasons bound up intimately with the anti-communist witch-hunts of the McCarthy years and the subsequent pressures of the Cold War.

When he arrived in America as a Russian-speaking Jew, everyone knew that Jakobson had narrowly escaped capture by the Nazis, who considered him a Bolshevik agent. He himself made no secret of the fact that he had begun to form his linguistic theories as a teenage avant-garde poet, his enthusiasm for futurism part of the intellectual, artistic and political ferment culminating in Russia's October Revolution. But in early post-war America, communist associations of this kind could only get him into trouble. Indeed, at one point he almost lost his job on those grounds.[1] There were good reasons why it made sense for Jakobson and his supporters, the young Noam Chomsky included, to keep quiet about his revolutionary past.

Although Jakobson was no Marxist, it is certainly true that during his early years as a futurist poet he was a Bolshevik sympathizer. Following the October revolution, he stayed on in Moscow and briefly joined the Bolsheviks, working for the Division of Fine Arts of the People's Commissariat of Enlightenment. In 1919, under the pseudonym Alyagrov, he delivered a typically fiery lecture entitled 'The Tasks of Artistic Propaganda'. In bold futurist spirit, he concluded with a call to intensify revolutionary confrontation with the 'old aesthetics':

At the moment, when the revolution in art is sharper, perhaps, than it has ever been, when the old aesthetics, which pervaded all aspects of life, is rotting away in every nook and cranny, interfering with a new, live aesthetics, it is not the task of an organization promoting artistic enlightenment to pour a conciliatory balm on the issue, but to reveal the growing conflicts – not to neutralize, but to sharpen the struggle of artistic trends, for in that struggle lies the life and development of art.[2]

Jakobson's uncompromising words were published in the journal *Iskusstvo* shortly before the 'Day of Soviet Propaganda' commemorating the second anniversary of the October insurrection. Vladimir Mayakovsky, Kazimir Malevich, Alexander Rodchenko and other revolutionary poets and artists were among the speakers at a meeting entitled 'The New Art and Soviet Power'. Unfortunately for Jakobson, their noisy endorsement of his call to 'sharpen the struggle of artistic trends' in fact led the authorities to close down the journal which had just published his speech; the issue of *Iskusstvo* dated 5 September 1919 was the last to appear.[3]

The revolutionary climate was beginning to cool and it seems to have been in response to these disappointing events that, in the spring of 1920, Jakobson suddenly decided he needed to leave Moscow and find work abroad. Aged 24, the brilliant linguist landed a job with one of the first groups of Soviet diplomats to be sent abroad, living first in Estonia and then moving to Prague. Competition for the Estonian post had not been too stiff, he was later told by the man who recruited him, because everyone else was afraid, convinced that the White army would blow them up the minute they crossed the border. 'But I wasn't afraid', recalls Jakobson.[4] When he took up residence in Prague, it was to be employed by the Bolsheviks as a translator, working in the young Soviet government's embassy there. Over the years, surviving largely on black coffee and the occasional bread roll, he established himself as a charismatic leader of the Linguistic Circle of Prague, whose members met regularly in Jakobson's favourite hang-out, the 'very comfortable and charming' Café Derby.[5]

Twentieth-century intellectual history owed much to those linguists as they deliberated late into the evenings in that crowded café. Frederick Newmeyer explains: 'it should be clear to anyone with the slightest familiarity with generative phonology how great a debt is owed to the Prague School phonologists. Roman Jakobson probably exerted a greater influence on transformational grammar than any other linguist.'[6]

In this light, it does seem odd that Chomsky should have so little to say about the distinctively Russian (and in many ways futurist and revolutionary) source of so many of his own most radical ideas.

In the years leading up to the 1917 revolutions, scientists, musicians, painters and poets were in a state of extraordinary ferment. In turn-of-the-century Geneva, Ferdinand de Saussure had spent the last years of his life developing the paradigm soon to be celebrated as 'structuralism', an approach based on the idea that a language system, like any other semiotic system, is a socially agreed arrangement of formal differences. Nowhere did his work have a deeper influence in the decade following its publication than in Russia, where it quickly became accepted as a manifestation of the 'formalism' then in vogue. But even before Saussure's ideas became known, modernist poets and theorists in Russia had already begun developing a subtly different notion of structure or form as the focus of interest in the study of language and the arts. Being revolutionaries, they hated the very idea of habit, convention or contract. Inspired by developments in physics – and especially by rumours that a young mathematician named Einstein was shattering all former conceptions of space, mass, energy and time – they seized on the idea that linguistic forms and meanings must be rooted somehow in nature's mathematical laws.

Although theirs was a pan-European cultural upheaval, it was in Russia that events came to a head. One manifestation, as everyone knows, was an uncompromising strand of political radicalism. Following the traumatic outbreak of the Great War, Lenin and other leading figures in Russia's Social Democratic and Labour Party strove to fan the flames of resistance to the horrors of war – their implacable hostility towards bellicose nationalism differentiating these dissidents from their social democratic counterparts elsewhere across Europe. Lenin's contribution to the great ferment that had seized Russian art, culture, politics and science in the immediate pre-war years soon became known to the world as 'Bolshevism' – anti-militarist defiance up to and including outright mutiny.

Running alongside Bolshevism, the most iconoclastic poets and painters formed an anarchic group and began calling themselves 'futurists' or 'futurians' – youthful rebels whose impatience with officialdom and with the bloodbath in the trenches led them to denounce the past in its entirety and to fight for a different world. Colluding with established reality, they insisted, was no part of the artist's job. Not all the futurists had opposed their country's war effort from the outset, but by 1916 the tide among them had turned. By 1917, they were voicing an overwhelming chorus of hostility directed against the war, against the regime that was held responsible for its horrors and against the capitalistic economic and social order that lay behind it. Everything had to be smashed and turned upside down.

The question these radicals asked was a fundamental one. How does creative art stop *mirroring* reality and instead transform it utterly? The answer, they decided, was to reverse the age-old priority given to artistic content, rather than form: instead form must be accorded priority. The artist's own paint must be allowed to speak. The bristles of the brush must make their mark. What matters now is not fidelity to some external 'meaning' or social use. The artist's materials and techniques must rise up in rebellion against all that. All settled, fixed or contractual understandings are suspect. Convention amounts to collusion. Meanings and reasons must go. Novel forms – animal cries, irrepressible laughter, splodges of brilliant colour, splintered fragments of metal or glass – must somehow create their own reasons, their own meanings. So-called 'logical' or 'rational' language is the warped, stifling instrument of state bureaucrats, landlords, executioners. It must be answered by a deeper logic of creativity, mathematics and science – of the world as it might be, if only we followed our dreams.

In these years, artists, poets, musicians and scientists were seized with such hopes, letting their imaginations run wild. Let form take priority over content! You dream, you play, you fantasize – and you fight to realize those dreams. Revolution transports you from the realm of necessity to that of freedom. When it came, these poets and artists instinctively embraced the 1917 February Revolution and subsequently its October insurrectionary climax, becoming in many ways that revolution's principal artistic expression.

From then on, Russia's 'formalists', as they came to be known, set out to establish a new branch of science – the technical study of 'spirit', by which they meant human artistic creativity. The futurists wanted the former servant of the artist to become master. That is, they wanted the various techniques used in a painting to escape their former instrumental status, becoming privileged in their own right as prominent subjects of the painting. Transforming the world meant empowering art's formal techniques to the point where they could be celebrated afresh as self-sufficient mechanisms of change. The 'device' would be 'laid bare' and fully 'realized', turning the tables on narrative meaning or content and seizing power in its place.

Kazimir Malevich's 1915 painting *Black Square* has today come to symbolize the thrust of this movement, stretching the idea of art to the point of self-annihilation. Obliterating all notion of 'content' – all notion of referential 'meaning' or 'usefulness' – this is Russian 'formalism' (Malevich coined the term 'suprematism') in its most daring and original form. If revolution is imminent, why not seize the moment? Why not defy the law – and, while you are about it, why not defy nature's laws as well? If Einstein is right and time is an illusion, why accept the inevitability of old age and death? Why not smash through all boundaries in time, swinging in giant leaps backwards and forwards at will?[7]

Velimir Khlebnikov was the Russian poetic visionary who soared highest with such dreams. By the time war broke out, he had already established himself as the ecstatic 'dervish' of Russian avant-garde poetry. Shy and inarticulate, he was popularly viewed as a wandering mystic or holy fool.[8] His friends eagerly appointed him to play the role of towering genius of their literary and artistic movement – 'futurism', in its distinctively Russian form. In 1912, Khlebnikov had coined the term *budetlyane* – 'futurians' or 'men of the future' – to distinguish Russian futurism from the Italian movement of the same name. His friend Mayakovsky attacked the Italian so-called 'futurists' as 'men of fist and fight'. According to Mayakovsky, Russian futurism had nothing whatsoever to do with its Western namesake.[9]

In January 1914, the Italians' right-wing leader, Filippo Marinetti, arrived in Moscow to begin a lecture tour. Khlebnikov tried to disrupt one of the first lectures, but was physically prevented – an incident which almost led to a fight. In anger, he told his friends that he was resigning from futurism. But 'Russian Futurism without Khlebnikov was like Bolshevism without Lenin',[10] and he was persuaded to change his mind. Unlike the war-mongering, machine-worshipping Marinetti, Khlebnikov dreamed of going *back* to the future, returning human language full-circle – via radio and electricity – to its preliterate vocal-auditory past.

Deep down, just as oil paintings are composed of visible colour, so sentences are arrangements of audible sound. Khlebnikov had the idea of a universal language of elemental sounds inseparable from equally elemental 'atoms' of meaning. Khlebnikov's notion of *intrinsic* meaningfulness was an attack on the Saussurean idea that speech sounds are only arbitrarily connected with their meanings. Arbitrary convention or habit was the enemy – it had to go. Language would be placed on deeper, more scientific and more natural foundations. Khlebnikov was not the first to discover and celebrate sound symbolism, but he was the first to trumpet it as a planet-wide legislative programme. Defying all convention, Khlebnikov resolved to reconstruct the universe around him, much as Einstein was reconstructing space and time.

In 1915, Khlebnikov announced that the initial consonant of a word root expresses a definite idea. To illustrate, he compiled a list of Russian words denoting various kinds of dwelling. All began with the letter *x* (*kh*), this sound indicating, he claimed, an outer surface protecting its contents from some external disturbance. Moving on to *l*, he wrote a poem using only words beginning with that consonant, which, he claimed, *naturally* signified a vertical movement, which then spread across a surface. Excited by what linguists nowadays term 'sound symbolism' or 'phonetic symbolism'[11] – a genuine phenomenon explored in depth by Jakobson, among others, and a basic resource for poets[12] – Khlebnikov concluded that each vowel or consonant has its own intrinsic meaning. His task was to restore those pristine meanings, in order to construct a futuristic universal language. Khlebnikov called this future tongue of all humanity *zaum* – 'transreason' or 'beyonsense'.

Jakobson recalls that in 1919 he was preparing with Khlebnikov a collection of the latter's work, unfortunately never published.[13] Jakobson's draft introduction, 'Approaches to Khlebnikov', traced two of the poet's characteristic innovations, both concerning technical devices such as metonymy and metaphor. Khlebnikov rejoiced in (i) 'laying bare the device' – emancipating it so that it was no longer a means to an end, but an autonomous actor in its own right; and (ii) 'realizing the device', so that a metaphor, for example, would jump out of the picture frame, unexpectedly gate-crashing the scene ostensibly being pictured. Instead of allowing metaphors to become conventionalized and eventually die, Khlebnikov would take a fossilized metaphor and encourage it to come back to life.

Jakobson then touched on the emotions experienced by Khlebnikov as he did such things. Khlebnikov had told Jakobson:

'When I was writing the beyonsense words of the dying Akhenaton in "Ka" – "Manch, Manch" – they almost hurt to look at; I couldn't read them. I kept seeing lightning bolts between them and myself. But now they don't move me at all. And I don't know why that is.'[14]

The poet recalled: 'Whenever I saw old lines of writing suddenly grow dim, and their hidden content become the present day, then I understood. The future is creation's homeland, and from it blows the word god's wind.'[15]

Years later, Jakobson noted that poets 'have made this confession from time immemorial'.[16] He quoted Shelley: 'Poets are the … mirrors of the gigantic shadows which futurity casts upon the present … the trumpets which sing to battle … Poets are the unacknowledged legislators of the world.'[17]

Khlebnikov, like Shelley, viewed the poet's job as legislation. The serious point is this: he refused to accept language's enslavement to the world in its current state. He insisted that, once a word has acquired some habitual, conventional meaning, it has become completely lifeless. Resurrecting words – freeing them to sparkle and speak to us afresh – meant scraping them clean of all such encrustations, 'laying bare' their intrinsic meanings and forms.

CHAPTER 11

INCANTATION BY LAUGHTER

Aside from the intrinsic value of his poems and other writings, Khlebnikov's importance is that he helped shape that particular strand of structural linguistics – associated with Roman Jakobson's Prague School – which (decades later and on a different continent) helped trigger the Chomskyan revolution.

Chomsky has always generously acknowledged Jakobson, singling him out as the one twentieth-century figure whose formalist approach to language provided a solid basis for his own. On other occasions, as we have seen, Chomsky has delved deeper into history, citing philosophers and theologians inspired by the idea of a perfect or universal language, Bishop Wilkins prominent among them.[1] But Chomsky has never shown the slightest interest in Khlebnikov.

In fairness, it would be far-fetched to claim any close relationship between Chomsky's rigorously formal theories and the playful dreamings of the Russian futurist poet. In Khlebnikov's world, the connection between sound and meaning is immediate and intimate, requiring no complex computations to bridge the gap. In sound symbolism, as in poetry, speech sounds are intelligible and meaningful in their own right. But for Zellig Harris, Yehoshua Bar-Hillel and Chomsky's other close associates of the 1950s – scientists working directly or indirectly on machine translation – no idea could possibly have been more useless. No machine could work out the difference between *John is easy to please* and *John is eager to please* by attending simply to the sounds. In the case of *Flying planes can be dangerous* – a sentence interpretable in two quite different ways – the task would be impossible, because the alternatives sound exactly the same. If sound symbolism worked, translation would not even be necessary. After all, anyone could comprehend a sentence in any language simply by listening properly, attending carefully to its sounds. In that case, translation in general – and MIT's project in particular – would be a pointless waste of time.

But, of course, Khlebnikov's notion of a universal language had nothing to do with computational processes in the first place. He was thinking in terms of sounds which – like musical notes – are heard as immediately and intrinsically meaningful, without need of complex computation. To recall all this is to be reminded of the difference between Chomsky's institutional and political situation when compared with that of Khlebnikov.

We are dealing with two acknowledged geniuses. In both situations – that of Khlebnikov in pre-revolutionary Russia and that of Chomsky in post-war America – a whole generation was crying out for a leader to inspire them in a period of turmoil and doubt. In both cases, the necessary genius was selected, assigned a special role and celebrated for answering a need of his times.

But, despite this commonality, one huge difference stands out. Anti-tsarist revolution on the one hand, Pentagon-funded counterrevolution on the other: the two situations were poles apart. Yet, once Chomsky had discovered Jakobson's 'distinctive features' theory and incorporated it into his own approach, the distance between these poles began to shrink. In contrast to American structural linguistics – a conservative paradigm if ever there was one – Jakobsonian structuralism emanated from the boldest and most optimistic social and political revolution the world has ever seen. Chomsky's own 'cognitive revolution' acquired its necessary aura of radicalism thanks partly to the fact that it was – or could be perceived as – a distant cultural aftershock of that earlier event, the excitement recalling 'the storming of the Winter Palace'.[2]

The following themes unite the thinking of Chomsky, Jakobson and Khlebnikov:

- Linguistics is *natural* science, on a par with physics, chemistry and astronomy.
- Social conventions are external accretions; the task is to uncover language's *internal* laws.
- Differences among languages are interesting only insofar as they enable us to glimpse the natural underlying basis of all the world's tongues.

Anticipating Chomsky, both Jakobson and Khlebnikov saw mathematical equations as perfect structures and viewed language, too, as following mathematical laws. Whether or not they were scientifically valid, Khlebnikov's ideas had about them at least the seductive aura of science. This rubbed off onto Jakobson, as he worked on his 'distinctive features' theory, and eventually from Jakobson onto Chomsky, too. Situated as he was in MIT's Research Laboratory of Electronics – at the heart of the US scientific research establishment – Chomsky's ability to benefit from something of the same revolutionary aura would prove a decisive advantage.

Jakobson was always adamant that the poet Khlebnikov was the source of his theory of 'distinctive features'. When Khlebnikov wrote a poem, he was developing in practice what would later become Jakobson's most celebrated theoretical idea.

The best-known early illustration was Khlebnikov's 'Incantation by Laughter', published in 1910. In it, the poet takes the Russian word for laughter – *smekh* – and from it elaborates a sequence of derivations invented in conformity with Russian morphological constraints. By adding prefixes, suffixes and diminutive endings, and by forcing the root *smekh-* to function in previously untried ways, Khlebnikov produced a poem, every word of which derives from this root. The result is untranslatable, even by Khlebnikov's impossible standards; but this attempt by translator Paul Schmidt conveys the spirit:

Hlahla! Uthlofan, lauflings!
Hlahla! Uthlofan, lauflings!
Who lawghen with lafe, who hlachen lewchly,
Hlahla! Uthlofan, hlouly!
Hlahla! Loufenish lauflings lafe, hlohan utlaufly!
Lawfen, lawfen,
Hloh, hlouh, hlou! luifekin, lufiekin,
Hlofeningum, hlofeningum.
Hlahla! Uthlofan, lauflings!
Hlahla! Uthlofan, lauflings![3]

Russian speakers find Khlebnikov's poem difficult to recite without bursting into laughter. Likening himself to an impish forest spirit, Khlebnikov succeeds in delving beneath the conventional forms of his native language in search of its innermost structure. In the beginning was the Word – actually a burst of irrepressible laughter – indicating the speaker's freedom from all the cares of the world.

Khlebnikov was forever searching for that isolable component of a speech sound which might switch the entire universe from its present state to an utterly different one. Little was to be gained, Khlebnikov argued, by documenting the accidents of this or that conventionally settled language. The scientist operates on a deeper level, exploring how one or other fundamental feature of language may transport the mind from a particular audible note to a visual sensation, or even to a characteristic smell. As he explained in 1904: 'There exist features that, by shifting continuously from one to another, bridging the gaps of which we humans have no premonition, would transform the blue color of a cornflower into the sound of cuckooing, or into a child's cry.'[4]

He sometimes found this difficult to explain:

> You still have not understood that my word
> Is a god howling in a cage.[5]

The cage was convention – those cruel forces of boredom which normally constrain language's infinite potential. The project to unlock the cage helps explain his particular fascination with the Russian word *sdvig*. This has no English counterpart, but means, roughly, 'sudden change of state', 'dislocation' or 'shift'. Khlebnikov was writing before electronic computers had been invented, but his fascination with abrupt changes of state did curiously anticipate the digital age. For Khlebnikov, a slight change in a sound could produce a sudden shift in meaning and so alter the structure of the universe. There is all the difference in the world between laughter and slaughter. His English translator Paul Schmidt explains:

> That *sword* becomes *word* when a consonant vanished gave him a vertiginous sense of the power of language to influence the natural world. The shift of a consonant was all that distinguished *inventors* from *investors* or *explorers* from *exploiters* – and suddenly there appears the image of a struggle between *N* and *S*, between *R* and *T*.[6]

Sdvig is a kind of death, the sudden abrogation of all that has gone before. To Khlebnikov, one meaning was revolution:

> Perhaps only on the threshold of death, in a single instant when everything would launch into flight for life, rush in panic, jump over all barriers . . . perhaps only in that instant will our mind overcome, with a horrendous speed, all crevices and ravines, smash all settled configurations and borderlines. Yet it is also possible that this is what routinely happens in everybody's mind, every time when a certain perception A shifts, with a horrible speed, into perception B.[7]

For Khlebnikov, the outbreak of the First World War was a gigantic shift of this kind, as was its sudden antithesis – the Bolshevik-led insurrection aimed at abrogating that war.

Khlebnikov perceived language – Russian first of all, but ultimately spoken language in general – as a vast field of sounds and meanings, linked to one another in a complex web. In principle, any unit of meaning might be magically transformed into any other, no matter how remote, by proceeding along an uninterrupted chain of minimal shifts between states. Just one shift – one minimal sound change – could switch in an instant from one meaning to the next. Khlebnikov offers a Russian illustration:

Ja videl
Vydel
Vësen
V osen',
Znaja
Znoi
Sinej
Soni.
['I saw the marking of springs into the fall, having learned about the glows of blue drowsiness.']⁸

In the words of one critic, this and other early poems of Khlebnikov 'sound as if they had been deliberately composed for the purpose of illustrating the basics of structural phonology'.⁹ Each couplet represents a perfect phonological opposition, illustrating what, in Jakobson's hands, would become known as a 'distinctive feature' – a minimal sound switch sufficient to turn one vowel or consonant *internally* into another. Khlebnikov was delighted when he found a pair of words differing by just one feature, as in the poem above, since such instances revealed how speech sounds and combinations can be handled at the deepest structural level.

Khlebnikov's central claim was to have discovered a lost alphabet of sounds – a universal language rooted not merely in habit or convention but in laws of nature yet to be fully understood:

We state that:
M [*m*] contains within it the disintegration of a whole into parts (a large entity into smaller ones).
Л [*l*] – the uncontrolled movement of a great force of freedom (time past).

K [*k*] – the conversion of a force of movement into a force of enduring stasis (from rush to rest).

T [*t*] – the subordination of movement to a greater force, a goal.

C [*s*] – the assembly of parts into a whole (a return).

H [*n*] – the conversion of something ponderable into nothing.

Б [*b*] – growth into something greater, the greatest point of the force of motion.

П [*p*] – lightweight bodies, the filling of an empty space by a supposedly empty body.

P [*r*] – unruly movement, insubordinate to the whole.

B [*v*] – the penetration of a large body by a smaller one.

Ж [*zh*] – increase caused by excess of force (*obzhog* [burned], *zhech* [to burn up] . . .)

Г [*g*] – too little as a result of a situation of insufficient force, hunger.

I – unites

A – against

O – increases size

E – decline, decay

U – submissiveness.[10]

The fundamental claim here is that every speech sound is *intrinsically* significant, retaining its meaning consistently across all the languages of the world.

This was the thinking that captivated Jakobson during his teenage years, when he idolized Khlebnikov, noting down his insights and seeking to integrate them into his own ideas about the nature of language and thought. Central to Khlebnikov's appeal was his view of mathematics as a source of magical power, prompting him to play with equations and formulae like a precocious child, or like a wizard with his spells. Following on from the Impressionists' well-known experiments with colour and light, and from the cubists' emphasis on geometrical forms, Khlebnikov wanted to turn science in all its forms into something tangible and accessible as a source of popular creativity. For Khlebnikov and the futurists, as for the cubists, science was *aesthetically* thrilling, its concepts and methods the explicit focus of their art. 'An Internationale of human beings', Khlebnikov jotted down in his notebook in 1921, 'is conceivable through an Internationale of scientific ideas.'[11]

Khlebnikov saw language not as habit or convention, but as itself part of nature. His aim was to delve beneath superficial differences in search of universal – quasi-mathematical – underlying laws. He dreamed of this in part for political reasons: if people are not talking – if they cannot establish communication – they are liable to start fighting one another instead. A universal language shorn of cultural accretions – one based on eternal mathematical forms – must bring humanity together. Here Khlebnikov describes the task he set himself:

> To find – without breaking the circle of roots – the magic touchstone of all Slavic words, the magic that transforms one into another, and so freely to fuse all Slavic words together: this was my first approach to language. This self-sufficient language stands outside historical fact and everyday utility. I observed that the roots of words are only phantoms behind which stand the strings of the alphabet, and so my second approach to language was to find the unity of the world's languages in general, built from units of the alphabet.[12]

It was Khlebnikov's 'strings of the alphabet' which would resurface in phonology as Jakobson's universal alphabet of 'distinctive features'.

From everything he wrote, it is clear that Khlebnikov's search for a universal alphabet was equally a search for certainty and security. In turbulent times, Khlebnikov yearned for constancy and reassurance, seeking this in astronomy and therefore in the night sky, whose faint patterns he associated with the intrinsic meanings of consonants and vowels, blotted out all too often by the sun:

> One can think of the word as concealing within itself both the reason of the starlit night and the reason of the sunlit day. This is because any single everyday meaning (*bytovoye znacheniye*) of a word also obscures from view all the word's remaining meanings, just as the daytime brings with it the disappearance of all the shining bodies of the starlit night.[13]

It was no accident that *Victory over the Sun* was the title of the futurists' first opera, premiered in Saint Petersburg in December 1913. In the story, those fighting for the future manage to stab the sun and finally capture it. The performance provoked hissing, booing and uproar, much to the authors' delight. Audience members jumped up, shook their fists and shouted: 'You're an ass yourself!' Someone hurled an apple, but generally the performance passed off 'in an atmosphere of good-natured laughter and mirth'.[14] Stage designer Kazimir Malevich depicted the sun's disc obscured by a rectangular black object – a premonition of his celebrated 1915 painting *Black Square*.[15] Khlebnikov wrote the prologue in exaggeratedly Russian-sounding nonsense-language,[16] its incomprehensibility underscoring the chasm between future and past. Here, as always, the idea was to get away from habit. 'For us,' wrote Khlebnikov in 1914, 'all freedoms have combined to form one fundamental freedom: freedom from the dead, i.e. from all these gentlemen who have lived before us.'[17] Combining primitivist longings with futuristic faith in mathematics, all Khlebnikov's writings in the years before 1917 gave eloquent voice to 'the cosmic mood of approaching revolution.'[18]

As early as 1904, before anyone had yet heard of Einstein, Khlebnikov was already dreaming of 'linking time and space'. Years later, shortly before his death, he returned to the theme, announcing that he had successfully established 'a quantitative link between the principles of time and space.'[19] In fact, Khlebnikov's ultimate aim was to transcend all dimensions of space by subordinating them to the deeper principles of time. He disliked 'space' because it meant fences and boundaries – cruel instruments of the territorial state. His universal language, he announced, would abolish all these, restoring the planet to its former unity as a sphere whose inhabitants would respect only borders in time – borders between such countries as night and day, dark moon and full, winter and summer. These frontiers would be maintained peacefully in accordance with certain mathematical formulae connecting Moon, Earth and Sun. Instead of relying on writing – the invention of despots and bureaucrats – language would return to its roots in laughter and song. Among the futuristic inventions promising this restoration of the voice, radio would occupy pride of place.[20]

Khlebnikov saw the written word as a bureaucratic instrument steeped in blood, its dominance testifying to the relatively recent political dominance of space (manifested as the territorial state) over time (humanity's tribal past and electronic, borderless future). Time, in his view, had become a 'Cinderella', a 'serf', a 'kitchen boy' rendered subservient to the divisive and bloody politics of space.[21] Determined to correct this cosmic imbalance, Khlebnikov glimpsed a parallel universe of laws governing the future 'state of time' – 'a vision of time in stone'.[22] In this futuristic universe, colours could be heard, sounds viewed in full colour. Widely separated historical events would now be closely aligned, visible to the eye at a glance. For the poet Osip Mandelstam, Khlebnikov was 'a kind of idiotic Einstein', unable to distinguish which was closer to him – a nearby railway bridge or *The Lay of Igor's Tale*, a saga describing events that occurred in the late twelfth century.[23]

During his student days, rocked by the upheavals of 1905, Khlebnikov began feverishly compiling dates of recent and ancient historical transitions, battles and other cataclysmic events. He was looking for the wavelength of history. Maybe it was 317? The idea seemed promising. In 1912, having presented a long list of dates, accompanied by careful calculations, he asked: 'Should we not therefore expect some state to fall in 1917?'[24] When, five years later, the tsar was duly overthrown on the predicted date, his friends triumphantly proclaimed him the 'King of Time'.

Having sensationally predicted the exact year of the revolution, Khlebnikov joyfully celebrated the event, viewing it as a cataclysm of cosmic significance. For him, it was the beginning of a worldwide insurrectionary rebirth destined to restore humanity's long-lost language of sounds. Wildly optimistic, Khlebnikov went into orbit. Without waiting for events, he announced during the spring of 1917: 'We alone have rolled up *your* three years of war into a single spiral, a terrifying trumpet, and now we sing and shout, we sing and shout, drunk with the audacity of this truth: The Government of Planet Earth already exists. We are It.'[25]

Khlebnikov anticipated supportive uprisings across the entire planet. The 'states of space' were about to be replaced by his electronic alternative – a state of global 'unrule' – operating within boundaries only of time.

In his 'October on the Neva', Khlebnikov recalls being caught up in the revolution, first in Petrograd, then in Moscow:

Nevsky Prospect was full of people, constantly crowded, and there was no shooting there whatsoever. . . . .

The situation in Moscow was entirely different; there the fighting was serious; we were holed up for a week . . .

The pitch dark was occasionally broken by passing armored cars; from time to time I heard shots.

And finally there was a truce.

We rushed outside. The cannons were silent. We ran through the hungry streets like kids after the first snowfall, looking at the frosty stars of bullet holes in windows, at the snowy flowers of tiny cracks; we walked through the shards of glass, clear as ice, that covered Tverskoy Boulevard. Pleasant, those first hours, when we picked up bullets that had smashed against walls, all bent and twisted, like the bodies of burnt-up butterflies.[26]

Wandering through the streets, Khlebnikov finally 'looked into the book of the dead'. Around the entrance to a morgue, a line of mourning relatives had arrived to claim their dead. Khlebnikov noted: 'The initial letter of a new age of freedom is often written with the ink of death.'[27]

As the young Soviet government came under siege from all sides, what had begun (at least in Petrograd) as a virtually bloodless revolution started descending into a terrible civil war. At first, Khlebnikov remained optimistic, anticipating opportunities for creative work in support of the new Soviet power. In Petrograd, his apartment became a gathering place for the city's literary and artistic avant-garde. 'Khlebnikov was the trunk of the age, and from it we were germinating branches', one critic wrote of himself and his friends then frequenting 'apartment no. 5'.[28]

Khlebnikov later went to Astrakhan, where he worked for a Bolshevik newspaper, *Red Soldier*, organ of the local military-political section. In the winter of 1919, he was in Kharkov, a city occupied alternately by Reds and Whites. Pronounced psychologically unfit for military service, he was confined for months in a psychiatric hospital, remaining in the besieged city until February 1920. On his release, he threw himself into publicity work for the army and navy on the southern front. Back in Kharkov, he lectured Red Army soldiers on subjects which included the cycles of time. In Baku, while writing propaganda posters for the Volga–Caspian fleet, he at last formulated to his own satisfaction his 'Laws of Time'. Shortly afterwards, he seized the opportunity to travel with the Red Army to Persia: in April 1921, he sailed for Enzeli – now Bandar-e Anzali in Iran. 'The banner of the Presidents of Planet Earth follows me wherever I go, and waves now over Persia', he wrote happily to his sister.[29]

On returning to Russia in the autumn of 1921, Khlebnikov stayed in the Caucasus, distressed by the experience of famine and starvation brought on by the civil war. By some heroic act of will, he defied reality sufficiently to write an optimistic essay on 'The Radio of the Future':

Let us try to imagine Radio's main station: in the air a spider's web of lines, a storm cloud of lightning bolts, some subsiding, some flaring up anew, crisscrossing the building from one end to the other. A bright blue ball of spherical lightning hanging in midair like a timid bird, guy wires stretched out at a slant.

From this point on Planet Earth, every day, like the flight of birds in springtime, a flock of news departs, news from the life of the spirit.

In the stream of lightning birds the spirit will prevail over force, good counsel over threats.

The activities of artists who work with pen and brush, the discoveries of artists who work with ideas (Mechnikov, Einstein) will instantly transport mankind to unknown shores.[30]

On arriving back in Moscow in December 1921, Khlebnikov spent some time attempting to publish his mathematical discoveries of the previous year. Weakened by hunger and illness, he finally decided to return to his parents in Astrakhan. Before setting out, he accepted a friend's invitation to recuperate in a small village near Novgorod. However, by the time he arrived there, in the spring of 1922, gangrene had spread to his legs. The local doctors were powerless. As Roman Jakobson writes: 'Khlebnikov knew he was dying. His body decomposed while he lived. He asked for flowers in his room so that the stench would not be noticed, and he kept writing to the end.'[31] He died in that village on 28 June.

The Russian formalists' vision of art as political creativity survived only for a brief period. As the revolutionary tide turned, succumbing to state tyranny and bureaucracy, Lenin died, Trotsky went into exile and Stalin consolidated his grip, proclaiming his new doctrine of 'socialism in one country' at the expense of earlier anticipations of planet-wide change. Khlebnikov's futurism and all other deviations from 'Socialist Realism' were now lumped together as 'formalism' – condemned as an elitist counter-revolutionary plot. One by one, the great poets and experimental artists of the revolutionary years were forced into exile, took their own lives or found themselves inexplicably imprisoned and then shot.

In 1930, responding to the recent tragic suicide of Vladimir Mayakovsky, Jakobson penned one of his finest essays, 'On a Generation that Squandered Its Poets'. He quotes these lines from the young Mayakovsky, who, like Khlebnikov, singled out *convention* as the main enemy:

> To be a bourgeois does not mean to own capital or squander gold. It means to be the heel of a corpse on the throat of the young. It means a mouth stopped up with fat. To be a proletarian doesn't mean to have a dirty face and work in a factory: it means to be in love with the future that's going to explode the filth of the cellars – believe me.[32]

But by the mid-1920s, according to Mayakovsky, boredom and habit – the Russian word is *byt* – had been restored to its dominance over the soul. 'A revolution of the spirit' – a new organization of life, art and the sciences – was therefore required. As Mayakovsky had proclaimed at the very start of the revolution:

> It's a small thing to build a locomotive:
> Wind up its wheels and off it goes.
> But if a song doesn't fill the railway station –
> Then why do we have alternating current?

These lines are from Mayakovsky's 1918 'Order to the Army of Art'. Recalling them in 1930, Jakobson notes sadly that 'to our generation has been allotted the morose feat of building without song. And even if new songs should ring out, they will belong to another generation and a different curve of time.' The hopes of an entire generation have been crushed, Jakobson continued – a tragedy which turned out to be that of his time, as if history itself had run out of breath. It was as if all those hopes had never been:

> We strained toward the future too impetuously and avidly to leave any past behind us. The connection of one period with another was broken. We lived too much for the future, thought about it, believed in it; the news of the day – sufficient unto itself – no longer existed for us. We lost a sense of the present.

Insufficiently appreciative of either the present or the past, Jakobson's extraordinary Russian generation had possessed 'only a single-minded, naked hatred for the ever more threadbare, ever more alien rubbish offered by the established order of things'. Jakobson concludes his essay:

> As for the future, it doesn't belong to us either. In a few decades we shall be cruelly labeled as products of the past millennium. All we had were compelling songs of the future; and suddenly these songs are no longer part of the dynamic of history, but have been transformed into historico-literary facts. When singers have been killed and their song has been dragged into a museum and pinned to the wall of the past, the generation that they represent is even more desolate, orphaned, and lost – impoverished in the most real sense of the word.[33]

Jakobson was to live for many more years, decisively influencing Claude Lévi-Strauss, Noam Chomsky and – through them – the course of twentieth-century philosophy and thought. But as a futurist and revolutionary, his song also belonged already in that museum, pinned sadly to that desolate wall.

Jakobson remained to the end of his life inspired by the hopes and dreams of Russia's heroic revolutionary years. Following his escape from the Nazis and wartime arrival in the United States, his protective friends saw no particular reason to advertise those earlier political sympathies. During the 1950s, when McCarthyite witch-hunts were still sweeping across America, coming out as a 'communist fellow-traveller' may not have seemed wise.

When he arrived on the scene, Chomsky may not have known much about Velimir Khlebnikov. He tended to be unimpressed by dreamy poets anyway. But Chomsky was most certainly familiar with – and decisively influenced by – Jakobson's theory of 'distinctive features'. As I have tried to show, this approach to language was in fact the mature scholarly realization of Khlebnikov's inspirational idea of a universal alphabet of sounds. As a theoretician, Khlebnikov was more dreamy and less organized than the relatively methodical Jakobson; not all his futuristic ideas made sufficient sense to be pinned down or rendered accessible to a later generation of academics and professionals. But, as a wordsmith, Khlebnikov was unequalled in his lifetime and has remained unequalled since, to this day recognized by Russians as one of their greatest poets.

In later life, Jakobson would often recall how his love affair with language had been inspired in the first instance by Khlebnikov.[34] Following the great scholar's death in 1982, a friend recalled: 'Futurism was innate in him. That is why Khlebnikov was so close to him, and why all his life Jakobson used to say that for him, Khlebnikov was the most important, the most fundamental poet of the twentieth century.'[35]

CHAPTER 12

TATLIN'S TOWER

Khlebnikov dreamed of rediscovering humanity's lost language of sounds. Jakobson inherited this idea, passing on his own version of it – his 'distinctive features' theory – to Chomsky. Chomsky embraced the extraordinary theory with enthusiasm, as we have seen, trying now this way and now that to extend its application from the level of phonetics over to semantics and ultimately to linguistics across the board. Jakobson's theory of universal distinctive features was the shining beacon inspiring Chomsky's still more breathtaking vision – Universal Grammar.

Chomsky's Universal Grammar was a sophisticated refinement of the central idea behind Warren Weaver's 'New Tower of Babel' project, which was designed to secure US state supremacy over 'communism' in the post-war world. The social and political forces inspiring Khlebnikov's work in an earlier age were anti-militarist, internationalist and totally unfunded. Yet the Tower of Babel provides a bridge between these two political worlds. Eclipsing whatever dreams Weaver may have entertained, Khlebnikov's prophetic vision culminated in an extraordinary engineering project – an immense tower reaching to the skies.

To understand Khlebnikov's tower, we need to recall his central project – which was to draw on mathematical and linguistic science to condense all history, uniting the future with the present and the past. Khlebnikov's mathematics, being poetic, was perhaps closer to numerology and the magic of numbers than to real science. But since this might equally be said of much modern linguistic theory, it need not trouble us. Khlebnikov's sensitivity to synaesthesia – to the colour of sounds, the tactile feel of numbers – extended to the universe's fundamental dimensions, dissolving even the boundary between space and time.

If time is really space in disguise, it should be possible to move through it in any direction. But to do this – to predict and master the course of events – it was necessary to elucidate time's mathematical periodicities, its ultimate physical wave-lengths. A Khlebnikovian character in an imagined conversation explains:

> Do you recall the seven heavens that the ancients wrote of? If we subtract 48 from 365 sequentially, until we reach 29 (29 days is the time it takes for the moon to revolve around the earth), we find we have created seven number-heavens: they surround the number of the moon and are contained within the number of the Earth. Here are the numbers: 29, 77, 125, 173, 221, 269, 317, 365 ... Now, why shouldn't events fly about beneath the firmament of these numbers?[1]

Khlebnikov published these strange words in 1913. Developing the theme, he became convinced that all history, life and experience, from microcosm to macrocosm, could be captured by a mathematical equation linking Moon, Earth and Sun. As he wrote in May 1914:

> I've put together the beginnings of a general law. (For example, the connection between our feelings and the summer and winter solstices.) You have to discover what relates to the moon and what to the sun. The equinoxes, sunsets, new moons, half-moons. That way it's possible to work out our stellar dispositions. Work out the exact curve of feelings in waves, rings, spirals, rotations, circles, declinations. I guarantee when it is all worked out MES will explain it – Moon, Earth, Sun.[2]

These gloriously optimistic ideas were not consistently followed, but seem to have morphed by degrees into a practical plan.

The plan was to build a tower. What kind of tower? To explain this, let me turn to one of Jakobson's most celebrated essays, in which he reminds the world of the importance of the Tower of Babel as a motif inspiring Russian poets and mystics in a line stretching back to the ninth century. When missionaries from Byzantium first introduced Christianity to the Moravians, they decided to abandon the traditionally sacred languages of Latin, Greek and Hebrew in favour of a liturgy newly translated into the local Slavic tongue. Jakobson shows how this sanctification of the regional vernacular led the locals to believe that their Slavic languages had been chosen by God. The language now known as Old Church Slavonic – designed to unite all Slavs – was believed to have originated in the Pentecostal miracle of the Gift of Tongues. This was, in turn, interpreted as God's decision to *reverse* the punishment he had meted out earlier to the Babylonian builders of the Tower of Babel.[3] In the Slav case, it was implied, restoring that pre-Babel language was a project not only permitted, but explicitly blessed by God. This idea exercised a profound influence on Russian shamanistic and mystical poetry – one that remained very much alive when, beginning with Russian and other Slavic tongues, the mystic Khlebnikov made it his mission to restore to all humanity its pre-Babel lost alphabet of sounds, unleashing enough mutual understanding to launch a revolution and establish heaven on Earth. All this helps to explain how Khlebnikov's Moon–Earth–Sun vision – complete with rings, spirals, rotations, circles and declinations – came to inspire one of the best-known icons of Bolshevik internationalism. The idea was to build, in real life, a seven-step ladder to heaven.

The suggestion that Khlebnikov was behind the startlingly beautiful Tatlin's Tower seems likely on various grounds. Khlebnikov imagined himself 'besieging' immense 'towers', central among them the 'tower of time'.[4] Apart from the fact that Khlebnikov's Moon–Earth–Sun motif was explicitly built into the ambitious project, we know that he himself was an intimate friend of the monument's designer, the former sailor, Vladimir Tatlin. 'If Tatlin loved one of his contemporaries with an inexhaustible and unreserved love,' recalls one student, 'it can only have been Khlebnikov. Mayakovsky remained for him a kind, magnanimous, brilliant friend, but Khlebnikov was his passion. Only *his* poems did Tatlin preserve in his memory. Only of *him* did Tatlin speak with reverence. He considered it the greatest fortune to have met him.'[5]

It was probably in May 1916, according to one eminent scholar, that 'Khlebnikov inspired Tatlin with the concept of a vast spirally based project of monumental architecture, one that after the revolution would become the famous "Tatlin's Tower" '.[6]

The chances of actually building the immense edifice looked promising following the October insurrection, when the young Soviet government – anticipating an imminent European extension of the revolution – recognized Khlebnikov, Mayakovsky, Tatlin and their circle as valuable cultural and artistic allies. As People's Commissar for Enlightenment, Anatoli Lunacharsky later recalled: 'The futurists were the first to come to the assistance of the Revolution. Amongst the intellectuals they most felt a kinship with it and were most sympathetic to it.'[7]

Following Red October, an Association of Art Activists was formed to help in creating 'new, free, popular forms of artistic life'. Filled with optimism, the futurists responded with 'Decree No. 1 on the Democratization of Art', subtitled 'The Hoarding of Literature and the Painting of Streets': 'From this day forward, with the abolition of Tsardom, the *domicile* of art in the closets and sheds of human genius – palaces, galleries, salons, libraries, theatres – is abrogated.'[8] 'The streets are our brushes, The squares are our palettes . . ', declared the poet Mayakovsky in his 'Order to the Army of Art.'[9]

It was in this spirit that Tatlin prepared detailed drawings and models for his planned tower. The project was officially commissioned in 1919 to commemorate the October Revolution, at that time already suffering under the weight of stultifying bureaucracy, civil war and terror from both sides.[10]

Striving to transcend everything, the edifice would rise 100 metres taller than the Eiffel Tower – at that time the tallest building in the world. An assistant of Tatlin's clarified that its two great arches would straddle the River Neva in Petrograd.[11] The vast structure, in the words of one admirer, 'employs a spine, legs, rib cage and vital organs that move'. All this suggested a living being – a dynamic embodiment of the collective. Its seven steps apparently emerged from the earth to a point in the distant future. Its forward-leaning outline brought to mind that mythic eternal moment when, on 25 October 1917, streams of proletarians crossed over the Neva from the Vyborg Side on their way to storm the Winter Palace.[12] In one diagram, a banner hangs from the tower, reading: 'The Soviet of Workers' and Peasants' Deputies of the World.'[13] Teeming with delegates from all countries, the heaving, twisting spirals would eternally enact and re-enact that mythic moment of origin. With neither end nor beginning, the tower suggested the gigantic screw thread of a tunnelling device emerging from the Earth's core and thrusting 400 metres into the sky. 'It seems as much related to the sky as to the earth as it screws out of one and into the other.'[14] To any observer from a later scientific generation – aware of the molecular structure of DNA – Tatlin's immense double helix would seem not only futuristic, but prophetic to an uncanny degree.

The first practical step was to build a scale model. Tatlin called it 'Monument to the Third International', exhibiting the wooden structure in Petrograd in November 1920. Writing of its full-scale projected realization, the contemporary critic Viktor Shklovsky commented: 'The Monument is made of glass, iron and Revolution.'[15] Of course, it was never built. The failure of the European revolution and the horrors of civil war and famine ruled out financing the project or giving any priority to its construction. Our only surviving evidence of the impossible dream consists of plan drawings and photographs of that iconic wooden model.

The tower, as passed down to us, appears never-ending, its summit unfinished, gradually disappearing as it rotates continuously into the clouds. Since its axial tilt is that of the Earth – exactly 23.5 degrees – it resembles a vast telescope aimed at the Pole Star, driving into the still point of the cosmos. Connecting its various levels is a seven-stepped flying buttress. Suspended within the double helix are three glass halls, each designed to rotate independently. The conference centre at the base is an immense cube that rotates slowly with the Sun – that is, one complete revolution a year. Above it dangles a pyramid designed to rotate more quickly with the Moon – one revolution a month. From the summit, topped by radio transmitters sending an endless stream of scientific and other vital information to all the proletarians of the world, a cylinder rotates with the Earth – one revolution a day.[16] Cylinder, pyramid, cube: there seems to be something missing. Has the designer forgotten to include a sphere? The mystery clears when we glimpse the immensity of the total concept. Tatlin did not need a sphere because he had already incorporated a sufficiently massive one – the spinning Earth, out of which the entire edifice emerged.[17]

Khlebnikov could hardly have asked for more: his friend's bridge to heaven would dynamically link Moon, Earth and Sun in a system of interconnected spirals. For that reason, the physical monument could have neither beginning nor end. 'The waves from the radio station situated at the very top of the spiral', noted Shklovsky, 'perpetuate the monument in the air.'[18] As we have seen, shortly before his death Khlebnikov would hail the 'Radio of the Future' as 'the central tree of our consciousness', destined to unite all humankind.[19]

Tatlin's Tower, then, pointed to a future in which all had at last come together, science now inseparable from art, poetry from mathematics, music from engineering. As an experienced sailor, Tatlin felt a special affection for the Pole Star – the one still point in a turbulent, suffering universe. For that reason, he aimed his tower at that point. Science, for these revolutionaries, was the piecing together of everything, condensing past, present and future onto a single plane. Khlebnikov and his friends hailed Picasso and Einstein as revolutionary prophets, challenging the prevailing order, as had Galileo in a previous age. A revolution not powered in this way would not be *their* revolution.

When Chomsky likewise puts simplicity and beauty first, he is not so much recalling the futurists as the contemporary physicist Paul Dirac:

> It seems that if one is working from the point of view of getting beauty in one's equations, and if one has really a sound insight, one is on a sure

line of progress. If there is not complete agreement between the results of one's work and experiment, one should not allow oneself to be too discouraged, because the discrepancy may well be due to minor features that are not properly taken into account and that will get cleared up with further developments of the theory.[20]

Despite his own self-image, Khlebnikov was not in fact a scientist. Rather he was a poet writing in an age of scientific revolutionary fervour, celebrating the aesthetics of science while giving it a political edge. He invoked such scientific geniuses as Mendeleev and Einstein to argue that perceived reality is entirely illusory, a different world being both possible and necessary. Emboldened by the conviction that his childlike optimism and delightful arrogance were scientifically justified, he and his circle were intolerant in the sense that all revolutionaries simply *must* be intolerant, refusing to fudge crucial issues or blur the lines, declining to compromise with common-sense resistance to their plans. In just these respects, Chomsky's leadership of the cognitive revolution, inspired essentially by modern physics, to my mind recalls Khlebnikov's supreme self-confidence during an earlier revolution. From a scientific standpoint, Khlebnikov's numerological theories and equations were delightful nonsense. But perhaps no more so than those of Chomsky four decades later. Under certain historical circumstances, when mountains must be moved, inspirational nonsense may be the best we can do.

For the futurists, stopping the sun in its tracks or reversing time's arrow were no more than minor difficulties. If the world disproves the theory, don't change that theory – the world must be wrong! Anyone who knows Chomsky will recognize the pattern. Experiments don't matter, observations don't matter and neither do awkward facts. It is certainly possible to interpret the physicist Paul Dirac as saying just this; but it is surely unfair. Dirac's statement that equations should be beautiful was made in the context of a discipline whose practitioners were steeped in experimental and peer-review procedures, endlessly testing their theories with sophisticated equipment to see if they worked. His advice to take courage from the beauty of an equation was not intended as licence to speculate without constraint. There is all the difference in the world between an Einstein and a Khlebnikov, between a giant of modern physics and (to paraphrase Chomsky's admirer Neil Smith) an endlessly creative Picasso of science.[21]

Emerging as it did from the Russian Revolution, Roman Jakobson's linguistics – and in turn Chomsky's – reflected faith in science, optimism about the future and confidence in the creative potential of the human mind.

CHAPTER 13

AN INSTINCT FOR FREEDOM

There was never any direct connection between Chomsky and Khlebnikov, who had about as much in common as matter and anti-matter. Yet any scholar tracing the history of ideas should be able to discern a link. Jakobson never tired of explaining how much he owed to Khlebnikov, while, as we have seen, Jakobson's subsequent influence on Chomsky is beyond doubt.

Although Jakobson loved Khlebnikov, he knew that his hero was more poetic visionary than sober scientist; before those inchoate notions about a universal language could be put to use, they needed time and work. In the years following Khlebnikov's tragic death on 28 June 1922, Jakobson worked hard, developing insights which would lead eventually to his recognition as the world's foremost linguist. But it would not be until 1938 that the theory which was destined to inspire Chomsky finally took shape in his mind.

The triggering event was a catastrophe – the Gestapo's confiscation and destruction of the only draft of Jakobson's close friend Nikolai Trubetzkoy's monumental book, *On the Pre-History of Slavic Languages*. Shaken to the core, Trubetzkoy died soon afterwards. 'Distinctive features' theory crystallized in Jakobson's mind during his final encounter with his friend, when the two men knew full well that they were meeting for the last time. Jakobson later remarked that life had by this stage confirmed to him that 'the most important ideas come in moments of catastrophe.'[1] Jakobson had by now come to *expect* his intellectual life to consist of one cataclysmic 'shift' (*sdvig*) after another.

As the German army invaded Czechoslovakia on 15 March 1939, looking in particular for Jews, Jakobson's friends warned him urgently to flee. Burning his precious archive of letters, books and papers, and leaving only 'nine piles of ashes', he fled to Prague, where he hid for a month in his father-in-law's wardrobe. It took him that time to obtain the necessary visa to enter Denmark. Anticipating that this country, too, would soon be invaded, he fled in September to Norway, which was in turn invaded on 9 April 1940. He escaped to Sweden thanks to a socialist from Oslo who managed to drive him to the far northern border in a cart. To avoid detection, Jakobson lay in a coffin in the back of the vehicle, while his wife sat in front with the driver, playing the role of grieving widow.² Jakobson was at this point journeying as a corpse between one defeated country and the next, between his previous life and a future unknown, hoping to escape from a battered, conquered continent to the shelter and safety of a beckoning New World. As during any rite of passage, his core experience was of irreparable loss and death. Yet, as he felt entitled to hope, death on this occasion might prove not a permanent state, but the prelude to new life.

After a brief stay in Sweden, Jakobson at last obtained a visa for America, crossing the Atlantic in May 1941. Once safely in New York, he joined the city's École Libre des Hautes Études, which had been set up by French refugees from the Nazi occupation and was now one of the few American scholarly institutions prepared to accept Jewish refugees. 'We were teachers and students of one another', Jakobson recalls. 'I introduced Claude Lévi-Strauss to linguistics and he introduced me to anthropology'.³ In 1946, Jakobson was appointed chair of Czechoslovak Studies at Columbia University, remaining there until 1949, despite being suspected of communist sympathies during the early McCarthyite years. Documents in Jakobson's archives at MIT show that he kept his job only thanks to the personal intervention of the university's president, Dwight D. Eisenhower.⁴ From Columbia, Jakobson went on to Harvard University and MIT, both institutions eager to pick the brains of the man known to everyone as the world's foremost linguist.

By 1955, Jakobson's Russian-speaking student Morris Halle was already an MIT insider, in a position to secure for his new friend Chomsky a job in the Humanities Department teaching French and German to scientists – despite the fact that Chomsky had never studied French and barely knew German.⁵ More significant was Chomsky's simultaneous appointment to a research position in MIT's famous Research Laboratory of Electronics.

As Chomsky quickly began rising to prominence, it was with the encouragement of a network of scientists and scholars in positions of influence, many of them – like Jakobson and Halle – Jewish refugees fortunate to have escaped the gas chambers. Chomsky's mathematical approach appealed to them because, as we have seen, it promised understanding beyond politics or ideology. Could Chomsky be the Galileo of our age, destined to revolutionize our understanding of language and mind? His teachers and friends gave him the benefit of the doubt, ensuring his extraordinarily rapid rise to fame. Revolutionaries usually have to fight their way up. In Chomsky's case, the gods seemed to be hoisting him aloft.

From all this we can see how Jakobson's escape from the Nazis and journey across the Atlantic helped make history. His story, like all magical fairy tales, boiled down to a transition from life to death and then back again to new life. Preserved intact inside one man's brain, ideas which had their origin in the ferment of revolutionary Russia ended up on a new continent, where they came to serve a rather different cause. A strand of formalism invented by revolutionary anarchists, communists and war-resisters ended up being sponsored by the US military in the hope of enhancing their systems of weapons command and control. Those hopes came to nothing, yet in Noam Chomsky's carefully crafted variant – meticulously shorn of either right-wing or left-wing political associations – the new science of language came to dominate much of Western intellectual life throughout the second half of the twentieth century.

Turning something revolutionary into something quite different – something perceived by many as deeply reactionary – was bound to cause internal anguish and pain. Towards the end of a 1971 debate pitting Chomsky against the left-wing French philosopher Michel Foucault, a questioner from the floor asked Chomsky to explain his involvement with the US military-industrial establishment. The transcript reads: '. . . how can you, with your very courageous attitude towards the war in Vietnam, survive in an institution like MIT, which is known here as one of the great war contractors and intellectual makers of this war?'[6]

Chomsky responded by invoking Karl Marx:

> There are people who argue, and I have never understood the logic of this, that a radical ought to dissociate himself from oppressive institutions. The logic of that argument is that Karl Marx shouldn't have studied in the British Museum which, if anything, was the symbol of the most vicious imperialism in the world, the place where all the treasures an empire had gathered from the rape of the colonies, were brought together.
>
> But I think Karl Marx was quite right in studying in the British Museum. He was right in using the resources and in fact the liberal values of the civilization that he was trying to overcome, against it. And I think the same applies in this case.[7]

Chomsky here seems to be easing his political conscience by claiming that Karl Marx's institutional environment was actually *more* 'oppressive' and indeed 'vicious' than the Pentagon-sponsored electronics lab in which he worked. Somehow, he manages to draw a favourable comparison between himself as a full-time salaried employee in one of the most advanced weapons research laboratories in the world and an impoverished Marx, taking notes for revolutionary purposes in a public library – the reading room of the British Museum.

In the late 1950s, while few people at MIT would have known or cared too much about Chomsky's politics, the mere fact that he was young, Jewish and a friend of Roman Jakobson would have led to the assumption that he must be a leftist of some kind. In those early years, he made no effort to signal his political commitments, concentrating mainly on his scientific work. But from the early 1960s, with President Kennedy's inauguration and a dramatic intensification of the Vietnam War, Chomsky decided to take a public stand. Showing unusual courage, he led demonstrations and helped organize civil disobedience in opposition to what he insisted was the United States' military invasion of a small country on the other side of the world. Those events were the beginning of Chomsky's lifelong, passionate and impressively effective public struggle to highlight the crimes of his own government and his own Pentagon-funded institutional milieu.

In his activist role, Chomsky's thinking is not idealist, but realistically materialist and broadly consistent with Marxism. As the political system is currently constituted, he argues, policies are determined by representatives of private economic power. In their institutional roles, these individuals 'will not be swayed by moral appeals', but can only be affected by the 'costs consequent upon the decisions they make'.[8] Chomsky and his allies seemed vindicated when, after the Tet offensive of 1968, the joint chiefs of staff pointed out that the deployment of additional troops to Vietnam was being hampered by the need to ensure that 'sufficient forces would still be available for civil disorder control' at home.[9] During these and subsequent years, no American public figure did more to put the record straight on the United States' invasion of Vietnam than Noam Chomsky. Other left-wing intellectuals may not have felt quite the same need to deny personal culpability for their country's actions around the world. Chomsky experienced this need as intimate, inescapable and so overwhelming that, as he told the *New York Times*, he felt 'guilty most of the time'.[10] This guilt was particularly evident when a long-standing friend asked Chomsky if he had any regrets about his life as an activist: 'His answer shocked me. Muttering more to himself than to me he said, "I didn't do nearly enough." '[11]

While speaking to audiences of anti-war activists, Chomsky was sometimes asked questions about the other side of his work – the stuff he was doing in an electronics laboratory sponsored by the military. His overall programme had to appear consistent. He could hardly afford to let his critics suggest that although his politics were progressive, his linguistic theories were clearly reactionary. His anarcho-syndicalism and anti-militarism had to be constructed as consistent with his linguistics. Somehow, the corporate-backed, militarily shaped and financed 'cognitive revolution' in psychology and related sciences had to be presented as intrinsically liberating and consistent with Chomsky's own political beliefs.

He did not have to look far for a solution. Central to anarchism is the celebration of spontaneity and self-organization. Chomsky found a way of presenting his computational 'device' as the secret of human creativity and freedom. Had the behaviourists been right in their claim that humans are blank slates from birth, there would have been no natural basis for revolution. Fortunately, however, the behaviourists are wrong. Humans, according to Chomsky, are resistant because they have a deep-rooted instinct to rebel against external control; they come into the world with the capacity to tell right from wrong. Most importantly, humans have innate knowledge of language. Children do not need to be taught grammar by external pressure or example because – thanks to the innately installed 'device' in their brains – they know the basics already. We 'can know so much', as Chomsky explains, 'because in a sense we already knew it'.[12]

If human mental nature is intricately structured and resistant, it must set limits on authoritarian control:

> If, indeed, human nature is governed by Bakunin's 'instinct for revolt' or the 'species character' on which Marx based his critique of alienated labor, then there must be continual struggle against authoritarian social forms that impose restrictions beyond those set by 'the laws of our own nature', as has long been advocated by authentic revolutionary thinkers and activists.[13]

Moving onto the offensive against his left-liberal critics, Chomsky explains:

> For intellectuals – that is, social, cultural, economic and political managers – it is very convenient to believe that people have 'no nature', that they are completely malleable. That eliminates any moral barrier to manipulation and control, an attractive idea for those who expect to conduct the manipulation, and to gain power, prestige and wealth thereby.[14]

You cannot be a revolutionary unless you set out with some notion of human nature:

Maybe the assumption is not explicit, in fact, it almost never is explicit. But the fact is that if there is any moral character to what we advocate, it is because we believe or are hoping that this change we are proposing is better for humans because of the way humans are. There is something about the way humans fundamentally are, about their fundamental nature, which requires that this change we are advocating take place.[15]

Indicating, in fact, quite an intimate link between his linguistic assumptions and those inspiring his politics, Chomsky makes the tentative suggestion that the

essential features of human nature involve a kind of creative urge, a need to control one's own productive, creative labor, to be free from authoritarian intrusions, a kind of instinct for liberty and creativity, a real human need to be able to work productively under conditions of one's own choosing and determination in voluntary association with others.[16]

Yet when asked whether this view qualifies as science, he makes it abundantly clear that it does not. His speculations about 'an instinct for freedom' are based, rather, on hope:

On this issue of human freedom, if you assume that there's no hope, you guarantee that there will be no hope. If you assume that there is an instinct for freedom, there are opportunities to change things, etc., there's a chance you may contribute to making a better world. That's your choice.[17]

Celebrating what he *hopes* might prove to be a rebellious and creative human nature, Chomsky repudiates the pessimistic view that humanity's 'passions and instincts' must forever prevent enjoyment of the 'scientific civilization' that reason might create. He speculates instead that 'human needs and capacities will find their fullest expression in a society of free and creative producers, working in a system of free association'. 'Success in this endeavour', he continues,

might reveal that these passions and instincts may yet succeed in bringing to a close what Marx called the 'prehistory of human society.' No longer repressed and distorted by competitive and authoritarian social structures, these passions and instincts may set the stage for a new scientific civilization in which 'animal nature' is transcended and human nature can truly flourish.[18]

To any socialist, these are surely inspiring words, perhaps among the finest that Chomsky has ever written.

Chomsky's dream of a 'new scientific civilization' reveals much about what might have been. On the one hand, he makes it clear that he does not consider his own views on political matters to be science. On the other, he expresses passionate opposition to any strand of anarchist libertarianism according to which ordinary people need not learn anything from scientific specialists but can think what they like. Instead of urging us to 'break free of the oppressive structures of scientific thinking', Chomsky recommends respecting and upholding precisely those 'structures'. The compatibility between anarchist politics and science, according to Chomsky, is proven by a multitude of precedents, among them the work of Peter Kropotkin, whose great book, *Mutual Aid* – a celebration of cooperative self-organization in nature – was 'perhaps the first major contribution to "sociobiology"'.[19]

Chomsky faced the daunting task of presenting his linguistic work, which he identified as 'science', as somehow consistent with his politics – identified as a set of common-sense beliefs and objectives falling outside the remit of science. To this end, he managed to turn his extraordinarily narrow intellectual commitment to science as rigorously 'natural' (rather than cultural or social) into a political asset. The difference between the humanities and the natural sciences, for Chomsky, is that real scientists must cooperate with one another across space and time, and therefore have little choice but to be honest. In the humanities, by contrast – as in ordinary life – people are free to ignore one another and can claim whatever they please. In the humanities, scholars tend to feel threatened by science precisely because of its unrestrictedly cooperative nature. Equally, they feel threatened by ideas that are genuinely new. Such defects may also afflict disciplines within natural science. But at least 'the sciences do instil habits of honesty, creativity and co-operation', features considered 'dangerous from the point of view of society'.[20] A student in a university mathematics or physics department will hardly survive without being questioning – whereas in the 'ideological disciplines', originality will just get you into trouble. Chomsky complains that in the 'domain of social criticism the normal attitudes of the scientist are feared and deplored as a form of subversion or as dangerous radicalism'.[21] According to him, the culture of genuine natural science is the real 'counter-culture' to the reigning oppressive ideology.[22]

During the 1970s and 1980s, historians of science began focusing on the social and political processes through which research agendas are set and 'facts' correspondingly selected and constructed.[23] Michel Foucault is perhaps the central figure here, although T.S. Kuhn – with his hugely influential book, *The Structure of Scientific Revolutions* – had arguably more impact on Chomsky and his circle. Under the influence of Foucault, Feyerabend and other philosophers and social theorists, the concept of a monolithic body of doctrine called 'science' was yielding during the 1980s to a more pluralistic vision of multiple sciences fashioned for diverse social purposes. Western science prevails over indigenous belief systems, so it was said, for no other reason than that its practitioners command disproportionate levels of economic and military power.[24]

Chomsky could never sympathize with such views. Since Copernicus and Galileo, we have known that the Earth is round and moves around the Sun – facts which remain true, regardless of anyone's tribal or religious beliefs to the contrary. For Chomsky, political pluralism does not license unqualified persons to intrude as they please into scientific debates. Those who have not mastered the relevant literature – internalizing its concepts and terms – have nothing of interest to contribute and should therefore expect to be excluded:

> Look, in the physical sciences there's by now a history of success, there's an accumulated record of achievement which simply is an intrinsic part of the field. You don't even have any right to enter the discussion unless you've mastered that. You could challenge it, it's not given by God, but nevertheless you have to at least understand it and understand why the theories have developed the way they have and what they're based on and so on. Otherwise, you're just not part of the discussion, and that's quite right.[25]

Since, according to Chomsky, the so-called 'social sciences' amount only to political ideology, it is right to exclude such perspectives from linguistic debate. Those who fail to understand this clearly have not mastered certain foundational concepts intrinsic to the field. For Chomsky, 'society' is not a valid scientific concept. No natural language should be conceptualized as belonging to a group. Neither can we say that in order to acquire linguistic competence, a child needs social relationships – science cannot say anything about 'obscure' matters of this kind.[26] 'Mind' has no necessary connection with 'society'. To study mental phenomena is to examine aspects of brain structure and function. Ignoring the so-called social sciences, Chomsky's dream is to unify the sciences by integrating linguistics into an expanded version of physics:

> The world has many aspects: mechanical, chemical, optical, electrical and so on. Among these are its mental aspects. The thesis is that all should be studied in the same way, whether we are considering the motion of the planets, fields of force, structural formulas for complex molecules, or computational properties of the language faculty.[27]

Consistently with this project, Chomsky defines language as 'an *individual* phenomenon, a system represented in the mind/brain of a particular individual',[28] contrasting this with the earlier view of language as 'a social phenomenon, a shared property of a community'. This earlier view is the one associated with the Swiss founder of general linguistics, Ferdinand de Saussure, who wrote of *langue*: 'It is the social side of speech, outside the individual who can never create nor modify it by himself; it exists only by virtue of a sort of contract signed by the members of a community.'[29]

Or again, still more emphatically: 'Contrary to what might appear to be the case, a language never exists even for a moment except as a social fact . . . Its social nature is one of its internal characteristics.'[30]

The problem with all such ideas, Chomsky complains, is that they involve 'obscure socio-political and normative factors' – about which science can have nothing to say.[31]

Chomsky, then, is a socialist who, in his scientific work, seemingly denies the existence of society. Even when considering language acquisition by the human child, he refuses to take social factors into account. The infant's linguistic knowledge, he insists, does not arise out of social engagement. Social factors can be ignored because 'they are relatively unimportant'.[32] Echoing ethologists such as Konrad Lorenz,[33] Chomsky views language acquisition as independent of experience:

> No one would take seriously a proposal that the human organism learns through experience to have arms rather than wings, or that the basic structure of particular organs results from accidental experience. Rather, it is taken for granted that the physical structure of the organism is genetically determined.[34]

Chomsky claims that human mental structures develop in exactly the same way. 'Acquisition of language', he concludes, 'is something that happens to you; it's not something that you do. Learning language is something like undergoing puberty. You don't learn to do it; you don't do it because you see other people doing it; you are just designed to do it at a certain time.'[35]

It would be hard to imagine a more extreme version of Lorenz-style genetic determinism, or a more extreme denial of the relevance of politics, culture and society to what makes people human beings. This makes all the more remarkable Chomsky's success in convincing his left-wing activist supporters that his science, although not directly relevant to their concerns, must somehow be politically liberating.

CHAPTER 14

THE LINGUISTICS WARS

Any theorist who flatly denies society is likely to have problems with those more traditional linguists who view language as a system of communication. Utterances are inherently ambiguous and interpreting their richness of meanings is a subtle undertaking. Which meaning you choose depends on a vast number of variables: the uncertain intention of the speaker, one's understanding of the context and each listener's personal range of interpretations. If you accept all this, it follows that the study of linguistic meaning must lead inevitably into the social sciences – into how people interact, how they express themselves socially, how they must collaborate in order to reach mutual understanding. Surely, it would be impossible to deal with such complex issues if you had to confine yourself rigidly within natural science.

Machines do not form social relationships with one another, do not have intentions and do not need to struggle to reach mutual understanding. Machines can collaborate, but in their case this is straightforward: they have no selfish or competitive instincts to overcome. When machines communicate, what passes between them is *information*, which in itself holds no referential meaning. The digital information used internally by a computer is utterly independent of what room the computer happens to be in, what day of the week it is, or who is making use of its operations.

In the immediate post-war period, the priority for the US military was to develop fast and reliable devices for encoding, transmitting and decoding *information*. In at least one critical way, this meant departing from behaviourism, since it meant acknowledging the existence of computational states, now conceptualized as variable states of mind. But 'mind' in the sense now intended was conceived as distinct from 'body' only as software is distinct from hardware. According to the instructions received from its software, a digital computing machine may be in various modes, metaphorically identified as alternative states of its mechanical mind. But while the priorities of the military were consistent with 'mind' in this new sense – mind now conceptualized as software – it is important to remember that meaning in any deep humanistic or philosophical sense was of no interest whatever. For the scientists involved, all that mattered was C3 – Communication, Command and Control. The top brass and their laboratory engineers were no more concerned with subtle questions of meaning than the behaviourists had been.

The behaviourist linguists among whom Chomsky learned his trade had long ago given up on 'meaning'. They had done so for a very good reason. For a behaviourist, the full meaning of a sentence depends on details of the context of utterance considered in stimulus–response terms. This makes things horribly complicated, as Bloomfield observed:

> The situations which prompt people to utter speech include every object and happening in their universe. In order to give a scientifically accurate definition of meaning for every form of language, we should have to have a scientifically accurate knowledge of everything in the speaker's world.[1]

So, to reconstruct the meaning of *John read the book*, you might have to know the temperature at the time of utterance, the speaker's state of health and a great deal more. Bloomfield concluded that the only realistic – the only *scientific* – course was to forget all this and restrict one's focus to the study of abstract form, leaving meaning to one side. Working in the same intellectual tradition, Chomsky's teacher Zellig Harris had been still more relentless in focusing on form to the exclusion of content or meaning. It was when Chomsky and his close colleagues began to allow questions of meaning to seep back into their debates, risking a Pandora's Box of disputes and complications, that the seeds of the linguistic wars were sown.

To demonstrate that you could do away with meaning while still studying grammar, Chomsky's *Syntactic Structures* had offered what was to become the best-known sample sentence in the history of linguistics: *Colorless green ideas sleep furiously.* Although nonsensical, this sentence does seem to be grammatical – illustrating, in Chomsky's words, that 'the notion "grammatical" cannot be identified with "meaningful" or "significant" in any semantic sense'.[2] 'Nevertheless', *Syntactic Structures* concluded, 'we do find many important correlations, quite naturally, between syntactic structure and meaning' – correlations which 'could form part of the subject matter for a more general theory of language concerned with syntax and semantics and their points of connection'.[3] With these carefully chosen words, Chomsky held out the promise of one day breaking into the forbidden citadel of meaning.

In fact, that day came sooner than he might have thought – and led to the period of academic bloodshed now known as the 'linguistics wars'. The struggles broke out in the late 1960s and rumbled on through the 1970s. They pitted Chomsky against many of his students and colleagues, who – while pursuing ideas which Chomsky himself had earlier suggested – arrived at conclusions that began to alarm him. The conflicts were all the more bitter for breaking out among former friends. As Chomsky publicly repudiated his own previous allies, they in turn fought back, the former band of comrades splintering into violently opposed factions prone to swearing at each other in public, staging walk-outs or even – on one occasion – switching off an opponent's microphone in mid-sentence. Something of the atmosphere can be gauged from the way in which, for several minutes during a plenary session of the 1969 Linguistic Society of America conference, contributors from either side 'hurled amplified obscenities at each other before 200 embarrassed onlookers'.[4]

The disputes centred on 'the semantic component' of the language device – that is, on questions of meaning. Beginning in the early 1960s, several of Chomsky's most brilliant students felt sufficiently inspired by their teacher to explore how semantics might be properly integrated for the first time into the new transformational paradigm.

Chomsky's position on 'meaning' had always been ambivalent. While insisting that grammar could be studied without bothering about meaning, he held out the hope that eventually his methods would explicate meanings as well. Contradicting Leonard Bloomfield's objection that meanings must always be horribly complicated, he argued that a sentence's meaning is in fact not external – not out there in the world – but internal, deep within language itself. In his first book, *Syntactic Structures*, he distinguished the outer layers of a sentence from its inner content or 'kernel'. Meaning was determined by this 'kernel'. Elaborating the idea in his 1965 book, *Aspects of the Theory of Syntax*, he suggested a still more ingenious way of dealing with what he termed 'the semantic component' of the language machine. The meaning of a sentence, he explained, comes from its 'deep structure' – the logical form underlying the surface arrangement of the words. Take, for example, these two sentences:

John hit me.
I was hit by John.

Both mean the same, despite the superficial reversal of word order. Harris had coined the term 'transformations' to describe such reversals and other logical operations. Retaining this term, Chomsky argued that a sentence's 'deep structure' was not tied to the vagaries of its surface structure – not tied to its optional 'transformations' – but determined meaning independently. Pursuing this logic to its limit, he insisted on the 'inadequacy' – even the complete 'irrelevance' – of surface structure to the fixing of meaning.[5]

Chomsky now had to find a persuasive way of making the strong claim that meanings are not affected by social life or culture, a necessary step if he wanted to define linguistics as a wholly non-cultural, wholly *natural* science. By the mid-1960s, for reasons we will explore later, he had disowned Harris, Bloomfield and just about every American who had ever taught him or influenced him, with the exception of one figure – Roman Jakobson. Jakobson had famously dissolved the phoneme – a cultural entity – into its allegedly natural components. But Chomsky needed to go further: the phoneme had to be eliminated altogether. In Jakobson's view, phonemes – the culturally determined vowels and consonants characteristic of a given tongue – were real elements of language. Admittedly, he had shown how to reduce them to universal phonetics. The world's seemingly infinite variety of phonemes, Jakobson had persuasively argued, could be decomposed into a much smaller set of fixed features which were natural. But that did not render the phonemic level of analysis invalid; rather, it allowed linguists to choose between levels – one for studying the historically determined phonological systems of particular languages, the other for elucidating the timeless principles of general phonology.

For Chomsky, however, this was not enough. Jakobson was insufficiently committed to natural science. After all, the renowned Russian–American linguist still allowed *some* role for cultural tradition. As far as Chomsky was concerned, conceding the validity of the phoneme in any shape or form might prove a slippery slope, suggesting in turn that other parts of language might also be culturally contaminated.

To guard against that danger, Chomsky took pre-emptive action. Invoking recent work by his friend Morris Halle,[6] Jakobson's favourite student, he decreed the complete *abolition* of the phoneme. It existed no more. Chomsky knew that he was trashing the centrepiece – the pride and glory – of Bloomfieldian structural linguistics; he must have anticipated the outrage this would cause. By now, however, he had sufficient institutional backing to simply legislate,[7] and the behaviourists were effectively cowed. He issued the decree – whereupon the phoneme was no more.[8] All that remained were species-wide *natural* entities – essentially Jakobson's universal alphabet of phonetic features. It was as if the English voiced consonant '*v*' could no longer be acknowledged as a unitary whole, its place taken by such instructions as '+ voicing', '– nasality' and so forth. In place of '*f*' was a similar set of instructions, with the single difference that the 'plus' sign before 'voicing' was now switched to 'minus'.

Applying all this to semantics was now the big challenge. As Chomsky explained in his 1965 book, *Aspects of the Theory of Syntax*,

> it is important to determine the universal, language-independent constraints on semantic features – in traditional terms, the system of possible concepts. The very notion of 'lexical entry' presupposes some sort of fixed, universal vocabulary in terms of which these objects are characterized, just as the notion of 'phonetic representation' presupposes some sort of universal phonetic theory.[9]

This idea, nowadays known as 'lexical decomposition', posits a fixed and innate repertoire of semantic features, just like the *phonetic* ones made famous by Jakobson. Take *bachelor*. Could this concept be abolished and replaced by separate instructions, for example '*plus* human, *plus* male, *minus* married'? Then *husband* might be the same set of instructions, with the single difference that in this case, the sign before 'married' has been switched to 'plus'.

Following up this idea, Chomsky invited his readers to consider the sample sentence *Sincerity may frighten the boy*. In this sentence, *frighten* was said to be:

> a complex symbol consisting of the features [+V, +–NP, + [+Abstract] . . . – . . . [+Animate]], and others. The rules of the grammar impose the dominance order [+V], [+–NP], [+[Abstract] . . . – . . . [+Animate]] . . .].[10]

Just as he had abolished the phoneme, Chomsky was in this way demoting word meanings to the status of epiphenomena. They should be replaced, for all scientific purposes, by more fundamental entities located deep inside the mind – binary-digital features such as 'animate/inanimate', 'abstract/ concrete', 'Noun Phrase/Verb Phrase' and so forth. A lexical item was reduced to a set of phonological features *together with a parallel set of semantic ones* emanating from a Jakobson-style 'distinctive features' set or 'universal alphabet'.[11]

But, although lexical items had now been reduced to underlying abstract features – features which a machine might be programmed to grasp – their syntactic combination in complex sentences remained a challenging problem. The linguistics wars were rooted in the attempt to address this problem. In this regard, an influential book was Jerrold Katz and Paul Postal's *An Integrated Theory of Linguistic Description*, which outlined what came to be known as the 'Katz–Postal Hypothesis'.[12] The claim was that transformations do not affect meanings – any transformation worked upon a kernel sentence will preserve the original meaning. So if you were to replace say, *Four languages were spoken by everyone in the room* with its active counterpart – *Everyone in the room spoke four languages* – the meaning should remain the same. In point of fact, it was quickly recognized that examples of this kind prove just the opposite: only the second of the two sentences specifies that each *individual* person spoke four languages. As so often, however, counterexamples did nothing to dampen the initial enthusiasm.

In 1964, Chomsky suggested that the 'syntactic description' of a sentence has two quite different 'aspects', one determining its 'surface structure' or 'phonetic form' and the other determining its 'semantic interpretation' or 'deep structure'.[13] In this early formulation, a sentence's deep structure *uniquely* fixed its meaning. Chomsky was clear on this point, emphasizing that a given deep structure 'incorporates all information relevant to a single interpretation of a particular sentence'.[14] If 'all information' is intended literally, it must mean that reference to adjacent sentences or to the external environment – to the context of utterance – is completely unnecessary. *All* the information relevant to each of a sentence's possible interpretations is incorporated within one of the 'deep structures' internal to the sentence itself.

Soon after beginning work on these ideas, Chomsky became anxious to avoid creating the impression that he was in any way departing from his foundational commitment to natural (as opposed to social) science. What better way to do this than to invoke, by way of authority, the philosopher who is widely celebrated as 'the father of natural science'? In this spirit, Chomsky wrote what to this day has remained his most popular and accessible scientific book, *Cartesian Linguistics*, published in 1966.

In point of fact, while René Descartes made some interesting comments on the subject of human linguistic ability, he had little to say on grammar. So Chomsky had to look elsewhere. He turned to a once-influential study published in 1660 by Antoine Arnauld and Claude Lancelot, both adherents of a heretical Catholic sect. According to the Port-Royal *Grammaire générale et raisonée*, the world's languages are only superficially diverse. External forms – arrangements of vocal sounds – vary widely, but their conceptual foundation is common to all languages, being (in Chomsky's paraphrase) 'a simple reflection of the forms of thought'.[15]

Chomsky quotes this sample sentence from the Port-Royal grammar: *Invisible God created the visible world.* The Port-Royal authors argue that three constituent propositions pass through the listener's mind:

(1) God is invisible.
(2) He created the world.
(3) The world is visible.

In Chomsky's reformulation, 'He created the world' is the essential or 'kernel' sentence, its content enriched by the accompanying ideas, which – because they are embedded inside the sentence – need not be spelled out in full.

The idea here is that the laws of thought – identical everywhere for all human beings – ensure that deep down, grammatical structure must everywhere be the same. Since Chomsky clearly endorses this idea, he is able to claim that his own approach 'expresses a view of the structure of language which is not at all new',[16] being 'essentially a modern and more explicit version of the Port-Royal theory.'[17] 'Deep structure', so it seemed, was not only pregnant with meaning but pan-human as well. It was only when he began to backtrack, insisting (to everyone's surprise) that 'deep structure' was not really 'deep' at all – and that it had nothing whatever to do with a hidden level of shared meanings – that divisions within his own camp began to break out into open civil war.

The strange fact is that in the mid-1960s, Chomsky's commitment to deep structure brought him popularity and success owing to a total misunderstanding. As one historian points out, 'deep structure' was taken to mean 'deep' in the sense of both 'profound' and 'universal':

> By the time Chomsky's ideas came onto the scene, nearly everyone under the age of 40 had imbibed the idea of a Freudian unconscious. All of this had predisposed them to believe that language does not necessarily mean what it appears to mean on the surface, but that there is a hidden, deeper level of meaning. When Chomsky began to write about a deep structure in language, connecting it with Universal Grammar, and these ideas were popularized, it was widely interpreted to mean that Chomsky had uncovered a universal level of real meaning that actual language distorts.[18]

Familiar as they were with Freud, people assumed this anyway, taking Chomsky to be confirming that scientific linguistics now endorsed their idea. When the activist Chomsky wrote books and articles explaining how politicians abuse language to cover up their crimes, the popular idea that the quest for 'deep structure' meant unmasking lies and revealing hidden truths took still firmer hold.

But such misunderstandings were not limited to the popular reception of Chomsky's ideas. 'Almost his whole first wave of students', writes the same historian, 'equated deep structure with a real and universal meaning out of which syntactic structures were generated.'[19] Although 'generative semantics' – the natural outcome of this calamitous misunderstanding – would soon be denounced and its defenders excommunicated, the irony is that it was precisely this misapprehension that seized the popular imagination and elevated Chomsky to worldwide fame.

Chomsky's initial concept of deep structure fell apart following the realization that a sentence taken in isolation rarely has just one correct meaning and that endless complexities arise if you try to attribute each new meaning to an additional 'deep structure'. It soon became obvious that the search for deep structures had been motivated to a large extent by Chomsky's ruthlessly consistent internalism – his reluctance to take account of what he considered 'external' factors, such as context. Yet, unless the context is known, it is often impossible to say which precise meaning of a sentence is the correct one – the meaning intended by the speaker. Take this sentence: *The missionaries are ready to eat.*[20]

Will the good people be the diners – or the meal? In real life, the ambiguity would usually vanish: which meaning was intended would be obvious to everyone. But this was the problem: for Chomsky to acknowledge context would mean stepping outside natural science to acknowledge real human beings conveying ideas under concrete conditions. You cannot study this kind of thing as a topic within natural science: it belongs in that department of linguistics known as 'pragmatics'. The problem for Chomsky has always been that this entire field is off limits, since it takes us immediately into the social and political world.

Here is another sentence: *He is barely keeping his head above the water.* We might *think* we know at least roughly what this means, but how can we be sure? Should we envisage a short person wading across a flooded stream? To illustrate the risk of assuming too much, Ronald Langacker writes:

imagine a race over the ocean by helicopter, where the contestants must transport a severed head, suspended by a rope from the helicopter, from the starting line to the finish; a contestant is disqualified if the head he is carrying ever dips below the water's surface.[21]

This curious scenario may not sound too plausible, yet the sentence – now unconnected with what we thought was its topic – would in this context make perfect sense.

In dealing with meaning, Chomsky flatly refused to take context – by definition an 'external' factor – into account. Defining 'deep structure' as an entirely *internal* component of *syntax*, he dealt with the problem of ambiguity by arguing that *two* deep structures must lurk beneath the surface of an ambiguous sentence such as *Flying planes can be dangerous* (or *The missionaries are ready to eat*).

It was not long before critics came up with sample sentences interpretable in so many different ways that the number of deep structures you would have to assume began to seem absurdly large.[22] The precise number depended on how many ingenious contextual meanings you could imagine. But then why posit deep structures in the first place? Far from *determining* sentence meaning, every structure which came to mind was itself evidently determined by the meaning you had just assumed. Imagine one more possible meaning – and you got one more deep structure! Clearly, 'deep structure' and 'meaning' were different names for the same thing. Back, then, to square one. You had sounds and you had corresponding meanings. 'Deep structure' had turned out to be a wholly redundant complication.

But if meanings are not determined by deep structure, there is only one alternative – they must be arrived at through negotiation and social agreement. To Chomsky in the late 1960s, this was a truly worrying thought. After all, those who became disenchanted with the Katz–Postal Hypothesis might well look for the sources of meaning in non-linguistic areas of human psychological, communicative and social life. George Lakoff's publicly announced direction of travel seemed particularly alarming. 'Nowadays,' observed Lakoff, 'students are interested in generative semantics because it is a way for them to investigate the nature of human thought and social interaction.'[23] Chomsky suspected, correctly as it turned out, that concerns of this kind could only drive linguists to re-engage with the social and behavioural sciences and, beyond that, with the wider world – posing a mortal threat to his attempt to confine 'scientific' linguistics wholly within natural science.

The more deeply the heretics explored meaning, the more they saw themselves in a landscape of interconnected humans with bodies as well as minds, striving through language to communicate their thoughts. Upset by Chomsky's retreat instead into more and more abstract formalisms, they plotted their escape. Historical accounts depict John Ross, Paul Postal, Jerrold Katz, George Lakoff and the other generative semanticists as Pilgrim Fathers persecuted at home and migrating in search of a New World.[24] Crossing an ocean, they discover fertile land – a continent whose indigenous philosophers do 'ordinary language'. Influenced by Wittgenstein, these thinkers – central among them the Oxford philosopher John Austin – are less interested in supposed intricacies inside the head than in the more down-to-earth practicalities of how language is actually used.

An influential insight was that, when people reach stable agreements as to the meanings of words, those convergences become 'facts' of an artificially legislated (as opposed to biological or 'natural') kind. Sound-meaning correlations are 'institutional facts'.[25] Contrary to Chomsky, then, sentences are not variable states of a biological object. Rather, they are institutionally generated digital abstractions, much like currency values. The exiles from Chomsky's circle do not agree on all points, but they do all agree that the *public* and *social* dimensions of language are intrinsic to its very nature. As a serious alternative to Chomsky's more narrowly focused paradigm, generative semantics was widely perceived as something of a fiasco. But in the new setting, the heretics succeed in making a fresh start, eventually producing lasting contributions to linguistic science.

Interviewed 20 years later, Ross recalls: 'It's funny, the Vietnam War, in its horror and nastiness, ran very much parallel with the horribleness of the syntax wars.'[26] Postal, who later drifted politically to the right, elaborated on a possible connection:

> One thing I think we should say is that the timing is at least correct. That the beginning of all this kind of explicit hassling significantly corresponds with Chomsky's public involvement in the anti-Vietnam War movement. There is a certain case to be made that the emotional fervor and sense of rightness that was involved in that political anti-war movement somehow migrated over to linguistics independent of the content. In other words, there is in his kind of linguistics . . . a kind of emotional and moral fervor, as if he were battling for the right.[27]

Postal is persuasive in suggesting a connection between Chomsky's activism and his moral fervour in combating linguistic heresy, but seems unable to explain what the connection was. He almost seems to be suggesting that Chomsky considered disputes over syntax relevant to the anti-war movement. This cannot be so.

In fact, the fervour of Chomsky's intolerance of heresy reflected his overriding need to keep questions of meaning, social interaction and politics *out of* his linguistics. This meant packaging his scientific work as politically neutral in one compartment of his life, while keeping his politics – labelled 'non-scientific' – in the other. Compartmentalizing things in this way began to assume an unprecedented urgency as he moved publicly and visibly onto the political stage.

CHAPTER 15

BETWEEN COLLIDING TECTONIC PLATES

If we accept that each side of Chomsky's output is intimately conditioned by the other, we have to explore his linguistics in order to understand his politics and vice versa. In order to understand the peculiarities of the science – this book's primary focus – we must understand the political commitments against which it has always been counterposed.

Chomsky was born in Philadelphia in 1928, during a time of great hardship; he describes himself as 'a child of the Depression'.[1] Many in his family were militant trade unionists: 'So you knew what a picket line was and what it meant for the forces of the employers to come in there swinging clubs and breaking it up.'[2]

'Some of my earliest memories,' Chomsky recalls, 'which are very vivid, are of people selling rags at our door, of violent police strikebreaking, and other Depression scenes.'[3] One incident made an indelible impression: 'I remember I was with my mother on a trolley car. I must have been five years old. There was a textile strike. Women workers were picketing. We just passed by and saw a very violent police attack on women strikers, picketers outside.'[4]

The women's response at first puzzled the five-year-old:

It was mostly women, and they were getting pretty brutally beaten up by the cops. I could see that much. Some of them were tearing off their clothes. I didn't understand that. The idea was to try to cut back the violence. It made quite an impression. I can't claim that I understood what was happening, but I sort of got the general idea. What I didn't understand was explained to me.

The women were in fact 'hoping the police would be embarrassed and back off. The police beat them up anyway.'[5]

Between the ages of two and twelve, Chomsky attended the Oak Lane Country Day School in Philadelphia. This was an experimental progressive institution, which sought to foster non-competitive creativity. Each child, Chomsky recalls, 'was regarded as somehow being a very successful student':

> It wasn't that they were a highly select group of students. In fact, it was the usual mixture in such a school, with some gifted students and some problem children who had dropped out of the public schools. But nevertheless, at least as a child, that was the sense that one had – that, if competing at all, you were competing with yourself. What can *I* do? But no sense of strain about it and certainly no sense of relative ranking.[6]

When, later, he entered a city high school, Chomsky was shocked to discover that none of this was considered normal. In other schools, apparently, competitive dynamics were encouraged and personal creativity suppressed. He comments:

> That's what schooling generally is, I suppose. It's a period of regimentation and control, part of which involves direct indoctrination, providing a system of false beliefs. But more important, I think, is the manner and style of preventing and blocking independent and creative thinking and imposing hierarchies and competitiveness and the need to excel – not in the sense of doing as well as you can, but doing better than the next person.[7]

He is here describing the educational philosophy he would denounce throughout his life.

Chomsky's real education, however, came less from school than from a lively intellectual culture, dominated by the radical Jewish intelligentsia of New York. It was, he recalls,

> [a] working class culture with working class values, solidarity, socialist values, etc. Within that it varied from Communist Party to radical semi-anarchist critique of Bolshevism . . . But that was only a part of it. People were having intensive debates about Stekel's version of Freudian theory, a lot of discussions about literature and music, what did you think of the latest Budapest String Quartet concert or Schnabel's version of a Beethoven sonata vs somebody else's version?[8]

At an early age, Chomsky was affected by the outcome of the Spanish Civil War: 'The first article I wrote was an editorial in the school newspaper on the fall of Barcelona, a few weeks after my 10th birthday.'[9] He describes the defeat as 'a big issue in my life at the time'.[10]

Referring to Germany and Italy after the First World War and 1936 in Spain, Chomsky comments:

> The anarcho-syndicalists, at least, took very seriously Bakunin's remark that the workers' organizations must create 'not only the ideas, but also the acts of the future itself' in the pre-revolutionary period. The accomplishments of the popular revolution in Spain, in particular, were based on the patient work of many years of organization and education, one component of a long tradition of commitment and militancy ... And workers' organizations existed with the structure, the experience and the understanding to undertake the task of social reconstruction when, with the Franco coup, the turmoil of early 1936 exploded into social revolution.[11]

By his twelfth birthday, Chomsky had already rejected the politics of the Communist Party. Inspired by Barcelona's anarchists, he adopted their defeated cause – and has never abandoned it in the years since.

Chomsky rejected not only Stalinism, but also Leninism, which he associated with elitist attempts at mass indoctrination. The Spanish anarchists, he felt, did not try to educate the masses by imposing a rigid ideology from above. They believed in self-organization and everyone's capacity – once personally and politically liberated – to contribute to the revolutionary cause.

When speaking as a scientist, Chomsky avoids any mention of the social dimensions of human nature. But in his activist role, needless to say, he celebrates these aspects with passion. 'I do not doubt', he writes, 'that it is a fundamental human need to take an active part in the democratic control of social institutions.'[12] The 'fundamental human capacity', in his view, 'is the capacity and the need for creative self-expression, for free control of all aspects of one's life and thought'. Contemporary capitalist society ensures rewards for the more selfish tendencies in human nature. 'A different society', however, 'might be organized in such a way that human feelings and emotions of other sorts – say solidarity, support, sympathy – become dominant'.[13] Chomsky observes:

> It is no wonder that 'fraternity' has traditionally been inscribed on the revolutionary banner alongside 'liberty' and 'equality'. Without bonds of

solidarity, sympathy and concern for others, a socialist society is unthinkable. We may only hope that human nature is so constituted that these elements of our essential nature may flourish and enrich our lives, once the social conditions that suppress them are overcome. Socialists are committed to the belief that we are not condemned to live in a society based on greed, envy and hate. I know of no way to prove that they are right, but there are also no grounds for the common belief that they must be wrong.[14]

These are moving and powerful words, invoking innate human capacities for sympathy and solidarity in support of the socialist cause. In them, we hear Chomsky the activist, still committed to his teenage ideals. This aspect of him takes human nature to be deeply social and cooperative. It is striking, therefore, to find Chomsky the scientist taking the most crucial component of our nature – the capacity for language – to be neither cooperative nor communicative at all.

But to understand the gulf, we need to understand fully the pressures Chomsky came under in Cambridge, Massachusetts, on taking up his post at MIT. The conflicts and contradictions involved were not just cognitive or institutional, but more difficult in being sometimes intimate and personal. Even Chomsky's most dedicated left-wing admirers find it hard to understand how he could possibly have formed a friendship with John Deutch.

Deutch was a brilliant MIT chemical engineer, destined to become director of the CIA. Chomsky recalls:

We were actually friends and got along fine, although we disagreed on about as many things as two human beings can disagree about. I liked him. We got along very well together. He's very honest, very direct. You know where you stand with him. We talked to each other. When we had disagreements, they were open, sharp, clear, honestly dealt with. I found it fine. I had no problem with him. I was one of the very few people on the faculty, I'm told, who was supporting his candidacy for the President of MIT.[15]

To appreciate the glaring contradiction involved here, it is worth remembering that, according to Chomsky, 'the CIA does what it wants', carrying out assassinations, systematic torture, bombings, invasions, mass murder of civilians and multiple other crimes.[16] In Indonesia in 1965, the CIA organized a military coup to prevent the Communist Party – described by Chomsky as the 'party of the poor' – from winning a key general election.

The ensuing repression resulted in a staggering mass slaughter of perhaps half a million people. Chomsky recalls: 'The CIA pointed out in its report, which has since come out, that the slaughter that took place ranks up with the Nazis and Stalin. They were very proud of it, of course, and said it was one of the most important events of the century.'[17] The massacre Chomsky describes is not in dispute and his outrage is clearly genuine.

So Chomsky was not just flitting between different worlds: as he settled into his MIT job, he found himself nudging between colliding tectonic plates. According to MIT's student activists in the 1990s, Deutch was conducting research into vacuum bombs, also known as fuel-air explosives, the most powerful non-nuclear weapons in existence.[18] Even before Deutch's appointment to run the CIA, Chomsky could hardly have introduced this particular friend of his to a meeting of anti-war activists. Conversely, while chatting with Deutch, he surely would not have welcomed the presence of any activists who wished to join the conversation. As he flitted between one side of his life and the other, Chomsky in fact needed – and did his best to construct – a veritable firewall to keep these two constituencies of his apart.

As an explanation for the extraordinary disconnect between Chomsky the scientist and Chomsky the activist, this firewall is surely enough. Let me emphasize: I am not criticizing Chomsky for being personally respectful towards his professional colleague, John Deutch. All of us have been in situations of this kind, where it simply makes no sense to get embroiled in painful interpersonal dynamics. My point is merely that it cannot have been easy. While actively exposing the criminality of his institutional milieu, Chomsky was rubbing shoulders with these people, working for them and apparently chatting with them as part of the same professional and scientific elite. How could he possibly cope?

Obviously, he had to divide himself in two. An interviewer once jokingly asked him about the two different incarnations of the famous Noam Chomsky. 'What do they say to each other when they meet?' Chomsky's reply was immediate: 'There is no connection, apart from some very tenuous relations at an abstract level.'[19] Apparently, then, the two Chomskys aren't on speaking terms.

When the activist speaks, his passions are engaged and he takes full personal responsibility for his words. When the scientist speaks, something quite different seems to occur. According to his own account, one modular component of his brain – 'the science-forming capacity' – functions autonomously as a computational device.[20] Chomsky the activist is not responsible for the science, which comes from a different region of his brain. 'The one talent that I have which I know many other friends don't seem to have',

Chomsky explains, 'is I've got some quirk in my brain which makes it work like separate buffers in a computer.'[21] These separate compartments can make Chomsky seem like separate people. The scientist must avoid interfering with the activist, while the activist must avoid interfering with his scientific alter ego.

This leads naturally to the idea that the mind consists of quite separate cognitive modules, each 'a special purpose computer with a proprietary database'.[22] In Chomsky's case, each of his two internal computers – the one for science and the one for activism – operates on the basis of its own pre-installed concepts and corresponding language, resistant to translation across the divide. 'Now exactly how one can maintain that sort of schizophrenic existence I am not sure', Chomsky admitted on another occasion. 'It is very difficult.'[23]

He manages the inconsistencies in an astonishing way. In the following passage, he is commenting mainly – but perhaps not entirely – on intellectuals other than himself:

> What happens is you begin to conform, you begin to get the privilege of conformity, you soon come to believe what you're saying because it's useful to believe it, and then you've internalized the system of indoctrination and distortion and deception, and then you're a willing member of the privileged élites that control thought and indoctrination. That happens all the time, all the way to the top. It's a very rare person, almost to the point of non-existence, who can tolerate what's called 'cognitive dissonance' – saying one thing and believing another. You start saying certain things because it's necessary to say them and pretty soon you believe them because you just have to.[24]

If this passage means anything, it is that most people who talk ideological rubbish actually believe the rubbish they are saying. Yet occasionally, Chomsky says, you may find the odd individual who can divide himself in two, saying one thing while, in another part of his brain, believing quite different things without suffering 'cognitive dissonance'. Quite who Chomsky has in mind is not made clear. But he cannot possibly be telling us that he himself belongs to that elite majority who talk ideological rubbish and conveniently believe what they are saying. On the other hand, he is surely not depicting himself as an example of that 'very rare person, almost to the point of non-existence' who can endure cognitive dissonance. The only alternative is that he sees himself as someone with two separate buffers in his brain, each committed to a quite different perspective on the world.

Chomsky's secret, then, is his extraordinary capacity to dissociate his politics from his science, made psychologically bearable by an accompanying belief that his talent is the consequence of a split brain. As a scientist working in an electronics laboratory, Chomsky believes – apparently sincerely – that language is a 'mechanism' or 'device' devoid of humour, metaphor, imagination, communicative intent, social meaning or anything else which the rest of us would associate with language. Meanwhile, Chomsky the activist has somehow managed to retain his critical faculties. He doesn't believe any of this nonsense – or, if he does, it certainly doesn't show. Chomsky the activist speaks with passion and flair. While the scientist says language is not for communication at all, the ordinary human Chomsky uses language precisely to communicate – to denounce his own state, his own government, his own employers, his own institutional milieu. Short of denouncing his own science, Chomsky the activist opposes just about everything which he embodies in his alternative role.

To describe Chomsky as someone caught up in intense contradiction would be an understatement. Since internal conflict is so central a feature of Chomsky's institutional position, and since explicit awareness might make his situation even more painful, it is perhaps not surprising that Chomsky 'reaches for his revolver' – his own phrase – on hearing the word 'dialectics'. He explains:

Dialectics is one that I've never understood, actually – I've just never understood what the word means . . . And if anybody can tell me what it is, I'll be happy. I mean, I've read all kinds of things which talk about 'dialectics' – I haven't the foggiest idea what it is.

I'll tell you the honest truth: I'm kind of simple-minded when it comes to these things. Whenever I hear a four-syllable word I get skeptical, because I want to make sure you can't say it in monosyllables . . . When words like 'dialectics' come along . . . like Goering, 'I reach for my revolver'.[25]

Chomsky may reach for his gun, but the relationship between his science and his politics remains nevertheless a *dialectical* one. There is no more appropriate adjective. Chomsky needed to speak out against the US military not *despite* his early institutional sponsorship by that same military, but precisely *because* of it. Under intense pressure, Chomsky's science and his politics crystallized out as polar opposite forms. In the words of Norbert Hornstein: 'The main thesis for the Chomsky problem is that he takes a split view of his work for, well, psychological reasons. He cannot tolerate the fact that he is actually a war scientist as this would crush his political self image.'[26]

In one sense, as a result, Chomsky's two interests appear to have nothing to do with each other. But, as Hegel always taught,[27] opposition is itself a relationship – and, at each pole, the terms were and remain shaped by that very tie.

Expressing the same point in less abstract philosophical terms, we can surely understand the very personal horror, almost individual responsibility, that Chomsky must have felt while working as a respected scientist in the belly of the beast. Denouncing other people's crimes, as he puts it, is all too easy. To sleep with an easy conscience, one must be ready to expose one's own crimes, the crimes of one's own government, one's own institutional milieu. Chomsky could reconcile his conscience with the job he loved only by publicly denouncing those funding his research.

Chomsky has always been, in effect, the conscience of America. He has retained that unique status to this day. That is why people come from far and wide to listen to him. It is not just his politics and it is not just his science. What attracts people – what carries conviction – is the painfully evident tension between the two.

CHAPTER 16

THE ESCAPOLOGIST

Chomsky freely admits that he felt uneasy on arriving in Harvard in 1951:

> Computers, electronics, acoustics, mathematical theory of communica-
> tion, cybernetics, all the technological approaches to human behavior
> enjoyed an extraordinary vogue . . . Some people, myself included, were
> rather concerned about these developments, in part for political reasons,
> at least as far as my motivations were concerned.[1]

Chomsky hastens to clarify that his political misgivings 'were irrelevant
to showing that all this was wrong, as I thought it was'. The theories were
defective anyway: 'As soon as they were analyzed carefully, they unraveled,
though not without leaving substantive and important contributions.'

We have seen how Chomsky regarded the Pentagon's laboratories as
'Departments of Death', devoted to 'destruction, murder and oppression'. We
can imagine that, in his eyes, this might even apply to MIT's Research
Laboratory of Electronics – the institutional setting in which he would
spend his working life – so that the word 'Electronics' in such descriptions
might even become something like 'Death Technology'. His point was that
if the work being conducted in such buildings was labelled correctly, it
might be disrupted by direct action. Alternatively, he mused, such relabel-
ling might suffice to make those doing such work aware of the moral
contempt in which they were held.[2]

It is probably owing to anxieties of this kind that his intensifying activism
through the 1960s came with an equal and opposite insistence on the exclusion of anything remotely political or even social in his scientific work. In
this ostensibly non-political, purely scientific capacity, he announced and

championed one arcane detail of linguistic doctrine after another as if battling for some moral cause.[3] To many within his own professional circle, the logic of all this was wholly unclear. To oppose America's invasion of a small country on the other side of the planet, why should it be necessary to adopt a position for or against, say, the 'A-over-A' principle in syntax? Why should resisting the US war effort in Vietnam have any bearing on whether, or how, Deep Structure determines the meaning of a sentence such as *John is easy to please*?

The reality was that Chomsky needed to ensure that *whatever his current theory*, it would be so constructed as to exclude the remotest possibility of engagement with the world – engagement which, given the present balance of forces, would inevitably be on the Pentagon's terms. For him, as previously for Descartes, escaping from grubby reality into the realm of eternal form met not just a cognitive need, but a pressing and inescapable *moral* one. This becomes particularly clear when we recall that he likened B.F. Skinner to a concentration camp guard:

> First you ask, is this science? No, it's fraud. And then you say, OK, then why the interest in it? Answer: because it tells any concentration camp guard that he can do what his instincts tell him to do, but pretend to be a scientist at the same time. So that makes it good, because science is good, or neutral, and so on.[4]

Chomsky likens the behaviourally regulated society championed by Skinner to 'a well-run concentration camp with inmates spying on one another and the gas ovens smoking in the distance'.[5]

When referring to 'behaviourists', moreover, Chomsky is not just thinking of Pavlov, Skinner and their like. Widening the net, he denounces *intellectuals in general* as 'an extremely corrupt and dangerous group'. To clarify that he is especially thinking of Marxist-influenced social scientists in the West, he explains that he does not mean 'somebody who is studying 14th-century manuscripts'. Rather his target is 'intellectuals in the sense of those who hold that they have something important to say about society and human life and human action and so on and so forth'. In a country like the United States, these are 'the commissars, the secular priesthood, the state ideologists, the people in charge of the system of indoctrination and control'.[6] Chomsky exempts natural scientists from such condemnation. The most 'corrupt' and 'dangerous' intellectuals do not include his own colleagues at MIT working on command-and-control systems or fuel-air explosives.

While accusing his behaviourist opponents of criminal fraud, Chomsky applies quite different standards to chemists and physicists conducting weapons research. Towards such scientists, he has adopted since the 1960s what he terms 'a pretty extreme position' – indeed, 'one that might be hard to defend had anyone ever criticized it'. He formulates it as follows: 'Nothing should be done to impede people from teaching and doing their research even if at the very moment it was being used to massacre and destroy.'[7] One should leave such choices to the conscience of the individual scientist.[8]

Given Chomsky's own institutional position, that may have been a message to himself: 'I would stop doing what I was doing if I discovered that I was engaged in an area of scientific research that I thought, under existing social conditions, would lead to, say, oppression, destruction, and pain.'[9] Yet even here, he followed his blunt declaration with a caveat. Asked point-blank whether he might have conducted research in nuclear physics back in 1929 – when it was already clear that it might lead to an atomic bomb – he replied: 'I don't think a glib answer is possible. Still, if you ask me specifically, I'm sure that my answer would have been yes. I would have done the work just out of interest and curiosity and with the hope that things would somehow work out.'[10]

One's research might lead to the death of millions, but that is not necessarily a good reason to stop. You just have to hope that things work out. In his own case, those sponsoring his research might hope for practical applications – they undoubtedly did – but he personally would do nothing to help. No institution should legislate which topics any scientist should be permitted to research. Instead, 'people have a responsibility for the foreseeable consequences of their actions, and therefore have the responsibility of thinking about the research they undertake and what it might lead to under existing conditions.'[11]

We have noted already how, from the mid-1960s onwards, Chomsky helped organize draft-card burning and other activities aimed at disrupting the US military onslaught against Vietnam, showing considerably more courage and integrity than other left-wing academics. Getting arrested and spending time in police cells amplified his public voice and profile, giving him a place high on Richard Nixon's 'enemy list' of extremely dangerous artists and intellectuals.[12] From Chomsky's standpoint, it was now more urgent than ever to clarify that his own scientific work was not part of the problem – not part of the thinking which he and his supporters were passionately denouncing. To do this successfully, he had to convince himself and his supporters that his linguistics had no connection with military priorities, nothing to do with the Pentagon-funded academic establishment and indeed no connection with any of those prominent American linguists – including Zellig Harris – whose unstinting support had previously propelled his extraordinary rise.

In the mid-1960s, Chomsky did two things simultaneously: he became politically active in the anti-war movement and he turned against key former colleagues and friends. Prior to becoming active, Chomsky would generously acknowledge the teacher who had introduced him to linguistics – noting, for example, that while working on the *Logical Structure of Linguistic Theory* (LSLT), 'I discussed all aspects of this material frequently and in great detail with Zellig Harris, whose influence is obvious throughout'.[13] There can be no doubt that this acknowledgement was more than a formality: Zellig Harris really did help Chomsky at this time. On a visit to Harris's home, for example, a student noticed Zellig and Noam 'going at it hammer and tongs, the LSLT manuscript spread out on the kitchen table'.[14] The help went beyond that manuscript. Here is Chomsky in his 1956 preface to *Syntactic Structures*:

> During the entire period of this research I have had the benefit of very frequent and lengthy discussions with Zellig S. Harris. So many of his ideas and suggestions are incorporated into the text below and in the research on which it is based that I will make no attempt to indicate them by special reference.[15]

Recommending *Syntactic Structures*, Zellig Harris warmly reciprocated: 'My many conversations with Chomsky have sharpened the work presented here, in addition to being a great pleasure in themselves.'[16]

However, following his turn to political activism, Chomsky began denying that Harris had helped him in any way. Referring to his 'Morphophonemics of Modern Hebrew', Chomsky claims: 'Harris never looked at it'.[17] Zellig Harris allegedly dismissed his entire approach as 'a private hobby', 'never paid the slightest attention' to it and 'probably thought [it] was crazy'.[18] On another occasion, Chomsky claims that 'Harris never looked at my 1949, 1951 work on generative grammar', adding 'it's next to inconceivable . . . that Harris looked at my Ph.D. dissertation or LSLT'.[19]

'These remarks', comments one historian of the period, 'speak to Chomsky's narrative of isolation, of pursuing a course so radical and difficult that it was incomprehensible to, and incapable of sustaining any interest from, any and all linguists, even his own supervisor.' Such allegations, continues this historian, are outrageous:

... if true, they suggest stunning professional negligence; if false, they suggest libel. Chomsky claims that his supervisor and acknowledged mentor never looked at his B.A. thesis, his M.A. thesis, or his Ph.D. dissertation – all of which Harris would have had to sign off that he had read and approved – or at LSLT, the huge draft of that dissertation which was in modest but noteworthy circulation (Yale and Harvard had library copies for instance), and which very prominently included Harris's innovation, the transformation.[20]

Leaving aside the fact that Chomsky's aspersions run counter to normal expectations of professionalism, ethics and scholarly responsibility for a teacher in Harris's position, they are also flatly contradicted by his own earlier remarks, as we have seen.

To explain such an extraordinary loss of memory, we need a reason beyond simple forgetfulness. By this stage, the explanation I favour will not seem unexpected. Quite simply, Chomsky felt under moral and political pressure. He needed to shake off the contamination implied by his MIT employment and institutional status. Dissociating himself from the Pentagon was not enough: he needed to burn his boats, cutting loose from all those who had previously had any connection with its funding priorities or agendas. Despite holding anti-capitalist views similar to those of Chomsky, Zellig Harris was a machine translation enthusiast, and had to be disowned by Chomsky for the same reason that he needed to disown his own previous involvement with machine translation. From the mid-1960s, Chomsky repudiated not only Zellig Harris, but also Leonard Bloomfield, Charles Hockett and most of the other US scholars whose work he had previously cited.

But before publicly announcing that break, he realized that he would need an alternative narrative. If his ideas about language had no American roots, what exactly was their ancestry? One possibility might have been to invoke Jakobson, the Prague School and an earlier epoch of revolutionary ferment. But that would have got him into political trouble of a different kind, risking the suspicion that communism somehow influenced his supposedly neutral, value-free science. To avoid all such problems in a single move, he jumped not only to another continent, but to a different age. He announced that his linguistics carried on the brilliant but neglected work of the seventeenth-century French philosopher René Descartes.

Celebrated as 'the father of modern scientific philosophy', Descartes was, of course, a towering European figure with immense intellectual authority. To align yourself with Descartes has the obvious advantage of suggesting that you are a philosopher of the highest order, while, at the same time, distracting attention away from politics of any kind. A further advantage was that Descartes treated uniquely human nature, expressed above all in the capacity for speech, not as social or political, but as God-given and bound up with the existence of the soul. To teach that language is a natural component of the human mind or soul is to elevate its study to a plane above politics.

Descartes never quite mentioned a language organ, but he came close. The pineal gland, Descartes wrote, is the 'principal seat' of 'the soul'.[21] From deep in the brain, this tiny organ animates the tongue and lips as we speak. When we give verbal expression to our thoughts, thanks to this gland, we have no need to be consciously aware of the complex tongue and lip movements involved:

> . . . when we speak, we think only of the meaning of what we want to say, and this makes us move our tongue and lips much more readily and effectively than if we thought of moving them in all the ways required for uttering the same words. For the habits acquired in learning to speak have made us join the action of the soul (which, by means of the gland, can move the tongue and lips) with the meaning of the words which follow upon these movements, rather than with the movements themselves.[22]

Anticipating the cognitive revolutionaries of the mid-twentieth century, Descartes was interested in whether a machine might be designed in such a way as to speak. Imagine, he writes, a collection of mechanical dolls replicating the appearance and behaviour of various beasts. In principle, these might be constructed so cleverly that no one would suspect that they were man-made. Such replicas were possible because animals *really are* machines, albeit complex ones. But mechanical human beings were another matter. Regardless of how cleverly they were made, writes Descartes, 'they could never use words, or put together other signs, as we do in order to declare our thoughts to others'. Descartes offers a simple explanation: animals don't have a soul.

Having highlighted the soul as the secret of language, Descartes felt tempted to explore the next question to arise – the nature of the interface between body and soul. Given the politics of his day, however, he feared that pronouncing on such matters might not be wise. From his correspondence, we can see why. In November 1633, he had been 'quite determined' to send his friend, Mersenne, a copy of his latest *Treatise on Man*:

But I have to say that in the meantime I took the trouble to inquire in Leiden and Amsterdam whether Galileo's *World System* was available, for I thought I had heard that it was published in Italy last year. I was told that it had indeed been published, but that all copies had immediately been burnt at Rome, and that Galileo had been convicted and fined. I was so astonished at this that I almost decided to burn all my papers or at least to let no-one see them. For I could not imagine that he – an Italian and, as I understand, in the good graces of the pope – could have been made a criminal for any other reason than that he tried, as he no doubt did, to establish that the earth moves.

With a moving Earth now deemed theologically false, the future facing Descartes seemed frightening:

I must admit that if the view is false, so too are the entire foundations of my philosophy, for it can be demonstrated from them quite clearly. And it is so closely interwoven in every part of my treatise that I could not remove it without rendering the whole work defective. But for all the world I did not want to publish a discourse in which a single word could be found that the church would have disapproved of; so I preferred to suppress it rather than to publish it in a mutilated form.[23]

Descartes, then, had no appetite for martyrdom. Excusing himself for holding back all extant copies of his treatise, he wrote that, if his views 'cannot be approved of without controversy, I have no desire ever to publish them'.

The suppressed treatise risked papal denunciation because it dealt with the body, with the soul and – most controversially of all – with their mysterious interconnection. Unless he took care, Descartes might be arrested for heresy. Suppose, for example, he declared man's soul to be under bodily influence. That might suggest that God's immortal realm could be corrupted and defiled – an obvious heresy. Against this background, we can appreciate why Descartes found it so tempting to construct soul-substance as utterly immune from contamination by bodily or corporeal substance – leading to the cracked mirror philosophical compromise known to the world as 'Cartesian dualism'.

Descartes wanted to know how the mind can affect the body, so that when, say, I decide to scratch my head, my hand obeys. He mused that possibly some unobserved, very fine material in the pineal gland brokers the interaction between body and soul. But he doubted whether any of his arguments were convincing, even to himself, and apparently settled despairingly for an unspoken concordat with the Church. In return for being left in peace to study man's body, he would leave the Vatican with its traditional monopoly on the soul.

When Descartes divided the cosmos into two categories – corporeal substance on the one hand and soul-substance on the other – his point was that only bodily entities could be weighed, measured, subjected to experiments and scientifically described. It seemed obvious to him that man's immortal *âme* could not conceivably be treated in this way. Such a thing cannot be scientifically studied. This makes Chomsky's invocation of Descartes in his 1966 book, *Cartesian Linguistics*, quite breathtakingly audacious. For throughout this work, as in subsequent elaborations, Chomsky takes Descartes's 'soul', translates it as 'mind' – and insists that this *mind* (Descartes's 'second substance') is the one part of human nature which *can* be scientifically studied! Far from following faithfully in the footsteps of Descartes, then, it turns out that Chomsky is doing something radically different, transmitting Cartesianism as in a *camera obscura*, turning the French philosopher-scientist upside down.

Chomsky's supporters apparently have not noticed this complication, but the facts are not in dispute. Chomsky himself admits to extending the definition of 'Cartesian' to encompass, as he puts it, 'a certain collection of ideas which were not expressed by Descartes, [were] rejected by followers of Descartes, and many first expressed by anti-Cartesians'.[24] That 'second substance' which, for Descartes, was needed as a concession to the religious authorities becomes, for Chomsky, the central point. 'Now I believe,' as Chomsky explains, 'and here I would differ a lot from my colleagues, that the move of Descartes to the postulation of a second substance was a very scientific move; it was not a metaphysical or an unscientific move.'[25]

Writing of Descartes's postulation of soul-substance, he continues: 'In fact, in many ways it was very much like Newton's intellectual move when he postulated action at a distance; he was moving into the domain of the occult, if you like.'[26]

Newton's occult notion of gravity exploded the whole notion of 'body', thereby getting rid of the age-old 'mind–body' problem:

Mind–body dualism is no longer tenable, because there is no notion of body. It is common in recent years to ridicule Descartes' 'ghost in the machine', and to speak of 'Descartes' error' in postulating a second substance: mind, distinct from body. It is true that Descartes was proven wrong, but not for those reasons. Newton exorcised the machine; he left the ghost intact. It was the first substance, extended matter, that dissolved into mysteries.[27]

Or, still more emphatically, 'it is important to recall that what collapsed was the Cartesian theory of matter; the theory of mind, such as it was, has undergone no fundamental critique'.[28] The fact that Descartes' 'theory of mind' has not been disproven, according to Chomsky, poses for us 'the task of carrying on and developing this, if you like, mathematical theory of mind'.[29] Chomsky, then, made it his life's mission to study mathematically the one thing which Descartes said could *not* be studied mathematically – man's immortal soul.

According to Chomsky's reverse-Cartesian philosophy, matter does not exist: there is only mind. Where Descartes opted to leave the study of soul-substance to the theologians, Chomsky chooses the opposite course. It is important to appreciate just how stark is the contrast between the two thinkers' approaches. Torn between fearless reason and worldly caution, Descartes presented man as a machine haunted by a ghost beyond the comprehension of science. In Chomsky's reversal, man's incomprehensible body is raised above the animal level, thanks to a comprehensible machine installed in the physical brain.

Once adopted, reverse Cartesianism had a logic of its own, triggering a cascade of distinctively Chomskyan doctrines. Primary among them was the exclusion of bodily performance, leaving only knowledge or 'competence' as his focus of study. This meant, Chomsky argued, that you did not need to study language use at all. 'In principle', as he puts it, 'one might have the cognitive structure that we call "knowledge of English", fully developed, with no capacity to use this structure'.[30] So a child might possess full command of English without once having spoken a word.

Excluding social use leads inevitably to further simplifying assumptions, such as that language exists for talking to oneself: 'Now, let's take language. What is its characteristic use? Well, probably 99.9 percent of its use is internal to the mind. You can't go a minute without talking to yourself.'[31]

But even this 'talking to oneself', according to Chomsky, is mostly in English, Swahili or whatever happens to be one's first language, acquired by the individual from outside.[32] This cannot be the focus of scientific interest. If we are doing *genuine natural science*, our focus must be on what all humans have in common – the internal language with which everyone is equipped from birth. If the components of this genetically fixed 'language of thought' exist at all, explains Chomsky, they are quite mysterious: 'We don't have any way of studying them – or very few ways, at least.'[33] They are hard to study because, as a child acquires its repertoire of concepts, they tend to be formulated in whatever happens to be its own language. Bringing to light the innate concepts in their pristine natural form is therefore difficult, if not impossible. And so, having insisted that linguistics should focus narrowly on our innate language of thought, Chomsky tells us that this cannot realistically be done.

Pursuing the logic still further, Chomsky arrives at yet another impasse, this time concerning referential meaning. He is forced to conclude that the whole idea of reference has no place in linguistics, being a fundamental misconception. The natural use of language is not reference to things in the world:

> It would work the same if there weren't any world. So you might as well put the brain in a vat, or whatever. And then, the question comes along, well look, we use [words] to talk about the world: how do we do it? Here, I think, philosophers and linguists and others who are in the modern intellectual tradition are caught in a kind of trap, namely, the trap that assumes that there is a reference relation.[34]

A comical response came when Peter Ludlow (normally a staunch supporter) was conducting an interview with Chomsky in front of a typically respectful invited audience. 'Well, you see,' Ludlow objected, 'one thing that I've never quite understood is exactly what the problem is with reference . . .' Chomsky replied that linguistic expressions do not refer to things, only people do that – and you can't study people if what you're doing is supposed to be natural science. Still protesting, Ludlow pronounced the name 'Noam Chomsky'. The transcript runs as follows:

PL: So the name 'Noam Chomsky', I can ask the audience here what person in this room does that term refer to . . .

NC: . . . When you ask what the term refers to, you are assuming that there is a relation between terms and things, and then have to explain what that relation is, what are the entities between which it holds, because that's a new concept. In this case, what do you mean by *person*. We don't have that concept of reference in ordinary language.[35]

Suppressing his evident exasperation, Ludlow decided to move on.

As an ordinary citizen, Chomsky knows full well that speaking involves reference. When we talk, we are likely to be talking *about* something or other. But while Chomsky the lay person can concede this, Chomsky the scientist cannot. The most likely explanation would seem to be as follows. Unlike his ordinary social intelligence, Chomsky's *scientific* intelligence had been shaped since his student days within a culture heavily dependent on state sponsorship, particularly by the US military.[36] Because these sponsors wanted practical applications, Chomsky had to take care to avoid violating his conscience. Taking no chances, he divorced the very concept of 'language' from *all possibility* of practical use. While accepting the funding, he would retreat into a space of his own in which language was not public, not social, not communicative – and not even capable of making reference to anything in the world, whether real or imagined.

In this rarefied space, language is nature, not nurture. It is silent, internal, private and unconscious. 'The language faculty is kind of like the digestive system, it grinds away and produces stuff that we use.'[37] Or again: 'It would work the same if there weren't any world.' The speaker might as well be a surgically amputated brain in a vat – cut off, isolated, disconnected from both society and the body. Not only is language isolated in space; it is also cut off in time. Language has no history, having emerged in an instant from no evolutionary precursor. Perfect in form, it resembles the work of a divine architect.[38]

This, then, was how Chomsky escaped from the political dilemmas he faced. Against an imperfect world – indeed, right inside the scientific brain of that world – he would construct its antithesis: language as timeless, perfect form. It was as if he had reached the following painful conclusion: only on condition that linguistics could be rendered quite meaningless could he feel comfortable working in that 'Department of Death' at MIT.

CHAPTER 17

THE SOUL MUTATION

Chomsky's insistence that language did not gradually evolve – his doctrine that it emerged in a single step – is so diametrically at odds with pretty much the whole of modern science that we are forced to ask why he feels driven to champion this strange idea. Since there cannot be any empirical justification for it, we are obliged to look elsewhere. I suspect that the explanation must lie in the pressures and contradictions of his institutional milieu.

Back in the seventeenth century, Descartes imagined himself an earthly clockmaker, concluding, naturally enough, that it was not possible, even in principle, to assemble a human soul in his workshop. He concluded that such a God-given entity – the secret of language – must forever remain a mystery. Chomsky is able to reverse Descartes's conclusion only by imagining himself in the reverse role, no longer as human clockmaker, but as God himself, the 'divine architect'. He then tries to imagine what God would have done.

Before telling us, Chomsky reminds us of the difficulties. 'Language', he states, 'is, at its core, a system that is both digital and infinite.' All known biological systems are the opposite, being graded and limited. Why the human language organ should be so utterly different 'is a problem, possibly even a mystery'. Since 'there is no other biological system with these properties,' Chomsky continues, we are left 'with the problem of how this capacity developed in humans and how a messy system such as the brain could have developed an infinite digital system in the first place'.[1]

Having posed the baffling question, Chomsky resists all serious attempts at an answer. During the 1950s and 1960s, he had not felt any strong need to come up with an answer to the evolutionary questions that must arise if you claim to be doing biology. He could simply say language is an organ and leave it at that. During the 1970s, however, scientists who had become aware of his claim to be doing biology began to wonder how such an organ might have evolved. If an organ exists, then logically it must have evolved – or so everyone thought. With that idea in mind, a number of evolutionary scholars invited the most prominent linguists of the day to contribute to an interdisciplinary conference on the origins and evolution of language.

Convened in Chicago in 1976, this was the first time an international academic event had been dedicated to this topic. Chomsky's contribution was entitled 'On the Nature of Language'. While talking at length about fixed principles, he said not a word about origins or evolution. As he outlined detail after detail of his latest linguistic thinking, a sense of frustration built up among the audience. Eventually, one of the organizers felt prompted to remind him of the topic: 'Let me just ask a question which everyone else who has been faithfully attending these sessions is surely burning to ask . . . how did language get that way?'[2] Chomsky's response was to stonewall: 'What is interesting to me is that the question should be asked. It seems to be a natural question; everyone asks it. And I think we should ask why people ask it.'[3]

Even when he was an invited speaker to a conference on the origins of language, then, Chomsky expressed impatient antipathy towards the organizers' aims, separating himself in a world apart. The experience prompted him to clarify his stance. 'There is a long history of study of origin of language', he commented during that same year, 'asking how it arose from calls of apes and so forth. That investigation in my view is a complete waste of time, because language is based on an entirely different principle than any animal communication system.'[4]

In recent years, he has been persuaded by supportive colleagues to soften that position, conceding that *peripheral aspects* of language ability may gradually have evolved.[5] But as to the essentials, his position has not changed. Descartes insisted that God cannot have created the soul by degrees: he must have implanted it instantaneously and in perfect form. Chomsky's variation on this theme clothes the ancient idea in the language of modern science. Without quite invoking divine intervention, Chomsky posits a 'strange cosmic ray shower' triggering a random mutation which implanted the language organ in one step:

> To tell a fairy story about it, it is almost as if there was some higher primate wandering around a long time ago and some random mutation took place, maybe after some strange cosmic ray shower, and it reorgan-

ized the brain, implanting a language organ in an otherwise primate brain.[6]

When I recount this story, I am often warned by Chomsky's supporters not to take it too seriously. It's just a metaphor, they say. Chomsky himself, after all, describes it as merely 'a fairy story'. But while this is true, we need to explain why Chomsky keeps returning to the same story. Here is a somewhat more elaborate version:

An elementary fact about the language faculty is that it is a system of discrete infinity. Any such system is based on a primitive operation that takes n objects already constructed, and constructs from them a new object: in the simplest case, the set of these n objects. Call that operation Merge. Either Merge or some equivalent is a minimal requirement. With Merge available, we instantly have an unbounded system of hierarchically structured expressions. The simplest account of the 'Great Leap Forward' in the evolution of humans would be that the brain was rewired, perhaps by some slight mutation, to provide the operation Merge, at once laying a core part of the basis for what is found at that dramatic 'moment' of human evolution.[7]

Merge, then, is the procedure central to any conceivable system of 'discrete infinity'. It means combining things, combining the combinations and combining these in turn – in principle to infinity. No matter how we imagine the detailed story of human evolution or the exact composition of the evolving human brain, writes Chomsky, the simplest (and hence best) theory is that Merge must have arisen in a single step:

There are speculations about the evolution of language that postulate a far more complex process: first some mutation that permits two-unit expressions (yielding selectional advantage in overcoming memory restrictions on lexical explosion), then mutations permitting larger expressions, and finally the Great Leap that yields Merge. Perhaps the earlier steps really took place, but a more parsimonious speculation is that they did not, and that the Great Leap was effectively instantaneous, in a single individual, who was instantly endowed with intellectual capacities far superior to those of others, transmitted to offspring and coming to predominate, perhaps linked as a secondary process to the SM [sensory-motor] system for externalization and interaction, including communication as a special case. At best a reasonable guess, as are all speculations about such matters, but about the simplest one imaginable, and not inconsistent with anything known or plausibly surmised.[8]

Having considered the alternatives, therefore, Chomsky really does seem to favour the theory that language emerged in a single step. Lest there be any doubt, he repeated another version to an appreciative audience in the Vatican early in 2014, lifting spirits by reassuring the assembled cardinals and theologians that his one-step theory was consistent with the latest developments in science.[9]

The problem with soul–body dualism is that it always falls apart. No one has ever found a way to explain how the two substances join up. In Chomsky's case, predictably, the 'cosmic ray shower' fairy story solved one problem only to pose another. To formulate matters using his own terminology, the old parts of the newly irradiated brain were still complex, analogue and in that sense 'messy'. By contrast, the recently installed new component was mathematically simple, digital and clean. So how did the two brain components talk to one another and connect up? Did the opposed systems – messy to one side, clean to the other – snap neatly into place? If so, how was that even theoretically possible?

Scientists these days tend to follow Darwin, rather than Plato or Descartes, and so questions of this kind tend not to arise. To explain his thinking, therefore, Chomsky feels obliged to make a special effort. He invites us to imagine an ancestor of today's gorillas being hit by just the right kind of cosmic ray shower – only to be equipped with a language organ that it couldn't use. There may have been insurmountable difficulties concerning 'interfaces' and corresponding 'legibility conditions'. Put simply, the unfortunate gorilla's messy old brain just couldn't register the messages being transmitted by its newly installed digital component. According to Chomsky, this would have been no surprise. After all, the relevant mutation had been triggered by a 'strange cosmic ray shower' – interpreted literally, an intervention from outer space – and was to that extent utterly unconnected with prior evolutionary developments on Earth. So we would hardly expect a convenient fit between old brain and new – or indeed any fit at all: 'In fact it is conceivable, it is an empirical possibility, though extremely unlikely, that higher primates, say, gorillas or whatever, actually have something like a human language faculty, but they just have no access to it. So, too bad, the legibility conditions are not satisfied.'[10]

It is the statistical improbability of *any fit at all* which makes Chomsky's account of what actually happened in the human case so astonishing and correspondingly inspiring. The fit between old brain and new, he says, by pure coincidence proved to be not just acceptable or workable – it was *perfect*.

I have mentioned Chomsky's refusal to discuss origins at the 1976 Chicago conference dedicated to that subject. The next major conference on language origins attended by Chomsky was held in Boston in 2002, organized by myself and Jim Hurford, among others. Chomsky gave the final plenary talk, using the occasion as an opportunity to spell out the evolutionary implications of his 'Minimalist' turn. Impressed by his contribution, biologists Marc Hauser and Tecumseh Fitch later persuaded Chomsky to collaborate with them in writing an article. The outcome – 'The Faculty of Language: What is It, Who Has It, and How Did It Evolve?', published in the journal *Science* – was destined to become one of the most widely cited publications in the language origins field.[11]

The joint authors argued that the Language Faculty as Broadly Construed (FLB) includes memory and various other things not unique either to humans or to language. The Faculty in the Narrow Sense (FLN) – the essential element unique to language and also to our species – is our ability to combine words and then combine the combinations, in principle without limit, a mechanism known as 'recursion'. FLN emerged suddenly and without any antecedent, co-opting ('exapting') perceptual, motor, cognitive and other capacities found across a wide range of animal species, many related only distantly to humans.

To qualify as FLN, recursion needed to be more than just an abstract principle. It had to perform a specific job. Inside the brain, it needed to mediate between *cognition* – the domain of thinking – and the *sensory-motor* interface, involving such moveable body parts as arms, hands, ears, eyes and tongue. In the authors' words:

> We assume, putting aside the precise mechanisms, that a key component of FLN is a computational system (narrow syntax) that generates internal representations and maps them into the sensory-motor interface by the phonological system, and into the conceptual-intentional interface by the (formal) semantic system.[12]

Critics soon began commenting that, despite its celebrated status, the article was in key respects bizarre. A consistent critic was Derek Bickerton, who observed that it 'stands out as being perhaps the only work on the evolution of language that includes not a single word about how humans evolved'. Bickerton asks us to imagine a paper on the evolution of dam-building among beavers without a word about how beavers themselves evolved.[13]

Worse than that, having discussed FLN's two supposed 'interfaces' –
thought to one side, external activity to the other – the authors make no
attempt to explain how either interface could possibly work.

A major problem concerned the evolutionary origin of words. Nothing
was said about this. Yet, if recursion involves combining things and then
combining the combinations, how can this work when there are, as yet, no
elements available to combine? The authors are quite clear that the necessary words needed to be quite unlike animal concepts in representing
abstract categories, not events or objects in the real world. Yet nothing
was said about where such a vast store of intricately complex lexical items
could possibly have come from. Comments Bickerton: 'Supporters of this
approach cannot afford even to look at the language-cognition interface,
since doing so would force them to admit this gaping hole in their theory.'[14]

A further problem was the Chomskyan assumption that neither manual
signing nor audible speech initially played any role. The new faculty enabled
only private thinking. During this early stage, there was no communication,
hence no need to 'externalize' at all. This meant, of course, that there was
neither a requirement nor even any possibility of a sensory-motor interface.

In short, we have no interface with internal cognition and equally no
interface with external reality. Recursion, then, seems to be a system for
mediating between nothing and nothing, an outcome which sheds
intriguing new light on Chomsky's celebrated claim that in accomplishing
this task, language turns out to be 'surprisingly perfect'.[15] Summing up the
situation, the linguist Denis Bouchard dismisses the entire theory as
vacuous: 'In its current state, the hypothesis of a mutation for recursion is
not very informative. It just says that language has recursion because it got
recursion.'[16]

No other biological adaptation has ever been explained in so tautolog-
ical a way.

Exempting 'biolinguistics' from the constraints operative elsewhere
across biological science, Chomsky's approach barricades *Homo sapiens* so
categorically from the rest of life on Earth as to shatter any hope of conceptual unification. In Chomsky's own words:

> There is no reason to suppose that the 'gaps' are bridgeable. There is no
> more of a basis for assuming an evolutionary development of 'higher'
> from 'lower' stages, in this case, than there is for assuming an evolu-
> tionary development from breathing to walking; the stages have no
> significant analogy, it appears, and seem to involve entirely different
> processes and principles.[17]

As if all this were not enough, Chomsky says that even if by some miracle we hit upon a good theory, it would make no difference to our scientific understanding of language. For all Chomsky cares, we might have been transported to Earth by aliens:

Similarly, suppose it were discovered that our ancestors had been constructed in an extraterrestrial laboratory and sent to Earth by space ship 30,000 years ago, so that natural selection played virtually no role in the formation of the kidney, visual system, arithmetical competence, or whatever. The technical sections of textbooks on the physiology of the kidney would not be modified, nor the actual theory of the functions computed by the retina or of other aspects of the human visual and other systems.[18]

Language is not bodily at all (despite the fact that he defines it as a component of the brain), so speculating about the possible role of the arms, hands or tongue in its evolution is pointless: 'If you look at the literature on the evolution of language, it's all about how language could have evolved from gesture, or from throwing, or something like chewing, or whatever. None of which makes any sense.'[19]

If Chomsky were dismissing this or that *specific* Darwinian theory, serious evolutionary scientists might be able to sympathize. Instead, however – at least where language in the 'narrow' sense is concerned – he repudiates all evolutionary approaches in advance, giving himself no choice but to proceed as if his story about visiting extra-terrestrials were true.

If you set out from an utterly false initial assumption, then you will find yourself driven to pile one impossibility on top of another in order to make each step of your argument work. No other species has anything resembling language, 'which means', according to Chomsky, 'that the language faculty appears to be biologically isolated in a curious and unexpected sense'.[20] You can explain its evolutionary emergence any way you like:

We can make up a lot of stories. It is quite easy . . . for example, take language as it is, break it up into fifty different things (syllable, word, putting things together, phrases and so on) and say: 'OK, I have the story: there was a mutation that gave syllables, there was another mutation that gave words, another one that gave phrases . . .' Another that (miraculously) yields the recursive property (actually, all the mutations are left as miracles). OK, maybe, or maybe something totally different . . . The story you choose is independent of the facts, pretty much.[21]

Note how, for each aspect of language which might need to be explained, Chomsky cannot envisage even the possibility of a theory beyond his repeated assumption of 'a mutation'. On grounds of simplicity, Chomsky prefers just one mutation. It would still be a miracle, but at least only one – better than dozens of miracles.

The 'simplest speculation' about language's evolution, he therefore proposes, is that 'within some small group from which we are all descended, a rewiring of the brain took place in some individual, call him Prometheus'.[22] Thanks to this singular event (which occurred, according to Chomsky, soon after the speciation of *Homo sapiens*), Prometheus and his descendants possessed modern human thought, with the language faculty fully installed. But now comes Chomsky's most eccentric idea, expressed as follows: 'The use of language, however, was delayed by maybe something like 50,000 years.'[23] For all that time, Chomsky seems to be saying, our species possessed language but never got round to using it. This did not matter: 'The capacity to think became well embedded. The use of it to communicate could have come later. Furthermore, it looks peripheral: as far as we can see from studying language, it doesn't seem to affect the structure of language very much.'[24]

For Chomsky, then, making oneself comprehensible to others is 'peripheral', having no effect on what language is: 'Language is not properly regarded as a system of communication. It is a system for expressing thought: something quite different. It can, of course, be used for communication, as can anything people do – manner of walking or style of clothes or hair, for example.'[25]

So language is no more designed for communicating your thoughts than are your legs, clothes or hair. Language exists for talking to just one person – *yourself*:

> Actually you can use language even if you are the only person in the universe with language, and in fact it would even have adaptive advantage. If one person suddenly got the language faculty, that person would have great advantages: the person could think, could articulate to itself its thoughts.[26]

So how 'perfect' was the new faculty? 'Recent work', Chomsky answers, 'suggests that language is surprisingly "perfect".'[27] In fact, it has the perfection – the mathematical elegance and simplicity – of a snowflake: 'Language is something like a snowflake, assuming its particular form by virtue of the laws of nature . . . once the basic mode of construction is available.'[28]

Individual snowflakes are actually very diverse in their detailed shapes, but at a gross level they all share a common pattern, hexagonal and symmetrical. They share this pattern not because each crystal comes with complex chromosome-like machinery encoding a hexagonal blueprint, but simply because that symmetry follows naturally from the physical chemistry of water molecules acting in accordance with the laws of physics. The same applies to cell division inside the body:

> Nobody thinks there are genes that tell a breaking cell to turn into spheres, just as you do not have a gene that tells you to fall if you walk off the roof of a building. That would be crazy, you just fall because physical laws are operating, and it is probably physical laws that are telling the cells to break up into two spheres.[29]

Extended from snowflakes and cells to Universal Grammar, this is the central idea behind Chomsky's 'Minimalist' approach. In the new theory, the complexities of genetics are abandoned – elegance and simplicity come first. 'Perhaps Descartes, even Plato, might have been pleased', observes Chomsky.[30]

As if in confirmation, here is Descartes on the subject of perfection: 'The substance which we understand to be supremely perfect, and in which we conceive absolutely nothing that implies any defect or limitation in that perfection, is called *god*.'[31]

In this light, the question arises whether Chomsky is doing science at all. Isn't it closer to philosophy, or perhaps even theology?

Like any theologian, Chomsky acknowledges that perfection in this world is not always easy to discern. Very little in language *looks like* a snowflake. In fact, he says, if you take morphology, phonology, pragmatics or pretty much anything else, what strikes you immediately is *imperfection*:

> Morphology is a very striking imperfection; at least, it is superficially an imperfection. If you were to design a system, you wouldn't put it in. It's not the only one, though; no formal language, for example, has a phonology or a pragmatics and things like dislocation . . . All of these are imperfections, in fact even the fact that there is more than one language is a kind of imperfection. Why should that be? All of these are at least *prima facie* imperfections, you would not put them into a system if you were trying to make it work simply.[32]

One problem is that the world's languages do not sound uniform to the ear: superficially, they appear to be quite different languages – an obvious imperfection. In particular:

The whole phonological system looks like a huge imperfection; it has every bad property you can think of . . . Probably the entire phonology is an imperfection. Furthermore the phonological system has, in a way, bad computational properties . . . If you look at the phonetics, it seems to violate every reasonable computational principle that you can think of. So, that raises a question: is the phonology just a kind of ugly system?[33]

Chomsky suggests that these bad, ugly, imperfect features stem from the unfortunate need for pure thought to interact with the sensory-motor system – the messy bodily system of arms, legs, ears, tongue and so forth which we use to communicate our thoughts.[34]

In order to convince us that language's imperfections are merely apparent – merely 'external' – Chomsky turns to a discussion of inflectional systems. Chinese, he notes, appears to lack such systems entirely, while Sanskrit and Latin have elaborate case systems. Chomsky warns us not to be fooled. Such alleged contrasts between different languages are apparent, not real. English, Latin, Chinese, Sanskrit – all in fact probably share one and the same case system. True, you can't always hear it: the system may be inaudible. But irrespective of whether the ear can detect anything, it's really there. 'Recent work', as Chomsky explains, indicates that 'Chinese and English . . . may have the same case system as Latin, but a different phonetic realization.'[35] If languages seem to vary, then, it is because speakers make audible sounds. If only we all remained quiet – all of us everywhere just talking to ourselves – such problems would disappear. As Chomsky reassuringly explains, 'a large range of imperfections may have to do with the need to "externalize" language. If we could communicate by telepathy, they would not arise.'[36]

Language, then, is perfect, universal and invariant – on the assumption that telepathy works.

Such mystifying statements capture the essence of Chomsky's recent thinking – his 'Minimalist Program'. They seem baffling, until we grasp the philosophical rationale. As we have seen, Chomsky readily admits that his aim from the outset has been to reinstate – albeit in updated, 'scientific' form – the ancient doctrine of the soul. It is not difficult to follow through the implications. By definition, the soul is perfect. Being immortal, it transcends all bodily limitations. It need not adjust itself to the body in any way. Occupying neither space nor time, it is impossible to cut up or divide. No creature can have, say, one half or one quarter of a soul. No baby sets out with an embryonic soul which then begins to grow: if the infant's soul exists at all, it must at each point be fully formed.

Pursuing the same theological thread, we arrive at the moment of creation. Darwinism as conventionally understood cannot apply. No soul can evolve by degrees. If an ancestral human did suddenly acquire a soul, it must from the outset have been perfect. Where did it come from? Who installed it? When? How? Why? There can be no scientific answer to such questions.

CHAPTER 18

CARBURETTOR AND OTHER INNATE CONCEPTS

One reason for investigating language origins is that it allows us to test the validity of theories about what language is. Denis Bouchard explains:

> For instance, if a theory assumes a dualist view in which Man is body and soul, as Descartes did, then, since the soul is not part of the mechanistic world and hence does not fall into the realm of scientific inquiry, the question of the origin of the soul cannot even be asked in science, nor the question of the origin of language, if language is a by-product of the soul.[1]

On the other hand, if you have a theory which says that humans are bodies with genes and a brain, the question of the origin of language can be raised. 'Whether a linguistic theory can address the question of the origin is a good test of its value', concludes Bouchard. 'We should be wary', he adds, 'of a general linguistic theory that cannot provide a reasonably good basis of explanation for the origins of language.'[2]

Chomsky has struggled to provide such a 'reasonably good basis'. In the early years, as we have seen, he poured contemptuous scorn on the very idea of investigating how language first evolved. As pressure built up during the 1990s, he responded with his 'Minimalist Program', which he hoped would deal once and for all with this embarrassing question. A single mutation for a hugely complex organ – Chomsky's initial theory – had always stretched credulity. On the other hand, the alternative scenario of a suite of simultaneous mutations would be 'improbable on an apocalyptic scale', to quote Pieter Seuren.[3] In response, Chomsky now made one of his U-turns, perhaps the most spectacular to date. He announced that no part of

Universal Grammar needed to be genetically installed apart from the simple mechanism of 'recursion', otherwise known as 'digital infinity', 'discrete infinity', or 'the infinite use of finite means'. Since it was so simple, Chomsky could now argue that one mutation might have been able to install it in a single step.

This was satisfyingly simple, but created a new difficulty. Complexity can be shifted from one part of your theory to another, but it won't go away. To arrive at Minimalism, Chomsky was forced to offload complexity from the biological language organ to the intricately varied lexical units – audible words or their corresponding concepts – on which recursion worked. At a stretch, perhaps, you could imagine a single mutation giving rise to recursion. But to invoke that same mutation to explain the innumerable complexities of words (lexical units) and their properties would be quite another matter.

Yet an explanation was required. After all, to be able to combine units, those units must exist. Chomsky himself admits as much, noting that, for the new combinatorial system to work, it needed concepts of a kind that animals do not have – concepts representing abstract categories.[4] So where did concepts of that special kind come from? Were they, too, installed by a single mutation coinciding with the origin of *Homo sapiens*? Not even Chomsky could make that idea sound plausible. As Bickerton put it: 'Another serendipitous mutation on top of the one for recursion is too much to swallow.'[5] But, mutation or no mutation, the lexicon still had to be innate.

According to Chomsky, *carburettor* is an innate concept. When I mention this to my left-wing activist friends, the usual reaction is disbelief. Surely I must be imagining things? Chomsky, they have heard, argues that certain fundamentals of *grammar* must be innate. But could any sane person possibly view individual words – *bureaucrat*, for example, or *carburettor* – as encoded somehow in the human genome? Has each and every word coined over the course of human history – together with all theoretically possible future words – been part of our nature since *Homo sapiens* first evolved?

As we will see, Chomsky does say that. Since the idea is so exotic, we need an explanation. Chomsky himself invokes conceptual necessity, by which he means logical consistency. He is forced to adopt that position in order to shore up his fundamental thesis that language is in essence *natural* and *biological*. If he conceded that, say, *carburettor* became installed in a young person's brain thanks to her socio-cultural environment, it would be hard to avoid logical extensions of that same idea. If *carburettor* needs to be externally acquired, why not *climb*, *house* and *book*? Indeed, why not the lexicon as a whole? Thoughts along these lines threatened to blow apart the entire biolinguistic paradigm.

'From the biolinguistic perspective,' explains Chomsky, 'we can think of language as, in essence, an "organ of the body", more or less on a par with the visual or digestive or immune systems.'[6]

But while in these words he defines language as an 'organ of the body', a few lines later, in the very same paragraph, he changes his mind, telling us: '. . . there is no longer a coherent concept of *body (matter, physical)*, a matter well understood in the eighteenth and nineteenth centuries.' The concept of 'body', then, has no place in modern science.

If there is no valid concept of 'body', it seems odd to describe language as a bodily organ. To get over the contradiction, Chomsky glides from the problematic realm of 'body' to what he regards as the solidly scientific domain of 'mind': 'We can think of language as a *mental organ*, where the term "mental" simply refers to certain aspects of the world, to be studied in the same way as chemical, optical, electrical, and other aspects.'[7]

Among other parallels, Chomsky likens the language organ to the immune system. His point here is that language and the immune system both seem to have prophetic powers, in that each is equipped to answer to any and all challenges likely to be thrown up by future events.

Long before it encounters a particular virus or dangerous antigen, according to Chomsky, the immune system has already registered that problem's potential to exist, anticipating it correctly and, when stimulated, producing the correct antibody in defence. It seems to be able to do this, he writes, even in the case of a completely novel challenge, which no previous immune system had ever encountered in the history of life on Earth.[8] Proceeding by analogy, Chomsky claims that when the human language organ was installed, it must have been connected in a similar way to an extraordinarily vast *lexicon* of potential words, enabling it to anticipate all future needs.

During the 1960s, when generative semantics was still in vogue, Chomsky invoked Jakobson's 'distinctive features' theory to argue that only the elemental *subcomponents* of words – abstract features such as 'animate'/'inanimate', 'masculine'/'feminine', etc. – needed to be genetically pre-installed. When that idea proved unworkable, he was obliged to switch instead to what seemed the only permissible alternative: each word had to be genetically encoded as a *whole entry*. That idea was developed in particular by Chomsky's close associate Jerry Fodor.[9] Chomsky has continued to defend it to this day.

Doing his best to make the argument sound plausible, Chomsky starts at the easy end, with concepts that might well sound 'natural'. *Climb* is a good example, since it is something that humans everywhere do. Chomsky wants us to think of *house* in the same way, presumably because it is natural to need somewhere to sleep and rest. Chomsky claims that every child comes into the world knowing already what a house is. As it grows up and acquires its natal tongue, it just has to connect that concept with the locally appropriate sound: 'There's a fixed and quite rich structure of understanding associated with the concept "house" and that's going to be cross-linguistic and it's going to arise independently of any evidence because it's just part of our nature.'[10]

If this applies to *house*, Chomsky reasons, it must apply in the same way to other concepts: 'There is overwhelming reason to believe that concepts like, say, *climb, chase, run, tree* and *book* and so on are fundamentally fixed.'[11]

Note the inclusion of *book* in this list. Needless to say, Chomsky knows that no actual books existed during the palaeolithic age, when humans were everywhere hunter-gatherers and writing had not been invented. Despite this, he says there is 'overwhelming reason' to believe that *book* had been installed already in these stone-age people's minds.

How can Chomsky make such a strange claim? As so often, he denies that he is offering a hypothesis.[12] If it were a testable hypothesis, his astonished critics might be tempted to cite counter-evidence. Not believing in empirical tests or experiments, Chomsky argues instead from conceptual necessity. Lexical concepts, he insists, 'have extremely complex properties when you look at them'. From this it logically follows 'that they've got to basically be there and then they get kind of triggered and you find out what sounds are associated with them'.[13]

So much for *climb, chase, run, tree* and *book*. But Chomsky knew he could not restrict himself to an arbitrarily chosen list of words. Was *house* natural, whereas *book* was cultural or artificial? Where exactly should one draw the line? For his thesis to have any merit, it needed to apply across the board. So what about, say, *carburettor*? Or *bureaucrat*? Or *quantum potential*? When the philosopher Hilary Putnam realized what Chomsky was claiming, he could hardly conceal his astonishment. Such nonsense, he complained, had nothing to do with any known branch of biology. To have installed in the ancestor of all of us a stock of the words which future generations might need, observed Putnam, 'evolution would have had to be able to anticipate all the contingencies of future physical and cultural environments. Obviously it didn't and couldn't do this.'[14]

But, to everyone's surprise, Chomsky did not flinch. Young children, he reaffirmed, acquire words so rapidly that learning cannot be what is happening: each child needs merely to discover which locally appropriate vocal label should be applied to a concept *already installed*.[15] After elaborating this idea with respect to relatively simple words such as 'table', Chomsky continued:

Furthermore, there is good reason to suppose that the argument is at least in substantial measure correct even for such words as *carburetor* and *bureaucrat*, which, in fact, pose the familiar problem of poverty of stimulus if we attend carefully to the enormous gap between what we know and the evidence on the basis of which we know it . . . However surprising the conclusion may be that nature has provided us with an innate stock of concepts, and that the child's task is to discover their labels, the empirical facts appear to leave open few other possibilities.[16]

'Thus Aristotle had the concept of an airplane in his brain, and also the concept of a bicycle – he just never had occasion to use them!', responded philosopher Daniel Dennett, adding that he and his colleagues find it hard not to burst out laughing at this point. Perhaps 'Aristotle had an innate airplane concept', Dennett continued, 'but did he also have a concept of *wide-bodied jumbo jet*? What about the concept of an *APEX fare Boston/London round trip*?'[17] Despite the hilarity, however, Chomsky has continued to defend the idea.

He is able to do this thanks to his special position. When speaking as a scientist, as we have seen, Chomsky is forever invoking a particular *kind* of authority – not worldly authority, not the say-so of this or that colleague or respected figure in his field, and certainly not the received wisdom of the moment – but something closer to *ancestral* authority, to the pantheon of science, to eternal truth as revealed by the likes of Plato, Descartes, Copernicus and Galileo. Chomsky falls back on what the French social anthropologist Pierre Bourdieu terms 'authorized language' – the kind you need when performing speech acts, such as declaring war, consecrating a church, naming a ship and so forth.[18]

Invoking the ancestors is a way of turning what might otherwise seem total nonsense into self-fulfilling prophecies or decrees.[19] You cannot legislate in your capacity as an individual; it only works if the community recognizes your right to act as its representative, conferring on you the necessary authority. The technique is thrown into sharp relief when we remember how, in his activist role, Chomsky repudiates all such nonsense, speaking modestly yet emphatically in a personal capacity, invoking nothing more than an open mind and ordinary common sense. In this ordinary domain he is not legislating or enacting – merely sharing his thoughts with us, like anyone else.

Given Chomsky's extraordinary determination to keep these two modes of thought and action apart, we need hardly be surprised that he splits his thinking about the origins of language into two distinct parts. One of these – which he considers the scientific one – is delivered with authority. It explains silent, invisible and eternal language. Meanwhile, the other theory, delivered without authority, touches on the faculty's subsequent externalization, by which Chomsky means the evolution of visible signing or audible speech. The versions are contrasted in this way:

Internal language	External language
Perfect	Imperfect
Suddenly installed	Historically evolved
Natural	Cultural
Unitary	Diverse
Can be studied scientifically	Cannot be studied scientifically

Only the left-hand column matters to Chomsky, and clearly it matters a lot. It specifies human nature, is elegant and simple, and must be defended at all costs; the right-hand column concerns complex details which are beyond the remit of natural science.

So different are these two narratives that Émile Durkheim's celebrated distinction between 'sacred' and 'profane' domains comes to mind.[20] Science, in this context, corresponds to the sacred. And so it is that in Chomsky's supposedly 'scientific' myth, language emerges suddenly as an invisible, inaudible mechanism in one individual's brain. Only in the 'profane' part of the story, conceived as a later development, does language become social and communicative. Chomsky helpfully summarizes the two stages: 'First people sort of learned how to think, and then later when there were enough of them they somehow tried to figure out a way to externalize it.'[21]

In this informal statement, the real Chomsky is talking to us. He is conceding that, at some point, evolving humans must have begun externalizing their new capacity, expressing their thoughts for the first time in words. When he suggests that these people 'somehow tried to figure out a way', he is conceding that the solution must have been a social and cultural one – a kind of invention. A small group of hunter-gatherers, he surmises, must have hit upon a particular method of externalization which worked. When this first hunter-gatherer band subsequently split, people found different solutions in different localities, leading quickly to Babel – to the regrettable cultural diversification of the world's actual languages.[22] Chomsky is not interested in any of this. His authority derives from the mythical version, which must be safeguarded from debate or contestation. Paradoxically, he does this by dismissing the sacred story – the one about a 'strange cosmic ray shower' – as only a myth.[23] It was almost as if Chomsky had long wondered how to make the most extraordinary event in the history of life on Earth – the emergence of language-speaking *Homo sapiens* – seem meaningless and politically irrelevant. His fairy tale brilliantly answered to that need.

In a 2008 interview, Chomsky stuck to his one-step mutation narrative under repeated questioning. Asked how we might test the speculative hypothesis, he replied that it was the wrong question because he knew of no such speculation. The published transcript runs as follows:

Q: *We are not clear in what sense the speculation you have just offered is testable.*

A: Don't quite understand the question. Which speculation do you have in mind?

Q: *We had in mind your whole speculative origins scenario.*

A: You'll have to explain to me what you mean by my 'speculative origins scenario'. In particular, can you identify what I've written about this that is even controversial enough to require empirical tests?

Q: *To count as scientific, a hypothesis surely has to be testable. Can you specify just one or two experimental results or archaeological finds or anything else that might in principle pose a problem for your hypothesis of instantaneous language evolution?*

A: If it is true that what I have suggested is not even controversial enough to require empirical test, is perfectly consistent with what is known about our ancestors, and is accepted, tacitly, by everyone who has a word to say on this topic, then I do not see how the question you are posing arises.[24]

In true minimalist spirit, Chomsky is here determined to establish certainty by cutting out anything which might be refuted by evidence. The less you say, the greater the likelihood that what you do say will be right, and if that means keeping to truisms, so be it. To explain any genetic change, you must assume at least one mutation, something which by definition must happen in an individual, not a group. Reduced to its bare essentials, therefore, any evolutionary account must highlight that singular event. The virtue of proceeding in this way is that, if you minimize your narrative to the very least you can possibly say, then what remains of your picture of the complex evolutionary process is not speculative at all. Your story is now true because it's conceptually unavoidable.

Let me repeat here one of Chomsky's best known quotes, from 1988: 'There is a long history of study of origin of language, asking how it arose from calls of apes and so forth. That investigation in my view is a complete waste of time, because language is based on an entirely different principle than any animal communication system.'[25] Confirming that this remained his position in 2009, he declared: 'We know almost nothing about the evolution of language, which is why people fill libraries with speculation about it'.[26] Two years later, he was even more blunt at a talk given at University College London: 'There's a field called "evolution of language" which has a burgeoning literature, most of which in my view is total nonsense'.[27] Yet this did not stop Chomsky from adding to this literature with his own book devoted entirely to language origins, written in collaboration with MIT computational linguist Robert Berwick.[28] More conscientiously than Chomsky, Berwick has sought to keep abreast of recent developments in language origins research, for example by participating from the beginning in precisely the interdisciplinary conferences whose published outcomes Chomsky rubbishes.[29]

Why Only Us: Language and Evolution packages the one-step mutation story in an attractive and erudite way, giving what Chomsky terms his 'fairy story'[30] an appearance of authority. Here is the key paragraph:

> In some completely unknown way, our ancestors developed human concepts. At some time in the recent past . . . a small group of hominids in East Africa underwent a minor biological change that provided the operation Merge – an operation that takes human concepts as computational atoms and yields structured expressions that, systematically interpreted by the conceptual system, provide a rich language of thought. These processes might be computationally perfect, or close to it, hence the result of physical laws independent of humans. The innovation had obvious advantages and took over the small group. At some later stage, the internal language of thought was connected to the sensorimotor system, a complex task that can be solved in many ways and at different times.[31]

Note the first step in the story: an event that occurs in 'some completely unknown way'. Turning to the second step – the mutation yielding Merge – notice how it gets us only to a 'language' of thought. Finally, note that the authors explain externalization – the crucial step to what the rest of us mean by 'language' – as the outcome of unspecified solutions said to have been arrived at 'in many ways and at different times'.

Strengthening the impression that such vague statements represent good science, Chomsky and his supporters have recently been able to claim Richard Dawkins as a convert to their cause. For much of his career, Dawkins has poured scorn on the so-called 'hopeful monster' alternative to natural selection – the idea that you can explain a complex adaptation by invoking one random mutation affecting a lucky individual. In the second volume of his autobiography, published in 2015, Dawkins acknowledges Chomsky's genius and, in the process, feels obliged to eat his former words, announcing: 'The origin of language may represent a rare example of the "hopeful monster" theory of evolution.'[32]

Berwick and Chomsky keep on safe ground by reducing the vastly complex process of human behavioural and cognitive origins to within a sliver of nothing. No competition, no cooperation, no social life or politics, no signal evolution or communication, no sex, no childcare, no trust or deception, no evolutionarily stable strategies, no reason for long-term selection pressures to drive evolution in this direction or that. Nor do we find anything about word coinage, metaphorical usage, grammaticalization or any of the other known processes through which languages actually undergo evolutionary and/or historical change. Although Berwick and Chomsky's writing can be seductive, in the end it comes down to the truism of a mutation having occurred. Despite Dawkins' recent surprising conversion to Chomsky's position, Maynard Smith was surely right when he remarked that, from a Darwinian standpoint, the whole idea of explaining language as the outcome of a single mutation is 'not so much an argument as a cop-out'.[33]

CHAPTER 19

A SCIENTIFIC REVOLUTION?

Chomsky's need to politically neutralize every aspect of linguistics led him to stranger and stranger places, excluding from his scientific work not just politics, but social life of any kind and even any use of language in communication. We have seen how this resulted in paradoxes and conundrums without end. But, even as it had these unfortunate effects, it enabled the most powerful military and corporate sponsors of science in the United States to give full backing to Chomsky's 'cognitive revolution' new paradigm, secure in the knowledge that no troubling politics were involved.

As we have seen, Chomsky is for his supporters 'the world-renowned leader of an intellectual revolution'[1] and 'the leading figure in contemporary linguistics'.[2] Almost single-handedly, he established linguistics on a scientific basis, triggering an intellectual earthquake – the 'second cognitive revolution' – echoing, according to his followers, the immense scientific revolution sparked by figures such as Galileo, Descartes and Newton three centuries earlier.

While wishing to appear modest, Chomsky does not always discourage such extravagant claims. 'The discovery of empty categories and the principles that govern them', he writes, 'may be compared with the discovery of waves, particles, genes and so on.' The same, he continues, is true of the principles of phrase structure, binding theory and other subsystems of Universal Grammar:

We are beginning to see into the hidden nature of the mind and to understand how it works, really for the first time in history ... It is possible that in the study of the mind/brain we are approaching a situation that is comparable with the physical sciences in the seventeenth century, when the great scientific revolution took place that laid the basis for the extraordinary accomplishments of subsequent years and determined much of the course of civilization since.[3]

Chomsky's words here indicate hope for a similar outcome – for a profoundly liberating scientific revolution setting Western civilization on a new course.

From early on, however, sceptics began expressing worries as to the real institutional source and agenda behind such claims. Michel Foucault's approach – linking elite structures of knowledge with politics and power – began gaining currency during the 1970s. For some, it seemed clear that there had been no 'Chomskyan revolution' at all – only a situation in which, 'as all admit, transformationalists have succeeded in capturing the organs of power'.[4] One critic wrote of 'an increasing number of linguists' who realized 'that this allegedly linguistic revolution was a social *coup d'état*'.[5] The sociologist Stephen Murray insisted that soon after the publication of *Syntactic Structures*, Chomsky and his associates successfully 'engineered a "palace coup"'.[6]

For many, a persistent worry was to explain how that first book, *Syntactic Structures*, managed to get so sensationally reviewed almost before the ink on its pages had dried. Having been given the manuscript long in advance, Robert Lees – a doctoral student working on machine translation in the same laboratory as Chomsky[7] – praised *Syntactic Structures* for applying to linguistics the proven methods of established sciences such as physics.[8] A critic observes:

> Robert Lees' widely acclaimed 'review' of *Syntactic Structures* was written and published while Lees was a close associate and, for all practical purposes, still a doctoral student of Chomsky's at MIT ... one may wonder if Lees was indeed the sole author of the 'review', considering his employment situation at the time.[9]

The allegation here is that Chomsky may possibly have played a role in actually writing that review of his own book. There is no reason to believe this, but the suggestion by a hostile critic conveys much about the perceived impact and significance of that review. Chomsky himself later wrote: 'I expect that there would have been little notice in the profession if it had not been for a provocative and extensive review article by Robert Lees that appeared almost simultaneously with the publication of *Syntactic Structures*'.[10] Before Lees' review, few scholars had heard of the recent paradigm change in linguistics. After its publication, Chomsky began to receive invitation after invitation to explain what it was all about.[11]

Needless to say, Chomsky himself dismisses allegations that this and other details of his 'revolution in linguistics' amounted to a state-sponsored coup. Over the years, however, he has come to concede that, even if it was intellectually autonomous and therefore genuine, it was not much of a revolution. In the event, not one of Chomsky's ideas has proved remotely comparable with 'the discovery of waves, particles, genes and so on' in the natural sciences.

Given all this, we might ask what a state-sponsored coup might actually look like. Suppose it happened that a figure of authority such as the pope or a head of state chose to intrude into a scientific debate on a highly technical subject. Suppose their goal was to transform the state of play regardless of past achievements, experimental results or consensually agreed facts. We might expect that influential figure to carry out the task by developing a complex argument according to which past achievements were not achievements at all, experimental results are not necessary and facts that contradict the new argument can legitimately be ignored.

And so it came to pass. When the facts contradict a theory, most scientists would agree that the theory needs to go. But not Chomsky. His early flat dismissal of the need for empirical tests or experiments[12] has been repeated in equally strong statements right up to the present day. When asked which possible facts might conceivably refute Strong Minimalism, he replied: 'All the phenomena of language appear to refute it, just as the phenomena of the world appeared to refute the Copernican thesis.'[13]

Galileo, according to Chomsky, was willing to say: 'Look, if the data refute the theory, the data are probably wrong.'[14] Chomsky calls this the 'Galilean move towards discarding recalcitrant phenomena.'[15] It allows you to ignore as many awkward facts as you like if your ideas 'look right': 'You just see that some ideas simply look right, and then you sort of put aside the data that refute them.'[16]

As one critic has observed, this leaves Chomsky 'free to ignore any kind of contrary or refractory evidence to uphold his theory ... One wonders what would become of science if everyone took similar liberties.'[17]

An interviewer once asked Chomsky whether his claims about child language acquisition were based on observations or experiments:

'Have you done any experiments with children to learn how they acquire language', I ask.

'No. I hate experiments.'

'Has your distaste for experiments played any part in your thinking about linguistics?'

'I think that even without doing any experiments you can deduce some pretty striking things about what must be happening in the infant's mind.'[18]

By thought alone, Chomsky felt able to conclude that a child supposedly 'learning' English must in fact *already know* that language:

After acquiring a rudimentary knowledge of a language, he observes the linguistic data and asks himself, 'Are the linguistic data that I'm presented with consistent with the hypothesis that this is, for example, English?' And then he says to himself, 'Yes, it is English. O.K., then, I know this language.'[19]

Keeping to this pattern, Chomsky later announced that language acquisition is not a gradual process which goes through various complex stages, but is best understood as a simple and instantaneous event.[20]

Real scientists cannot afford to be arrogant. But if you are in the business of engineering a coup – or a cultural revolution – a certain level of arrogance and intolerance is required. Chomsky's qualifications in this particular respect have become part of the legend. For example, he once gave a talk followed by questions and discussion. The event was a conference in Texas in 1969. John Ross – a former student of Jakobson and Chomsky – began presenting counterexamples to Chomsky's claims. George Lakoff recalls: 'At each point, Chomsky cut him off and refused to let him finish, saying that no individual linguistic examples could possibly be counterexamples to his proposals.'[21]

Ross himself recalls: 'Noam wouldn't let me say the end of the question. He started drowning me out from the podium. He had a microphone and I didn't. So I had to sit down.'[22] Lakoff adds: 'We saw his treatment of Ross as scandalous and aggressive behavior.'[23]

Chomsky argues that it is misguided even to *expect* a theory to be evidentially true. He claims that, like any scientist, he is endowed with a dedicated faculty for spinning theories in a vacuum, regardless of the outside world: 'Call this the science-forming capacity . . . we may assume it to be fixed, in the manner of the language faculty.'

This mental module, he says, 'is supplemented with certain background assumptions, determined by the state of current scientific understanding'. The faculty then works like a demon or robot, independently of the scientist as human being. Confronted with a query, Chomsky explains, 'the science-forming capacity then seeks to construct a theoretical explanation that will respond to this query. Its own internal criteria will determine whether the task has been successfully accomplished.'[24]

'Successfully accomplishing' the task does not mean getting to the truth. 'Sometimes the theories produced may be in the neighbourhood of the truth', notes Chomsky. But since it is all happening in a vacuum, do not expect adjustment to anything in the external world. According to Chomsky, the unpredictable output of the faculty is no more constrained by evidence than is a lottery wheel: 'Notice that it is just blind luck if the human science-forming capacity, a particular component of the human biological endowment, happens to yield a result that conforms more or less to the truth about the world.'[25]

In his own case, Chomsky has been forced to admit that results which only yesterday somehow 'looked right' to his science-forming capacity have a disconcerting tendency to look quite wrong today.

While remaining true to his core mentalist assumptions, Chomsky has come close to disowning each and every detailed claim he has ever made:

> My own view is that almost everything is subject to question, especially if you look at it from a minimalist perspective . . . So, if you had asked me ten years ago, I would have said government is a unifying concept, X-bar theory is a unifying concept, the head parameter is an obvious parameter, ECP, etc., but now none of these looks obvious. X-bar theory, I think, is probably wrong, government maybe does not exist.[26]

In view of all this, does he still believe that his own work triggered a scientific revolution? During the pre-Minimalist years, he began entertaining second thoughts: 'my own sense of the field is that contrary to what is often said, it has not undergone any intellectual or conceptual revolution.'[27]

Still more emphatically, he declared that linguistics has 'not even reached anything like a Galilean revolution'.[28] At best, his life's work might just possibly qualify as a 'preliminary to a future conceptual revolution which I think we can begin to speculate the vague outlines of'.[29] When Chomsky was making these remarks during the early 1980s, no more than the haziest outlines of Minimalism were coming into view.

Shortly afterwards, the following exchange occurred:

> *How would you assess your own contribution to linguistics?*
> They seem sort of pre-Galilean.
> *Like physics before the scientific revolution in the 17th century?*

Yes. In the pre-Galilean period, people were beginning to formulate problems in physics in the right way. The answers weren't there, but the problems were finally being framed in a way that in retrospect we can see was right.

How 'pre-' do you mean? Are you saying that linguistics is about where physics was in the 16th century? Or are we going back still further, to Aristotle and to other Greek ideas about physics?

We don't know. It depends, you see, on when the breakthrough comes. But my feeling is that someday someone is going to come along and say, 'Look, you guys, you're on the right track, but you went wrong here. It should have been done this way.' Well, that will be it. Suddenly things will fall into place.[30]

With each new disappointment, then, Chomsky turns with undimmed optimism toward the future – to a moment of revelation when, quite suddenly, 'things will fall into place'. The announcement of Minimalism in the early 1990s was supposed to be that moment.

A full list of the zigzags and rethinks in Chomsky's intellectual Odyssey would be a long one, but the following examples stand out.

Phrase structure rules

Initially, Chomsky proposed that 'phrase structure rules' are fundamental to Transformational Grammar;[31] subsequently, we were informed that these same rules 'may be completely superfluous'.[32]

Transformational rules

Chomsky initially proposed that 'transformational rules' are either optional or obligatory;[33] subsequently, he announced that all are in fact optional[34] – a proposal then abandoned in its turn.[35]

Grammar

In his early work, Chomsky insisted that the notion of a 'well formed' sentence is fundamental to linguistics.[36] On converting to Minimalism, however, he announced that 'we have no notion of "well formed sentence"',[37] dismissing even 'grammatical' as a notion 'without characterization or known empirical justification'.[38]

Deep structure

One of Chomsky's most dazzling early ideas was his insistence that the meaning of a sentence is determined not by the superficial clustering of its words – its 'surface structure' – but by its underlying 'logical kernel', later termed its *deep structure*. In 1967, Chomsky insisted on the 'inadequacy' and, indeed, 'irrelevance' of surface structure to the specification of meaning.[39] Shortly afterwards, this idea was reversed: in 1976, we find him arguing that surface structure is by no means 'inadequate' or 'irrelevant'. Quite the opposite: it now transpires that 'a suitably enriched notion of surface structure suffices to determine the meaning of sentences'.[40] It turned out, then, that Chomsky's previous position – central to Transformational Grammar as proclaimed from the rooftops – was so wrong that it would be improved by saying the opposite.

Was Chomsky the new Copernicus? If so, the great astronomer now seemed to be clarifying that, after all, the sun circled around a motionless earth – or at least around a 'suitably enriched notion' of the earth. Shortly afterwards, Chomsky had retracted even this retraction. Having replaced 'deep structure' and 'surface structure' by the more abstract and obscure 'D-structure' and 'S-structure', he now explained that it made no difference whether you saw a difference between these levels:

> A theory that postulates a D-structure or LF level distinct from S-structure is only subtly different from one that does not, and it is not at all clear that the theories, when properly understood at the appropriate level of abstraction, will prove to be empirically distinguishable.[41]

The slippage continued until nothing was left. Deep structure – once celebrated as fundamental[42] – finally bit the dust in 1995. From a minimalist perspective, the world was informed, 'D-Structure disappears, along with the problems it raised.'[43] Since that moment, no more has been heard of D-structure, S-structure or any of the supposed levels of structure in between.

The nature of the language organ

Since meeting Eric Lenneberg in the 1960s, Chomsky has always insisted that language is biology. He presents it as a biological organ, which would make it – like the heart, for example – something with average weight, distribution in the body, physical dimensions and so forth. All along, however, this idea has been accompanied by a strikingly incompatible one, expressed with particular clarity in a journal article co-authored by Chomsky among others in 2014: 'Language, in all aspects, consists of abstract units of information.'[44] The problem here is that 'abstract units of information' do not weigh anything, do not take up space and – in short – do not exist in the physical world. If the fabled language faculty is in fact just a set of abstractions, we need to know. If that really is the Chomskyan position, we can at long last set aside the former insistence that the term 'language' refers to a biological organ.

Complex or simple?

Initially, Chomsky described the 'language organ' as hugely complex. As late as 1979, he was still describing it as 'among the most complex structures in the universe.'[45] During a debate with Jean Piaget in 1975, he implied that it was a physical object with a specific size and location in the brain, describing it as 'that little part of the left hemisphere that is responsible for the very specific structures of human language'.[46] On most occasions, however, he has tended to be more circumspect. While insisting that we should study the language organ in essentially the way biologists study the liver or heart, he hastens to warn that 'of course it is not an organ in the sense that we can delimit it physically'.[47]

With the adoption of 'Minimalism' in the 1990s, the once wildly complex portion of the left cerebral hemisphere morphed into a mysterious object lacking weight, size, location or any other measurable properties or physical dimensions. Far from being almost unimaginably complex, the strange organ's most astonishing, marvellous and inspiring feature now turned out to be its stunning simplicity. Its beauty and perfection lay in the curious fact that it was reducible in essence to just one thing: 'recursion'.[48] What exactly did that mean? Following a decade of arcane controversies and technical debates, Chomsky announced in 2009 that he had had enough: 'Here, some clarification should be made: there's a lot of talk about recursion and it's not a mystical notion: all it means is discrete infinity. If you've got discrete infinity, you've got recursion.'[49]

So that was that. Apart from noting the obvious and familiar fact that in language we arrange and rearrange words to produce sentences without limit – 'the infinite use of finite means' – Chomsky apparently had nothing to say.[50]

Universal Grammar

Chomsky's aim, from the mid-1960s, was to elucidate the properties of 'Universal Grammar' or UG, defined as follows: 'UG may be regarded as a characterization of the genetically determined language faculty. One may think of this faculty as a "language acquisition device," an innate component of the human mind.[51]

UG, then, is another name for the language organ. More specifically:

> It is the sum total of all the immutable principles that heredity builds into the language organ ... We now assume that universal grammar consists of a collection of preprogrammed subsystems that include, for example, one responsible for meaning, another responsible for stringing together phrases in a sentence, a third one that deals, among other things, with the kinds of relationships between nouns and pronouns ... And there are a number of others.[52]

Quite a lot, then – rather more than you would expect evolution to produce all at once from a single mutation. But those rich conceptions of Universal Grammar were formulated during the 1980s, prior to the dramatic announcement of 'Minimalism' a decade later, when the vast complexities of the relevant organ were suddenly repudiated in favour of that same entity's astonishing simplicity. In 2012, a sympathetic interviewer felt bold enough to ask the following question: what exactly *is* UG at this stage? Here is Chomsky's reply: 'Well, what's Universal Grammar? It's anybody's best theory about what language is at this point. I can make my own guesses.'[53]

Summing up decades of intensive work, then, Chomsky can only tell us that it's anyone's guess.

Creativity

Early in his career, Chomsky asserted that 'mathematical developments' enabling 'the full-scale study of generative grammars' made it possible for the first time 'to attempt an explicit formulation of the "creative" processes of language'.[54] This was widely perceived to be the whole point of Chomsky's early work. 'By focusing on syntax,' confirms Frederick Newmeyer, 'Chomsky was able to lay the groundwork for an explanation of the most distinctive aspect of human language: its creativity. The revolutionary importance of the centrality of syntax cannot be overstated.'[55]

When Chomsky celebrates the 'creative aspect of language', he means 'the ability of normal persons to produce speech that is appropriate to situations though perhaps quite novel, and to understand when others do so.'[56] 'It is clear', Chomsky explained in 1964, 'that a theory of language that neglects this "creative" aspect of language is of only marginal interest.'[57]

But then came the inevitable U-turn, as he announced to the world that this ability of speakers is as much a mystery to him as to anyone else: 'We do not understand, and for all we know, we may never come to understand what makes it possible for a normal human intelligence to use language as an instrument for the free expression of thought and feeling.'[58]

So it now transpired that the 'creative processes of language' were unlikely *ever* to receive that 'explicit formulation' which had been heralded to the sound of trumpets as the revolution began.

Merge

Chomsky has often been asked where lexical concepts come from. 'That's a huge issue', he concedes. He doesn't know. The one idea he insists on is that, somehow, the operation known as 'Merge' (i.e. putting words together) presupposes lexical items, wherever these come from and whatever they might be:

> Among the properties of lexical items, I suspect there are parameters. So they're probably lexical, and probably in a small part of the lexicon. Apart from that there's the construction of expressions. It looks more and more as if you can eliminate everything except just for the constraint of Merge. Then you go on to sharpen it. It's a fact – a clear fact – that the syntactic objects you construct have some information in them relevant for further computation . . .[59]

Translating all this, Chomsky seems to be explaining that you combine words in ways permitted by their properties, whatever these might be.

Remarkably few of Chomsky's concrete proposals, then, have managed to survive for long. One is reminded of a man on the doorstep fumbling with his key in the half-light. He twists it clockwise, then anticlockwise, turning it this way and that. Despite all his fumblings, the lock just will not yield. To those watching, the most likely explanation is that he's got the wrong key.

Chomsky's staunch admirer Neil Smith, however, justifies these fumblings as *artistic licence*:

> The closest, if unlikely, parallel comes from the 'relentlessly revolu-tionary' artistic development of Picasso, who evoked admiration, bemusement, and incomprehension in equal measure. For those who appreciated the paintings of the Blue and Pink periods, cubism seemed

like an assault; for those who grew to appreciate cubist pictures, the excursions into surrealism and welded sculpture were baffling. But even those who don't like it agree that *Les Demoiselles d'Avignon* changed art forever, and those who banned his work during the war appreciated the awesome power of *Guernica*. Chomsky has several times overthrown the system he has developed, confusing many, alienating some, inspiring a few.[60]

It is an intriguing parallel and, if you perceive Chomsky's theories as alluringly beautiful, possibly all too accurate. But, while we might agree with Smith about Chomsky's resemblance to Picasso, a still better parallel is the Russian suprematist painter Kazimir Malevich. Chomsky's Minimalism has the uncompromising audacity of Malevich's *Black Square*. In that extreme example of revolutionary nihilism, light, colour, pattern – all the diverse materials of traditional visual art – have been reduced not just to a minimum but to zero. Nothingness is being, being is nothingness. If the aesthetically alluring object appears to us simple, perfect, certain and absolute, the explanation is that there is nothing there.

Referring to Chomsky's creative exploits over the years, Peter Matthews – another staunch supporter – comments generously: 'They have changed as the clouds change in the sky, or as the key changes in a classical sonata movement.'[61] We may agree that constant change relieves boredom. Yet another advantage, if we are talking about art, may be psychological benefit for the artist. While conceding that his recent thinking might be completely wrong, Chomsky himself suggests that 'the Minimalist Program, right or wrong, has a certain therapeutic value'.[62]

But is this not supposed to be science? 'I don't think it is good', comments Steven Pinker:

Because Chomsky has such an outsize influence in the field of linguistics, when he has an intuition as to what a theory ought to look like, an army of people go out and reanalyze everything to conform to that intuition. To have a whole field turn on its heels every time one person wakes up with a revelation can't be healthy. It leads to a lack of cumulativeness, and an unhealthy fractiousness. It's an Orwellian situation where today Oceania is the ally and Eurasia is the enemy, and tomorrow it's the other way around.[63]

Chomsky can impose such reversals because, thanks to his massive institutional backing, he *is* 'modern linguistics'. It therefore seems that he can do what he likes. Conceding this point while attempting to justify it, loyal follower Robert Fiengo counters the allegation that Chomsky violates every norm of scientific procedure with the strangest of arguments. Because Chomsky is special, claims Fiengo, we can exempt him from the normal rules: 'Let us remember we are talking about someone who tries to reinvent the field every time he sits down to write. Why should we expect Chomsky to follow normal scientific procedure?'[64]

Or as yet another admirer has explained: 'in no other contemporary cognitive science is the work of a single individual so key and irreplaceable. In a nontrivial sense, the history of modern linguistics *is* the history of Chomsky's ideas and of the diverse reactions to them on the part of the community.'[65]

This is an extraordinary statement – so bizarre that something must surely be wrong.

According to Frederick Newmeyer, the proof that Chomsky really did accomplish a scientific revolution lies not in any evidence that his theories actually worked. Nor need we expect him to have united his discipline around him – that certainly never happened. What made it a genuine revolution was the indisputable fact that Chomsky's intervention proved impossible to ignore: 'There was a Chomskyan revolution because anyone who hopes to win general acceptance for a new theory of language is obligated to show how the theory is better than Chomsky's. Indeed, the perceived need to outdo Chomsky has led him to be the most attacked linguist in history.'[66]

The implications of this assertion are worth pondering. Taken literally, it would mean that if, say, the pope provoked controversy by intervening to serious effect in modern science, we would have to hail this on the grounds that no one could possibly ignore the results.

Newmeyer's strange claim is the reverse of the truth. The evidence that Galileo accomplished a scientific revolution lies in the fact that in the end, despite the Vatican's best efforts, a hard-headed scientific community endorsed his once heretical views. The difference with Chomsky is that, despite vast sums of Defense Department money and consistent institutional backing over half a century, his other-worldly doctrinal pronouncements have produced no sign of consensus or agreement, but instead unending controversy, uproar and incredulity at the implausibility of it all.

In the period since the announcement of Minimalism, Newmeyer – once one of Chomsky's staunchest supporters – has become a disillusioned critic: 'The empirical results of the minimalist program that Chomsky can point to . . . do not come close to approaching his level of rhetoric about the program's degree of success.'[67] This is significant because – more effectively than anyone else – it was once Newmeyer who, as an accomplished linguist and historian of linguistics, championed Chomsky as the scientific revolutionary of our age.

Although Chomsky's admirers may try to keep abreast, in the final analysis his position remains unassailable, cut off – and always one step ahead. 'At least as I look back over my own relation to the field,' he explains, 'at every point it has been completely isolated, or almost completely isolated. I do not see that the situation is very different now.'[68]

Massive institutional endorsement and corresponding influence, then, sit paradoxically alongside utter solitude. Admittedly, we might interpret this positively, in principle at least. Genius on the scale of an Einstein or Galileo must inevitably bring with it a deep sense of personal mission and isolation – that comes with the job. But neither Galileo nor Einstein was *ultimately* alone. When their theories were taken up by others and tested, it turned out that they worked. Chomsky's situation is quite different: he *really is* ultimately isolated. Shortly after he proposes each new theoretical edifice, cracks in it begin to appear. The critic who first discovers this, very often, is none other than Chomsky himself.[69]

Designing and manufacturing his own telescope, Galileo arrived at precision and simplicity through careful measurements and detailed observations, basing his findings on these. Chomsky does something very different. Relying on personal intuition, he sets aside *all* observations concerning particular languages. As he puts it, studying 'the distinctive properties of language – what really makes it different from the digestive system . . . means abstracting away from the whole mass of data that interests the linguist who wants to work on a particular language'.[70] Perhaps the most infamous statement of this position was Chomsky's 1980 comment that you can deduce general principles concerning all the world's languages even if your knowledge is restricted to just one, perhaps your own: 'I have not hesitated to propose a general principle of linguistic structure on the basis of observation of a single language. The inference is legitimate, on the assumption that humans are not specifically adapted to learn one rather than another human language.'[71]

To be fair, Chomsky immediately qualified that statement: 'To test such a conclusion, we will naturally want to investigate other languages in comparable detail. We may find that our inference is refuted by such investigation.'[72] To say that Chomsky's conclusions have indeed been refuted time and again would be an understatement.[73] In fact, although Chomsky never admits it, the story of his dramatic reversals of earlier positions can mostly be explained in this way, as experts in particular languages – for example in

Amazonia or Aboriginal Australia – have come up with counterrexamples to Chomsky's initial assumptions, which were based heavily on English. No one can deny that much has been learned in this way. Without Chomsky's provocative claims, spurring others to disprove them, the development of linguistics over the decades would have been very much poorer.

But this does not alter the fact that Chomsky's insistence on 'abstracting away from the whole mass of data' on particular languages was misguided from the start. It is as if Galileo had tried to understand the solar system by ignoring detailed records of the actual paths taken by each *particular* planet. As the philosopher Christina Behme points out:

> If, when studying L_1, one should abstract away from the whole mass of data of interest to the linguist about L_1, the same logic would hold for $L_2 \ldots L_n$. So one would have to abstract away from everything of linguistic interest about all languages to uncover the nature of language and explain how it differs from digestion.[74]

Apart from a few truisms – such as that all languages have words which can be combined to produce sentences – one would then end up with approximately nothing. As we have seen, Chomsky's current position on Merge or Universal Grammar amounts to little more. The difference between him and Galileo is that the astronomer's initial claims about a moving Earth survived every test. Convinced that he was on the right track, Galileo saw no need to keep changing his mind on key issues during his lifetime, as Chomsky has repeatedly done.

Time after time, Chomsky has laid claim to a grammatical constraint supposedly operative in all the world's languages. Time after time, his colleagues have checked things out and found some example that contradicts the claim.[75] Rudolf Botha recalls, for example, how Chomsky's Government and Binding (GB) theory led to the prediction that a native English speaker should find the following sentences ungrammatical:

a. *They read each other's books.*
b. *They heard stories about each other.*
c. *They heard the stories about each other that had been published last year.*

The fact that native English speakers find these sentences perfectly acceptable clearly violated Chomsky's theoretical prediction.[76] Instead of gracefully admitting this, however, the Lord of the Labyrinth (Botha's name for Chomsky)[77] claimed that an important distinction needed to be made between 'core' aspects of grammar and 'peripheral' ones. Those sentence structures which did not fit his theory were said to be 'marked' – meaning conveniently released from the constraints of the 'core grammar'. Where structures belong only to the 'periphery', he explained, we need not expect them to conform to the rule. On this basis, Chomsky sought to justify his claim that 'the predictions of the GB system are in fact correct . . .'[78] He went on to observe that, anyway, sentences of this particular kind 'appear to be rare' in the languages of the world.[79]

The transparent foul play here caused such uproar that Chomsky – to his credit – quickly realized that he had no choice but to own up. Referring to those inconvenient sentence types and his reaction to them, he admitted in 1982: 'I've always assumed they were a little odd in their behavior, but they really just didn't fall into the theory I outlined there at all, so I just had to say they're totally marked. I gave a half-baked argument about that, and there was some bad conscience, I must concede.'[80]

That was refreshing. Botha points out, however, that Chomsky's admission on this particular occasion was 'a statement about the past' and held 'no promise for the future'.[81] It was unlikely to divert Chomsky from his habitual response to accusations of foul play, which was to avoid admitting anything and brazen things out. Faced with counter-evidence, Chomsky's usual tactic is to claim that refutation helps by offering 'the possibility of a deeper understanding of the real principles involved'.[82] To be wrong, then, is – given a 'deeper understanding' – to be right.

In this way, Chomsky keeps hope alive. *Syntactic Structures* was acclaimed by his disciples as 'The Old Testament', superseded in key respects by 'The New Testament' – *Aspects of the Theory of Syntax*.[83] *The Minimalist Program* would then be the fifth – or perhaps sixth – in a sequence of Testaments; there have been more revelations since. Not many linguists have needed to clarify their human, mortal status by insisting '*I am not God*', but in 1992 Chomsky felt obliged to say exactly that.[84] Paul Postal likens him to a millenarian preacher who keeps confidently predicting the end of the world:

Then the day would come, the world would not end, and one might figure that the movement would collapse, right? But no, quite the contrary. The fervor of the group members became even greater. They would go out and proselytize, passionately trying to get more members. A new date would be set. When that date would arrive, the prediction would again obviously be falsified and one would assume that the movement would this time surely collapse. No. Again, there was increased proselytizing, increased fervour . . .[85]

Having abandoned one theory, Chomsky turns to his next, while vigorously denouncing anyone who dares to question this latest idea.

But if the assumptions of the project were a fiasco from start to finish, why did they gain such extraordinary institutional support? Here is one possible explanation, which may be consistent with Chomsky's own analysis of the workings of power. The Pentagon is the Vatican of our times. It is a state within a state, an apparatus wielding vast resources, shaping the sponsorship and funding of research projects in virtually every branch of science, enforcing a regime of subtle censorship and patronage sanctioned by loss of income or worse – and cloaking its self-serving activities in a veneer of piety and concern for the welfare of all.

The following exchange is from an interview conducted in 1995:

Q: One of the questions you are often asked after your talks is the one about, How can you work at MIT? You've never had any interference with your work, have you?

A: Quite the contrary. MIT has been very supportive. I don't know the figures now, but in 1969, when the only serious faculty/student inquiry was undertaken, into funding, there was a commission set up at the time of local ferment about military labs, and I was on it, and at that time MIT funding was almost entirely the Pentagon. About half the Institute's budget was coming from two major military laboratories that they administered, and of the rest, the academic side, it would have been something like 90% or so from the Pentagon. Something like that. Very high. So it was a Pentagon-based university. And I was at a military-funded lab.

'But', added Chomsky, 'I never had the slightest interference with anything I did.'[86]

Chomsky's activist admirers tend to express bewilderment at this point. No interference? Yet Chomsky has always been an anarchist? How can that possibly be?

In fact, however, Chomsky's scientific work – unlike his activism – did not trouble the authorities at all. In the wider scheme of things, even his left-wing politics may not have posed much of a threat. The contradiction is resolved when we remember that institutions like the Vatican require not only sinners, but also a sprinkling of saints. To enhance their public relations, they need genuinely idealistic individuals whose expressions of political dissidence may strike a chord with key sectors of the public – in particular, with those who might otherwise lead revolts from below. Behind the scenes, the string-pullers and fixers need real science – the Vatican's instruments of torture must actually work, its gunpowder properly explode – but equally they need figures who can provide stained-glass windows, painted ceilings, comforting hymns and persuasive myths.

In 2014, as I mentioned in Chapter 1, the Roman Catholic newspaper *The Tablet* reported on an address delivered by Chomsky to a Vatican audience in Rome:

> Just what might one of the world's great atheist scientists and philosophers have in common with the Church? Quite a lot, as became clear when Noam Chomsky addressed a Vatican foundation that aims to promote dialogue between science and religion.
>
> On the face of it, Noam Chomsky and the Catholic Church seem unlikely bedfellows . . . A few days ago, however, he found considerable common ground with the Church.[87]

The reporter tells us that their atheist guest speaker 'was certainly made to feel at home'. In his introductory remarks, Cardinal Gianfranco Ravasi, president of the Pontifical Council for Culture, welcomed Chomsky warmly as 'one of the princes of linguistics'.[88] The event was the brainchild of the Science, Theology and the Ontological Quest Foundation, an offshoot of the commission set up by Pope John Paul II to investigate what the theologians were still calling 'the Galileo affair'.[89]

In Vatican circles, Galileo still remains something of a worry. To the evident relief of his audience, however, Chomsky acted less like Galileo and more like a present-day pope. In contrast to Galileo's fearless scientific stance – which shook the Church to the core – Chomsky went out of his way to express intellectual humility and deference to mystery. *The Tablet* reminded its readers of Chomsky's core belief 'that all human beings have an innate capacity for language', and continued:

> It is this belief that provides Chomsky with grounds for mystery, as he explained to his audience on Saturday . . . Humans are limited by the biological blueprint that nature has given them. Crucially, he believes, these inherent constraints extend to our 'higher mental faculties'. We simply don't have the capacity to understand everything, not least the workings of our own minds. This position has earned Chomsky . . . the label 'new mysterian' . . . he underlined the limits of human knowledge, arguing that a fundamental understanding of the world is likely to remain forever beyond the reach of scientists. Indeed, he urged his audience to embrace mystery.[90]

Journalists are not always reliable. But an examination of the full version of Chomsky's talk, as published subsequently on Chomsky's own website, revealed that *The Tablet*'s account was accurate, as far as I could see.[91] To be absolutely sure, I consulted a more extensively referenced version, entitled 'The Mystery of Language Evolution', co-authored by Chomsky with Marc Hauser and others, and published as a journal article the same year.[92] As you would expect in a scientific paper, the terminology here is more opaque and scientific sounding, but, yes, the punch-line is the same – a call to embrace mystery. The authors pour scorn on all efforts to explain the origins of language, predicting that, on current evidence, the baffling problem is likely to remain 'one of the great mysteries of our time'.[93]

The Tablet summed up the invited speaker's message to the Vatican as follows: 'Chomsky says that the limits of science are caused simply by mankind's finite brainpower.'[94] While natural selection can explain certain things, the creative use of language is certainly not one of them. Questions about human free choice, whether moral or political, are off limits. Far from promising revolutionary new insight into human linguistic creativity – the whole point of the 'Chomskyan revolution' as announced from the rooftops in the 1960s – Chomsky reassured the Vatican that we know scarcely more about these things than we did 500 years ago. Unless or until this bleak situation unexpectedly changes, we should be honest about our God-given limitations and embrace mystery, consigning such difficult questions to 'that obscurity in which they ever did and ever will remain'.[95]

CHAPTER 20

MINDLESS ACTIVISM, TONGUE-TIED SCIENCE

During a 1971 interview, Chomsky frankly admitted that his antipathy towards social science might be interpreted as a personal failing hampering his activism:

> I am skeptical as to whether the fundamental problems of man and society can be studied in any very profound manner, at least in ways resembling scientific inquiry, perhaps because of temporary gaps in our understanding, or perhaps because of deeper limitations of human intelligence. These personal tendencies and beliefs probably lead me to underestimate the potentialities of activism or perhaps even social criticism and analysis, as well as to restrict, no doubt improperly, my own personal involvement.[1]

Chomsky, then, is acutely aware that his own activist involvement might suffer from the fact that it has no basis in science. Justifying his personal doubts about a science of 'man and society', he wavers between 'temporary gaps in our understanding' and 'deeper limitations of human intelligence'. It is easy to see that such philosophical suggestions do little to explain his real anxieties. His rejection of the claims made by self-styled 'social scientists' go far beyond intellectual doubt. Chomsky has always had good reason to suspect that any Harvard or MIT anthropologists or sociologists in receipt of Pentagon funding were pursuing something closer to state-sponsored deception, propaganda and criminal fraud than to honest scholarship.

In 1956, the Office of the Chief of Psychological Warfare, the section of the US army responsible for all aspects of unconventional warfare, set up the Special Operations Research Office (SORO). While most Pentagon research programmes focused on the bombs, tanks and missiles needed to wage war, SORO's work centred on ideas. Its researchers produced 'area studies, reports, analyses of the causes of revolution, descriptions of communist underground movements, and assessments of psychological strategy in dozens of foreign nations'.[2] Armed with this arsenal of social scientific information, the US army hoped to make the complexities of Third World anti-colonial and national liberation movements comprehensible and manageable. SORO's task, as stipulated in its contract with the army, was to provide the army with 'scientific bases for decision and action' in the battle for 'hearts and minds'.[3]

This involved a relentless new focus on the anthropological study of 'human nature'. In 1962, psychologist Leonard Doob explained to an audience of military officials and academics that the armed services had no use for '200 monographs on the 200 tribes of Nigeria'. Rather, they required to have explained to them the 'basic concepts involved in Nigeria or any place in the world because everywhere there are human beings'. Instead of remaining satisfied with a growing database of area knowledge about strategically important peoples, SORO would seek to make intelligible the complicated psychology, politics and sociology of Cold War geopolitics. As MIT political scientist Ithiel de Sola Pool explained, the job of anthropologists and social scientists was to bring 'the benefits of formalization and systematization to the experience-derived wisdom' of military officers.[4]

Initiated in 1960, one of SORO's most significant projects was Project Revolt, a series of studies designed to anticipate and prevent communist revolutions. This required intellectual advances in the study of revolution itself, an area of work considered to be as yet in its infancy. A 1962 symposium held at the American University in Washington, DC, brought social scientists and military men together in an effort to address this problem. The atmosphere of the time is conveyed by Lieutenant General Arthur G. Trudeau's welcoming address:

> I want to say right here and now . . . that our whole civilization is on trial today. Forces are loose in this world that would destroy all that we hold dear. These forces stem from a malignant organism that grows and thrives on human misery – which reaches out its long tendrils in every field of human endeavor, seeking to strangle and destroy.[5]

'You know as well as I', he continued, 'that in our effort to remove this communist cancer from the world society we have relied principally on our superior advantages in the physical sciences . . . Science and technology have given us the tools by which we can meet and master the communist challenge.'[6]

US superiority in physics and hardware, however, would prove fruitless 'without concurrent improved understanding of peoples and their societies, particularly in the underdeveloped areas of Asia, Africa, the Middle East and South America'.[7] The United States was faced with 'a multidimensional communist challenge – in paramilitary warfare, in psychological warfare, and in the conventional and nuclear fields – in short, from zero to infinity across the military spectrum of force'.[8] All this had led the lieutenant general 'to look to the social sciences for assistance in our efforts', the aim being to develop concepts and techniques 'for successful organization and control of guerrillas and indigenous peoples by external friendly forces'.[9]

Princeton sociologist Harry Eckstein was one among many academics eager to help, urging that 'we desperately need knowledge of how to turn revolutionary forces to our own account, how to use revolutionary ferment'. As he explained: 'The final knowledge we need is knowledge of the causes of revolutionary ferment in order to be able to repress it at its source, or for that matter to induce it at the source'.[10]

Eckstein's basic complaint was that the communistic enemy 'are immeasurably farther ahead of us in revolutionary theory'.[11] Eager to catch up in this respect, much of the US social science establishment now followed Eckstein in abandoning any last pretence at dispassionate scholarship in the rush to benefit from the military funding now expected to pour in. Only two years later, in 1964, Eckstein published an edited volume – entitled *Internal War* – boasting contributions from Talcott Parsons, William Kornhauser, Lucian Pye, Sidney Verba, Gabriel Almond and Seymour Martin Lipset among other sociological luminaries of the time.[12]

It is against this background that we need to view Chomsky's ferocious hostility to so-called 'social science'. Once we take all this into account, moreover, it becomes easier to understand his categorical insistence that his political analyses are completely devoid of any theoretical basis and have nothing to do with science. He says that 'one should be careful not to link the analysis of social issues with scientific topics'.[13] Keeping the two in isolation is a constant theme of his. So utterly separate is his own scientific work, he tells us, that his political output 'could have been written by someone else'.[14]

Many of Chomsky's most ardent supporters on the left have expressed their bewilderment at this strange self-distancing of Chomsky from himself. Why produce a painstaking analysis of an episode in recent history – say, US involvement in the Vietnam War – only to spoil the effect by saying that there is nothing scientific or even remotely theoretical about the analysis? The problem is so puzzling that one sympathetic writer has devoted an entire book to it, albeit without coming to any firm conclusion.[15]

Context is everything. From Chomsky's perspective, serious scientific work is something which nowadays requires a period of training, a supportive institutional framework, ongoing opportunities for collaboration between experts and – most often – substantial corporate funding as well. As a scientific linguist, he himself has always enjoyed such favourable conditions. But that has applied only to his scientific work. When he turned to politics, he was always – and for obvious reasons – on his own. At no time did Chomsky receive institutional or financial backing for his sustained critique of US foreign policy. He would have refused such support even if it were offered.

We have seen how, during the late 1950s and early 1960s, a growing number of sociologists, anthropologists and political scientists began benefiting from state and corporate support much as Chomsky had been receiving it as a linguist. From where he was standing, *inside* this framework of institutional support, Chomsky could not afford to allow any perception that his own work amounted to collusion. He knew that large sections of the social science community were working hand in glove with their corporate-military paymasters in a sustained effort to find new ways to control people in the Third World, subvert and defeat their wars of 'national liberation', spy on resistance networks and install friendly dictatorships. To maintain his integrity, Chomsky would have none of this, situating himself categorically outside this entire specialist and expert framework. That may be sufficient explanation for his decision to identify as an activist no different from the ordinary citizen, speaking in his own voice outside the state-sponsored juggernaut of science. This surely explains why his writings on political issues are so proudly declared non-theoretical and independent of his science.

For the US establishment, the ultimate intellectual enemy was, of course, Karl Marx. By way of example, Chomsky cites Stanford University's housing of and support for 'the Hoover Institution on War, Revolution and Peace', which was directed by its private benefactor as follows: 'The purpose of this institution must be, by its research and publications, to demonstrate the evils of the doctrines of Karl Marx – whether Communism, Socialism, economic materialism, or atheism – thus to protect the American way of life from such ideologies, their conspiracies, and to reaffirm the validity of the American system.'[16] One problem for the recruited sociologists was that Marx was an ancestral figure internal to their own discipline – no area was free from his influence. Combating his ideas, therefore, inevitably threw them into an 'internal war' of their own.

From early in the twentieth century, sociologists and historians across the West had come to accept Marx's insight that the dominant ideas in any epoch are best understood as those of the dominant socio-economic class. His general formulation is well known:

> The ideas of the ruling class are in every epoch the ruling ideas, i.e. the class which is the ruling material force of society is at the same time its ruling intellectual force . . . The ruling ideas are nothing more than the ideal expression of the dominant material relationships, the dominant relationships grasped as ideas.[17]

Having internalized this approach, scholars in search of explanatory theories began looking beyond the ideas and agendas of prominent individuals, exploring instead how developments in technology, production, distribution and exchange determine the balance of power between forces at a deeper level. 'Morality, religion, metaphysics, all the rest of ideology and their corresponding forms of consciousness', as Marx and Engels wrote in 1845, 'have no history, no development; but men, developing their material production and their material intercourse, alter, along with this their real existence, their thinking and the products of their thinking. Life is not determined by consciousness, but consciousness by life.'[18]

By the middle of the twentieth century, this materialist philosophical cornerstone of Marxism had become widely influential, producing repercussions across the humanities and social sciences as a whole.

Marx and Engels had aimed to launch a worldwide social and political revolution based intellectually on a scientific one – on what philosophers of science would nowadays term a 'paradigm change'.[19] It hardly needs stressing here that the two thinkers failed during their lifetime, while early in the next century their anticipated revolution was stopped in its tracks and subsequently forced into reverse by the most brutal of material forces.

The first catastrophe was the crushing of the international labour movement that Marx had helped found, when – from August 1914 – the dream of working-class solidarity was poisoned by nationalism and shattered in the mud and blood of trenches dug right across Europe. The Great War's closing years saw mounting waves of desertion and mutiny that culminated in the February 1917 overthrow of the Russian tsar and – in November 1918 – of the German kaiser as well. Europe as a whole appeared momentarily to be teetering on the brink of revolution. But it was not to be. Victorious only in a very limited way in Russia, the long-anticipated revolution was strangled almost at birth. Subsequent waves of counterrevolution, nationalism and defeat culminated in 'Socialism in One Country' in the Soviet Union and National Socialism in Germany.

The second catastrophe was a continuation of the first, as, within two decades, the world began descending helplessly into yet another war. After five years of slaughter, genocide and war crimes against civilians perpetrated by both sides, Nazi Germany and its allies were finally crushed. In the immediate post-war years, popular unrest across much of liberated Europe and Asia rekindled sparks of revolutionary hope. Once again, however (as Chomsky among others has well documented), the resistance was derailed, drowned in blood or successfully contained, any immediate prospects for revolution yielding to a carve-up between power blocs and half a century of superpower rivalry and the Cold War.

After the Second World War, the United States of America faced a daunting task. The intellectual, scientific, economic and military energies that it had successfully deployed against Nazi Germany needed to be preserved intact for redeployment against the new official enemy, 'world communism'. Following Fidel Castro's victory in the 1959 Cuban revolution, 'Better dead than red' – a slogan anticipated, ironically, by the German Nazis during the Second World War (*Lieber tot als rot*) – became a popular right-wing mantra.[20] The fight against communism was not only a military and economic challenge, but was equally a political and intellectual one. Much of the artistic, literary and academic intelligentsia across the Western world still tended to be not only anti-fascist, but in many ways sympathetic towards Marx. The question was how to break this spell.

In the end, it was broken most decisively by the cognitive revolution. While it would be an exaggeration to describe this as the main factor, it was certainly a primary one as far as intellectual life was concerned. The new cognitive paradigm placed mind over matter, consciousness over life, theory over practice – in such a way that the fundamental premises of Marxist materialism were almost imperceptibly undermined, dissolved and eventually dismissed as no more than old-fashioned dogma.

The intellectual most closely associated with this molecular process was Noam Chomsky. He did not achieve this extraordinary result – the overthrow of Marxism – by indulging in the kind of sectarian disputes for which the left is infamous. He did not achieve it by explicitly counterposing his own particular version of anarcho-syndicalism to the more Marxist-inspired alternatives then in circulation. That kind of thing would have been pointless, as it always is. Chomsky did not even need to mention Marx, because in the early Cold War years the job could be done far more powerfully by redefining linguistics as natural science, by insisting that social science is essentially fraudulent and by attacking Marxism at its roots – setting up a firewall designed to separate 'science' from any kind of social or political activism, with his own success in both spheres serving as a model.

David Golumbia is a historian of the period, with a special focus on the cognitive revolution. Golumbia is interested in how someone with Chomsky's deeply held anarchist convictions managed to retain the trust and endorsement of the American corporate and intellectual establishment. While acknowledging Chomsky's political courage and sincerity, Golumbia views his success as part of 'a deliberate and also largely covert effort to resist the possibility of communist/Marxist encroachment on the US *conceptual* establishment'.[21] The entire intellectual landscape was decisively reshaped, as 'individuals, government entities including military and intelligence bodies, and private foundations like the RAND Corporation, promoted values like objectivity and rationalism over against [*sic*] subjectivity, collectivity, and shared social responsibility'.[22] In this way, Marx's ideas were successfully discredited on every level – not only intellectually, but also politically and morally.

Central to Marxism is the unity of theory and practice. 'The philosophers have only interpreted the world, in various ways', wrote Marx in 1845. 'The point is to change it.'[23] Or again: 'All social life is essentially *practical*. All the mysteries which lead theory towards mysticism find their rational solution in human practice and in the comprehension of this practice.'[24]

The previous year Marx had written: 'History itself is a *real* part of *natural history*, of the development of Nature into man. Natural science will one day incorporate the science of man, just as the science of man will incorporate natural science; there will be a *single* science.'[25]

Marx's 'single science' would bring together past and present, theory and practice, subject and object, experience and knowledge. Crucially, it would not remain *purely* theoretical: in building class confidence and corresponding action, it would inspire the unification of humanity, bringing together the whole world.

To destroy Marxism, therefore, it was necessary to strike at this point, shattering the all-important junction between theory and practice. Chomsky's intellectual status, perceived moral integrity and impeccable left-wing credentials made him the perfect candidate for this job. Going to the heart of the matter, he insisted that 'there is no relationship at all between what is humanly interesting and what is intellectually interesting'.[26] The challenges of science have no bearing on the problems we face in our lives: 'Naturalistic inquiry is a particular human enterprise that seeks a special kind of understanding, attainable for humans in some few domains when problems can be simplified enough. Meanwhile, we live our lives, facing as best we can problems of radically different kinds.'[27]

Science and life just pass one another by: 'The search for theoretical understanding pursues its own paths, leading to a completely different picture of the world, which neither vindicates nor eliminates our ordinary ways of talking and thinking.'[28]

Again and again, Chomsky has been asked by activists to clarify the connection between his politics and his science. Again and again, he replies in the same deflationary way: 'There is no connection, apart from some very tenuous relations at an abstract level.'[29] Chomsky does not deny that 'those who wish to change the world should have the best possible understanding of the world, including what is revealed by the sciences', but seems anxious not to elaborate, particularly where his own work is concerned. 'Might we yet see a pamphlet by Noam Chomsky, linking your scientific and your political thinking for a popular audience?', an interviewer asked in 2008. As always, Chomsky poured cold water on the idea:

> I have written occasionally on links between my scientific work and political thinking, but not much, because the links seem to me abstract and speculative. Others believe the links to be closer, and have written more about them (Carlos Otero, James McGilvray, Neil Smith, and others). If I can be convinced that the links are significant, I'll be happy to write about them.[30]

The interviewer queried whether scientists, worried about the effects of unregulated capitalism in driving dangerous climate change, should be encouraged to become collectively self-organized and consciously activist. While agreeing that this might apply in certain cases, Chomsky seemed generally dismissive of the idea: 'If scientists and scholars were to become "collectively self-organized and consciously activist" today, they would probably devote themselves to service to state and private power.'[31] His words at this point suggest that the very concept of science conducted autonomously and collectively, beyond reach of the state, is for Chomsky almost a contradiction in terms. Asked whether he would encourage socialists or anarchists to view science – specifically, his own linguistic science – as having revolutionary potential, he replied: 'I don't encourage socialists or anarchists to accept falsehoods, in particular, to see revolutionary potential where there is none.'[32]

Science in one compartment, politics and life in another. The two should be kept apart. But how to ensure this? To Marx, no project could possibly have seemed more alien and unforgivable. '*One* basis for life and another for *science* is *a priori* a falsehood', Marx insisted.[33] For the founders of scientific socialism, it went without saying that workers could educate themselves and become scientifically aware. It went without saying that revolutionaries should keep abreast of the very latest, most exciting developments in science. 'The more ruthlessly and disinterestedly science proceeds,' wrote Engels in 1886, 'the more it finds itself in harmony with the interests of the workers.'[34] As humanity's only universal, international, unifying form of knowledge, science had always to be placed first. As the discussion of Russian futurism showed us, much the same idea at a later date inspired Velimir Khlebnikov: 'An Internationale of human beings is conceivable through an Internationale of scientific ideas.'[35]

For Marx, social science – including his own – was as much a product of class relationships as any other form of knowledge. For this reason, Marx thought it impossible to establish a real science of society without simultaneously combating those forces responsible for fragmenting and distorting science. This is what he meant when he wrote:

> The resolution of theoretical contradictions is possible only through practical means, only through the practical energy of man. Their resolution is by no means, therefore, the task only of the understanding, but is a real task of life, a task which philosophy was unable to accomplish precisely because it saw there a purely theoretical problem.[36]

Marx and Engels concluded, therefore, that in order to remain true to the interests of science – to solve its internal contradictions – they had no choice but to become politically active. Their idea was not that science is inadequate, needing guidance from politicians to keep it on track. On the contrary, they believed that, when true to itself, science is *intrinsically* revolutionary and must acknowledge no politics but its own.

Marx and Engels believed that for science to free itself politically, it needed a mass social constituency – a class which was not really a class at all but 'the dissolution of all classes'. 'There must be formed', as Marx explained,

> a sphere of society which claims no *traditional* status but only a *human* status, a sphere which is not opposed to particular consequences but is totally opposed to the assumptions of the ... political system, a sphere finally which cannot emancipate itself without emancipating itself from all other spheres of society, without therefore emancipating all these other spheres, which is, in short, a *total loss* of humanity and which can only redeem itself by a *total redemption of humanity*.[37]

This, then, was Marx's most thrilling idea: science itself was generating the social conditions for its own emancipation and fulfilment. In driving industrialization, science was replacing the former complex patchwork of castes and classes with a new and simplified polarity – capitalists on the one hand, the rest of us on the other. That second category – everyone else – Marx called 'the proletariat' and its allies. Lacking property and therefore any stake in the prevailing system, 'our class' alone could view the world objectively, as if from outside. 'Here,' wrote Engels, 'there is no concern for careers, for profit-making or for gracious patronage from above.'[38] Only in this kind of environment, free from corrupting material incentives and institutional pressures, could science at last be true to itself.

The spread of Marxist ideas in the period following the Second World War seemed unsettling and dangerous to the world's dominant superpower. But the McCarthyite witch-hunts of the early 1950s were too crude and erratic to qualify as an appropriate establishment response. Something far more sophisticated, subtle and intellectually defensible was required. Let me quote Golumbia:

> Scholars have offered any number of plausible explanations for Chomsky's rise to prominence, not least his own personal brilliance, and the incisiveness of his linguistic theories. Yet it seems reasonable to set aside some of these explanations and to think carefully about just what the times were and just what was the content of Chomsky's writing that made it seem not merely compelling but revolutionary. In what sense was the world ready and waiting for this particular Chomsky to emerge?[39]

Golumbia describes how recent scholarship has begun tracing certain political movements within the English-speaking academy in the 1950s and early 1960s, all suggesting 'a directed search for a particular ideological view that would help guide intellectual work toward a goal we have now come to recognize as neoliberalism', a project that profoundly shaped the intellectual climate established in all the leading universities.[40]

This was where Chomsky came in: his revolution in linguistics – crown jewels of the cognitive revolution – satisfied a deep social need. By splitting apart mind and body, the new paradigm provided intellectual justification for the urgent job of disconnecting revolutionary theory from corresponding practice, quarantining the best of science in a world of its own.

Chomsky was not alone in defending this institutional split, but his status as standard-bearer for the cognitive revolution gave him unique prominence and authority. 'More than any other figure,' to repeat

Golumbia's observation, 'Noam Chomsky defined the intellectual climate in the English-speaking world in the second half of the 20th century.'[41]

Central to this climate was an unprecedented belief in the value of scholarly detachment. Chomsky's insistence on divorcing his activism from his science became in many ways a model for the rest of us. For Chomsky it was crucial that his scientific work was *not* perceived as partisan in moral or political terms. During the student upheavals at MIT in the late 1960s, Chomsky endorsed the MIT management line that development of weapons of mass destruction – research into their design – was perfectly acceptable, provided it was kept separate from subsequent deployment of such weapons. This distinction – which to my mind uncannily recalls Chomsky's distinction between 'competence' and 'performance' – met with considerable opposition from colleagues on the political left. An eloquent critic of this management line was Howard Zinn, political scientist at Boston University and close friend of Chomsky. Zinn complained that 'the call to disinterested scholarship is one of the greatest deceptions of our time, because scholarship may be disinterested but no one else around us is disinterested. And when you have a disinterested academy operating in a very interested world, you have disaster.'[42]

Had it been management alone that equated professionalism with neutrality, the left might have found it easier to expose what Zinn termed the great deception. But Chomsky's separation of his own science from his activism set an example from the left which seemed morally defensible. For this reason among others, an increasingly sanctified ideal of individualism and professionalism took hold, denigrating emotion, subjectivity, collectivity and notions of shared social responsibility – values increasingly viewed as incompatible with the objectivity and rationalism of science.[43] As we have seen, Chomsky decreed that his own scientific work was non-political. By the same token, his political opinions were declared non-scientific and non-theoretical. At any given moment, according to this model, you are either a scientist or an activist; you cannot play both roles at the same time. A climate scientist, for example, will be respected for reporting worrying findings, but condemned for resorting to direct action to avert the consequences. Those who do confuse roles in this way risk being accused of betraying their vocation.

The result has been tongue-tied science and correspondingly mindless activism. It is a situation in which science and activism are incapable of sharing the same language or of collaborating in pursuit of any shared purpose at all. If Golumbia is right, *that* is our global situation today, *that* is a major reason why socialists have felt helpless and paralysed – and *that* is

why the US establishment has always felt comfortable with such a division
of intellectual labour. Building on Chomsky's success and the inspiration to
dissidents he was able to provide, this exercise in self-censorship – a variation on the theme of 'divide and rule' – became intellectually fashionable,
morally respectable and in countless ways institutionally endorsed.

It was certainly no accident that this academic consensus became
entrenched following the defeat of fascism, just as democracy began
sweeping the world as the dominant political ideology. Prior to this moment,
the establishment did not urgently need to separate science from activism
– people could not act on their scientific discoveries anyway. After it, there
was a danger that science could become a hugely potent force for social and
political change. That had to be stopped.

The moment came in the late 1950s, by which time much of the world
had signed up to American-style democracy, with its ballot boxes and
ostensibly free press. When people have human rights and can vote, explains
Chomsky, it is less easy to keep them down. If you are a dictator, it hardly
matters what people think – you can simply kill them or lock them up. But
with democracy, it does begin to matter. To retain control under democratic
conditions, you need a sophisticated system of thought control – preferably
one so subtle and clever that even its own managers are likely to convince
themselves. Chomsky explains the paradox:

> I mean, if there is a society that is free and open but has a highly class
> conscious dominating elite, ruling group, such a society is going to be
> forced to have a very effective system of indoctrination, precisely
> because it cannot rely on force and violence to ensure obedience. It is
> going to rely on a very sophisticated indoctrination and thought control.
>
> So it is not at all paradoxical that in the most free and open society,
> you should have the most sophisticated, well-grafted and effective
> system of indoctrination and thought control.[44]

The 'new science of the mind', as it was called, undoubtedly met at least
some of the requirements which Chomsky here sets out. Focused on the
mind rather than the body, the new 'cognitive' orthodoxy was, to adopt
Chomsky's terms, 'sophisticated, well-grafted and effective'. It distorted
social and political thought not directly, through crude propaganda, but
almost unnoticeably, by isolating science in a world of its own, as if it really
were a sphere unconnected with people's ordinary lives. If science (as Marx
believed) is communal intelligence – the most sophisticated intellectual
activity of which we humans are capable – then severing its connection

with life amounts to a kind of decapitation, severing human knowledge and understanding from our capacity to act.

In his activist role, Chomsky has written at length on what he terms 'the manufacture of consent'. His position is straightforward: the ruling élite do this by means of propaganda, which they can direct at the populace thanks to their ownership and control of the mass media. This culminates, according to Chomsky, in brainwashing more subtle, potent and all-pervasive than anything previously imagined:

> Propaganda is to democracy what violence is to totalitarianism. The techniques have been honed to a high art, far beyond anything that Orwell dreamt of. The device of feigned dissent, incorporating the doctrines of the state religion and eliminating rational critical discussion, is one of the more subtle means, though more crude techniques are also widely used and are highly effective in protecting us from seeing what we observe, from knowledge and understanding of the world in which we live.[45]

Many on the left would agree with these words, yet the precise mechanisms of control remain somewhat obscure.

Although Chomsky does not mention it, one possible control mechanism is to separate mind from body – the secret of the 'cognitive revolution'. If this is accepted, we have the deeply ironic possibility that Chomsky's own work served to meet the new need for a sophisticated system of control.

The 1950s cognitive revolution was ostensibly directed against behaviourism. Quite clearly, however, it went deeper than that. Chomsky uses the terms 'behaviourism' and 'empiricism' more or less interchangeably.[46] For him, 'behaviourism' has always served as a sweeping catch-all which included not just Pavlov and Skinner, but also Durkheim, Foucault and vast swathes of materialist philosophy and social theory.

Chomsky has described how, by the time he had worked out his mature political philosophy, he had 'passed through the various stage of Trotskyism and gone on to Marxist-Anarchist ideas'.[47] He has also described his standpoint as 'anarchist and left anti-Bolshevik and anti-Marxist'.[48] So far is he to the left of, say, Lenin or Trotsky that such figures appear to him little different from fascists.[49] In keeping with such perceptions, he tends to single out Marxist intellectuals as particularly dangerous advocates of behaviourism. Discussing 'the contemporary intelligentsia', he accuses them of aspiring to exercise power as 'either ideological managers or state managers', continuing: 'Well, I think that this is part of the strain of thinking that is very central to Marxism, and expresses itself in its clearest form in the

Leninist variety of Marxism and also in fascism, which is in many respects not a dissimilar position.'[50]

While applauding the Russian democratic revolution of February 1917, Chomsky found nothing to celebrate in the subsequent October insurrection, condemning it as essentially a military coup. Before, during and after 1917, according to Chomsky, Bolshevism was an oppressive ideology with similarities to the virulent fascism soon to be unleashed upon the world by Hitler and Mussolini.

This is not the place to debate in any depth the merits of Chomsky's critique of Bolshevism. In his defence, few would deny that, once in power, Russia's Bolsheviks faced a situation of economic collapse, hunger, international isolation and descent into civil war that encouraged the more authoritarian tendencies in Bolshevism. Even among Marxists, it is a rare historian who today feels able to defend or excuse, for example, Trotsky's strained attempts at self-justification when, with Lenin's support, he deployed the Red Army to crush the March 1921 sailors' rebellion in Kronstadt.[51]

But whatever the moral passion fuelling Chomsky's critique, it should also be remembered that it was in some ways expedient, too, meshing in key respects with the thrust of US policy in his time. From the onset of the Cold War, the world's major superpower needed to switch attention from Germany to Russia, with Bolshevism – now demonized as 'world communism' – constructed as a straightforward extension of the previous Nazi threat. If Bolshevism was fascism and Stalin the new Hitler, shouldn't the defenders of freedom remain on a war footing until victory was won?

In all fairness, no one could say that Chomsky wanted any of this. As a self-proclaimed revolutionary socialist, he passionately advocated just the opposite – his country's renunciation of imperialism, allowing other parts of the world to pursue their own chosen paths. But personal motives are one thing, outcomes another. The US elite were not greatly inconvenienced by Chomsky's moral conscience or his relentless exposure of their undoubted crimes. His role as the conscience of America may even in subtle ways have helped deflect criticism, indicating as it did that America was, despite everything, a 'free country'. Chomsky's early move in splitting politics so categorically from science must certainly have seemed helpful. The outcome – tongue-tied, politically inarticulate science alongside mindless, scientifically illiterate activism – would prove a disaster for the global revolutionary left. To the extent that science and activism are disconnected, humanity as a whole is deprived of a shared language through which to pursue any common purpose at all.

CHAPTER 21

CHOMSKY'S TOWER

We have been left without even a language. It was not just Chomsky's incommunicado theory – his idea that language is for talking to yourself – that bitterly divided linguists and met with incomprehension everywhere else. Equally baffling was the way Chomsky relentlessly switched codes, as if determined *not* to be understood by anyone except those closest to him at the time. It is as if he actively needed his two separate constituencies – fellow scientists on the one hand, fellow activists on the other – to speak mutually incomprehensible tongues.

A key strategy for making himself incomprehensible was to resort to so-called 'technical' terms. According to Chomsky, for example, the 'technical term "language" has no relation at all to the pre-theoretical term "language"'.[1] He justifies this by claiming that the 'rule system is something real, it is in your head, it is in my head, it is physically represented in some fashion'. Meanwhile, 'what is now called "language" does not need any term at all, because it is a totally useless concept . . . It does not fit with linguistic theory, it has no existence.'[2]

Note Chomsky's reasoning here: what the rest of us term 'language' has no existence because it fails to 'fit with linguistic theory' – by which he means his own particular theory. Chomsky proceeds to swap the terms 'grammar' and 'language' around, justifying this by stating that he has no wish to mislead:

Now, the technical term for a rule system is 'grammar'. So we have this odd and very misleading situation, where the technical term 'grammar' is very close to the pre-theoretical intuitive term 'language', whereas the technical term 'language' has no relation at all to the pre-theoretical term 'language'.[3]

His ideal solution would be to avoid the use of English when talking about language:

> It is only in the last couple of years that I have come to realize how much people are misled by this and have begun to suggest that we simply over-throw the whole terminology and start over again, now using the term 'language' to refer to the system of rules and principles, what has previously been termed grammar, and dismissing entirely what has previously been called language, because it is a concept with no use, corresponding to nothing in the physical world.[4]

So 'what has previously been called language' falls outside linguistics. It cannot be studied. It does not exist. 'Some people', concedes Chomsky, 'are surprised when you say that linguistics is about grammar and not language. But that just means that linguistics is not about what we mean by "language".'[5]

If you think linguistics is about language, Chomsky advises, you are making a big mistake. Linguistics is about something previously unknown – an elusive object thought to exist in the head. It's impossible to study it without special methods – those of science. Not only is special training required, your brain must be equipped with the necessary compu-tational module, 'the science-forming capacity'.[6] When scientists produce science, their work is the output of that faculty. If you don't happen to possess it, then that's tough – obviously you can't do science.[7] Likewise if you lack the necessary scientific training or background, it's important that you don't try to intervene.[8]

But it gets worse. Even if by some chance you were 'part of the discus-sion', no enlightenment would ensue. A constant Chomskyan theme is the ultimate unintelligibility of the world. Writing of Sir Isaac Newton's discovery of gravity, Chomsky says:

> It was assumed by Galileo and Descartes ... that the world would be intelligible to us ... Newton disproved them. He showed that the world is not intelligible to us ... And by the time that sank in, which was quite some time, it just changed the conception of science. Instead of trying to show that the world is intelligible to us, we recognized that it's not intelligible to us ... And then the aim of science is reduced from trying to show that the world is intelligible to us, which it is not, to trying to show that there are theories of the world which are intelligible to us.[9]

So the best science can do is to make certain *theories* intelligible – theories devoid of human meaning or interest. It was Albert Einstein who famously wrote: 'The most incomprehensible thing about the universe is that it is comprehensible.'[10] It took some audacity to turn that idea on its head while still calling it science.

Naturally, Chomsky faces problems when attempting to explain all this to his anarchist supporters and friends. Insisting that scientists must possess a special cognitive module, must restrict themselves to specialist terminology and must undergo specialist training does not sound too democratic or inclusive. Adding that science is of no human interest and that the world is unintelligible anyway tends to make matters worse.

Chomsky's solution is the one we have come to know: depending on the audience, he says different things. And so it is that when he speaks as an activist, all is reversed. The arrangement whereby authorized 'experts' speak in specialist, meaningless jargon is recognized for what it is – a time-honoured strategy of the ruling managerial elite.

It may seem hard to believe, but it's a fact that Chomsky recommends standards for others which are the reverse of those he applies to himself: 'There are things that we ought to be able to talk about in ordinary, simple words . . . We ought to be able to talk about these things in simple, straightforward words and sentences without evasion and without going to some expert to try to make it look complicated.'[11]

So it does seem extraordinary that Chomsky immediately exempts himself from this prescription. Why is language – the capacity which defines our very humanity – *not* one of those 'things that we ought to be able to talk about in ordinary, simple words'?

Chomsky justifies himself by arguing that his own work, unlike that of his opponents, is genuinely scientific because it is natural rather than social science. Expressed in terms of the contrast between Harvard University and his own MIT, Chomsky's formula is simple: 'Harvard is humanities based: it's the place where people are trained to rule the world', whereas 'MIT is science-based: it's the place where people are trained to make the world work.'[12] Dominating the whole world naturally involves deception and corruption; by contrast, making the world work enforces respect for nature's laws. As we saw earlier, Chomsky's colleagues at MIT were in fact 'making the world work' by designing fuel–air explosives and guidance systems for nuclear missiles.

To break the spell of Marxism, then, the world's dominant superpower needed to sever all connection between political activism and science. Science as a global force had to be deprived of political will; meanwhile, any internationalist activism which remained had to be deprived of guidance or inspiration from science. To be clear, I am not claiming that there was any conscious plan here: that seems unlikely. It's just that certain approaches seemed helpful; meanwhile, others appeared politically inconvenient, hence probably wrong. Chomsky himself has a phrase to describe this kind of semi-conscious political and ideological process which might be mistaken for conspiracy: he calls it 'the manufacture of consent'.[13]

The US military-industrial elite's exaggerated anxieties about a world communist conspiracy led quite naturally to fears regarding the still widely influential theories of Engels and Marx. It is now widely recognized that the doctrines promulgated by Stalin and those like him bore little relationship to anything which Marx or Engels actually wrote. It seems unlikely, for example, that either thinker would have had a high opinion of the disastrous genetics of Trofim Lysenko. We also know that Marx hated the very idea of a body of doctrine known as 'Marxism'; still more, he would have poured scorn on the idea that such a doctrine could be confused with genuine science. Basing one's politics on science, however, is always a good idea. Thanks partly to Marx – and despite official Soviet canonization of his ideas – the notion of 'science' continued to hold immense intellectual authority in both East and West right up until the Second World War. Indeed, throughout those decades, 'science' tended to be linked with atheism, social criticism and the left.

Influenced by Marx and other socialist thinkers, the inter-war years had witnessed a surge of passion for popular science, working-class intellectuals forming educational associations, teaching that all science is within popular reach and writing books with titles such as *Teach Yourself Physics* or *Mathematics for the Million*. There was a surge of interest in Esperanto as a means of breaking down borders and forming a global community, in which knowledge of every kind could be shared. The idea was simple: knowledge is power and everyone should have it. Or as Leon Trotsky put it: 'Science is knowledge that endows us with power.'[14]

During the 1930s and 1940s, revolutionary Marxism appeared to many US intellectuals – including some of the most prominent figures on the right – to be 'the inevitable, if not yet victorious, structural principle that eventually would govern world affairs'.[15] This widespread assumption lasted right up until the end of the Second World War. One incident captures the turning point. On 22 February 1946, George Kennan, then American chargé d'affaires in Moscow, sent an 8,000-word telegram to the Department of State urging 'containment' of communism by all ideological, economic, military and other means short of nuclear war. Kennan's message quickly caused a sensation and provided one of the most influential underpinnings for America's Cold War policy over subsequent decades. From that date until the final collapse of the Soviet Union in December 1991, the 'central organizing principle' for America and much of the West was the Cold War effort to contain the spread of 'communist' ideas.[16]

The urgent need was to tackle the threat not only militarily and economically, but also intellectually and ideologically. We have already seen how projects such as SORO were sponsored to fight 'communism' on the level of ideas. But any survey of supposed achievements will confirm the impression that this and related projects fell far short of an effective intellectual contribution to the battle of ideas within American academia itself. Much more effective was a model of academic professionalism designed to divorce scientific and other scholarship from any direct connection with radical political activism or life.

A key figure here was Raymond B. Allen – president of the University of Washington. It was Allen who, in 1950, wrote a letter to the American Philosophical Association (APA) supporting the firing of a philosopher, Henry Phillips of the same university, for being a self-confessed Marxist. Allen's intervention, according to historian John McCumber, was effective, in that 'it took the APA out of the business of defending philosophers attacked by McCarthyites'.[17] From that moment, Allen went on to become the foremost articulator of academic McCarthyism. In carefully chosen language, he warned that, to avoid being suspected of communist leanings, academics must henceforth restrict themselves to the dispassionate pursuit of truth: 'If a University ever loses its dispassionate objectivity and incites or leads parades, it will have lost its integrity as an institution and abandoned the timeless, selfless pursuit of truth.'[18]

Long after the witch-hunts had been abandoned, virtually all US academics continued to devise ways to avoid any suspicion that they might conceivably incite or lead parades. Philosophers had always felt especially threatened owing to the nature of their discipline: 'It had disciplinary links to subversion. For what was Marxism, basically, but a kind of philosophy? . . . Philosophy was intrinsically open to suspicion.'[19]

The way to stay safe was to make absolutely clear that post-war philosophy did not threaten to change the world. It was restricted to logic, the business of which was to evaluate statements. Here is W.V.O. Quine: 'Logic, like any science, has as its business the pursuit of truth. What is true are certain statements; and the pursuit of truth is the endeavor to sort out the true statements from the others, which are false.'[20]

Referring to scholarship in general, the philosopher John Searle expressed the prevailing ideal with unusual clarity in 1971:

The basic actions of the faculty member, the core of his professional activity, so to speak, lie in teaching students and conducting and publishing research. In each case he seeks to impart the truth or as nearly what is the truth as he can get according to professional standards of evidence and reason . . . He could regard it correctly as a violation of professional ethics if he made his utterances for the purpose of achieving some practical effects rather than for the purpose of communicating the truth. Not only does he not consider the consequences of his actions when making moral utterances but he would consider it somewhat immoral to do so.[21]

So if you happen to be a climate scientist, on Searle's account, you should *not* produce utterances or join politically with colleagues for the purpose of, say, averting climate catastrophe – that would be 'somewhat immoral'.

Academics' quite peculiar commitment to this self-denying ordinance was one of the permanent achievements of the witch-hunts of the McCarthy years. Historian John McCumber interprets Searle's prescription as follows:

True to the spirit of Raymond B. Allen, it denies the teacher and scholar any other goal than finding and communicating the truth. If someone should say that his or her purpose as a teacher is to produce educated Americans, or moral people, or a fuller appreciation of art – for Searle that person is, apparently, abandoning his or her vocation.[22]

The schools of analytic philosophy which now became dominant invoked 'scientific truth' as timeless and abstract, cut off from the actual lives of scientists, as from any social and cultural goals. A key aim of policy makers during the McCarthy years had been to detach the concept of 'science' once and for all from previous popular understandings concerning progress, enlightenment, secularism or associated political radicalism. An increasingly educated populace needed somehow to be alienated from science, persuaded on the one hand of its incomprehensibility and, on the other, of its ultimate irrelevance to ordinary people's lives. In many ways, the witch-hunts of the 1950s succeeded in this.

Anyone who has looked carefully at post-1950 developments in linguistics, psychology, anthropology and philosophy will recognize the pattern: intellectual life became steadily professionalized, rarefied and institutionally atomized into mutually incompatible sub-disciplines. The process was so powerful that even European intellectuals who once imagined themselves as dangerous dissidents – Sartrean existentialists, Althusserian Marxists, Foucauldian social historians, postmodernist critical theorists – found the ground moving beneath their feet, sweeping them steadily along. The widening of access to university education did nothing to reverse this process, as students from working-class backgrounds became convinced by their 'Marxist' teachers that, in shaping history, ideas gleaned during seminars exert causal primacy over practical action in the world. Over the years, Chomsky has been one of the few prominent intellectuals sufficiently aware of this process to denounce it explicitly, describing Western education in general as 'a period of regimentation and control, part of which involves direct indoctrination, providing a system of false beliefs'.[23] To illustrate, Chomsky cites the damaging belief that ordinary people must defer to so-called 'experts' to have any chance of comprehending how the supposedly complex system of modern economics, social life and politics works.

Chomsky's words are forthright and appealing. Yet when we recall his extraordinary claims about the complexities of linguistics – his claim, for example, that linguistics is not really about language at all – it is impossible to miss the irony. It is also noticeable that when Chomsky denounces the culture of deference to 'experts', he is actually quite selective, typically singling out *Marxists* as the people he has particularly in mind. Marxists, according to him, do not believe in human nature, their 'blank slate' ideology licensing them to manipulate people as they please. When Chomsky denounces what he terms 'externalism', he does so on those grounds. Marxism is 'externalism' carried to extremes, in essence no different from Skinner's behaviourism.

Even to this day, you do not have to endorse or even mention Marx to find yourself attacked in this way, as a recent somewhat bizarre example illustrates. Derek Bickerton – at one time a committed Chomskyan – is a specialist in Creole languages. In a 2009 book entitled *Adam's Tongue*, he suggested that humans may first have begun using words as a means of coordinating cooperative foraging. Hardly a scandalous idea, one might have thought. To his surprise, however, Bickerton found himself virulently attacked for showing Marxist tendencies. A review by supporters of Chomsky in the journal *Biolinguistics* denounced his proposal as 'plain radical externalism' – a 'scenario not too different from the one suggested by Marx and Engels more than 150 years ago'. Citing a short passage from *The German Ideology* – in which Marx and Engels attribute the origins of language to the emergence of cooperative labour – the reviewers observe that 'Bickerton needed a whole book to say more or less the same thing.'

Bickerton's approach, according to these authors, is 'poor and uncouth biology' – poor and uncouth *because* it takes environmental factors and topics such as social cooperation into account. They conclude that it 'is not just bad evolutionary linguistics, it's bad science, very bad science. If it is

science at all. It's something not even able to match the pre-Darwinian envi-
ronmentalism of Marx and Engels.'[24]

Note that Marxism is here dismissed as 'pre-Darwinian environmen-
talism' – a nineteenth-century doctrine, according to which the environment bears down on human beings who are powerless to transform that
environment in turn. Needless to say, it would be hard to find any idea
further removed from what Marx himself actually wrote and passionately
believed. What mattered, however, was that the cognitive revolution was
supposed to have buried Marxist and all other manifestations of 'externalism' once and for all. Bickerton – once a signed-up Chomskyan – had
apparently defected to the enemy camp and so had to be shot down.

In one sense, Chomsky's 'cognitive revolution' was a liberating event
because it brought together a grand coalition of scholars from multiple disciplines to detonate behaviourism and begin seriously investigating problems
of meaning and mind. But if this was scientific revolution, it was counter-
revolution, too. With so much money invested and so many brilliant minds
recruited as a result, it was inevitable that major progress in key areas of
cognitive science would be made. Everyone recognizes that linguistics today
is a vastly richer, more sophisticated and more convincingly scientific discipline than it was in the pre-Chomskyan era. But, wherever lasting insights
have been gained, these results have surely occurred *despite* the bizarre ideological premises of Chomsky and his followers, not because of them.

What some called revolution, others considered a state-sponsored coup,
and a decisive one at that. According to Chomsky's former supporter George
Lakoff, writing in 1971, the newly triumphant linguistics was 'as much part
of the intellectual establishment as General Motors is a part of the military-
industrial establishment'.[25] Cut off from anthropology and equally from the
rest of science, Chomskyan linguistics became isolated in its own disciplinary compartment, almost as if the dials, levers and electronic buttons of a
hugely sensitive command-and-control centre needed to be kept under
lock and key. To deter intruders, arbitrarily assigned entry codes would be
periodically changed. Linguistics, in this its most authoritative and prestigious form, was dehumanized, mystified and to a large extent rendered
incomprehensible. But, even as it became popularly inaccessible, the study
of language remained no less central and indispensable to us all. Research
on the faculty which makes us human – which allows us to communicate
and share our dreams – was placed beyond reach of all but an inner circle
of self-proclaimed specialists speaking to one another in code.

How can the rest of us proceed if the very word 'language' has been with-
drawn from circulation? Chomsky puts it well: 'If it's impossible to talk

about anything, then you've got them under control.'[26] His institutional backing enabled him to legislate language away. As the concept of language vanished from the scene, all that remained was internal, incommunicable thought. Warren Weaver once proclaimed that the new scientific linguistics would rebuild the Tower of Babel, restoring to humanity the primordial universal language shattered by a vengeful God. Something similar had also been the dream of Roman Jakobson, inspired as he was by Khlebnikov's poetry and by Tatlin's Tower. In Chomsky's hands, the project for a universal tongue ended up serving the exact opposite purpose, rendering language incomprehensible even to ourselves.

My argument throughout has been that, as the Russian Revolution played itself out, it produced echoes across Europe and the world, the tremors of that seismic event eventually surfacing as a wealth of novel insights into the nature of language and mind. In that sense, the post-war US cognitive revolution was genuinely revolutionary. Although Chomsky's contribution to it was lit up by the sparks of those earlier events, the dominance of his Cartesian paradigm reflected a quite different agenda. Sponsored by the US military, it was one more top-down project to combat egalitarianism and communism, wrenching the mind from the body, divorcing heaven from Earth, and preventing that tower of Tatlin's from ever reaching the sky.

It hardly needs saying that Chomsky the activist was no ivory-tower intellectual, aloof from the world. On the contrary, it would be difficult to think of any prominent academic who has done more to take to the streets, risk arrest, measure up to the events of the day, speak truth to power and, in the process, endure ferocious political hostility matched only by passionate grass-roots support. Meanwhile, the scientist in Chomsky took a different and quieter path, retreating to his own safe space – his own lofty tower – as if conversing with power in its own esoteric tongue.

Chomsky's tower was not physically high. It was a modest, somewhat shabby office in Building No. 20 of the Massachusetts Institute of Technology – nerve-centre of the US scientific establishment for several decades after the Second World War. Despite his lifetime commitment to activism – or more likely, as I have argued here, precisely *because* of that activist commitment – Chomsky's tower offered an escape from his political conscience and from the unspeakable horrors of this world. The escape was always a search for the unattainable – for a mysterious, long-hidden, mathematically perfect Universal Grammar. In this, he has echoed the message of mystics down through the ages, receiving and transmitting messages from the sky. Ultimately, Chomsky reminds us, there is just one tongue – Human – lying unfortunately silent within us all.

CHAPTER 22

BEFORE LANGUAGE

Chomsky is adamant that the problem of the origin of language can be explained without reference to any precursor. All you need is a mutation. Although this idea remains influential, it today faces stiff competition from a paradigm which conforms to Darwinian principles. The American comparative psychologist Michael Tomasello is the only scientist in the world to have devoted an equal amount of time to (a) the study of children's language acquisition and (b) the communicative skills of apes. This cross-species perspective has enabled him to inspire a wide-ranging body of research by primatologists, anthropologists, archaeologists and others who have concentrated on precisely those topics which Chomsky dismisses. Since it is impossible to do justice to everyone in this camp I will single out the biological anthropologist Sarah Hrdy, who has shone brilliant new light on the likely social life of our species in the prelinguistic period. Whereas Chomsky dismisses the very idea of precursors, Hrdy talks of empathetic cooperation and mutual understanding, illuminating a long chain of developments without which language could not possibly have evolved.

For both Tomasello and Hrdy, it is a kind of madness to imagine that language can be explained without reference to previous Darwinian evolution, to cumulative cultural evolution or to our species' unique capacities for collective imagination and collaboratively agreed action in the world. Even if we restrict our focus to innate cognitive capacities, the *social* brain lying behind our species' unique *social* intelligence is surely impossible to ignore. Unlike chimpanzees, all of us – right back to our earliest hunter-gatherer ancestors – are constantly putting ourselves imaginatively in each other's shoes.[1] Language would be impossible without what is termed 'egocentric perspective reversal'.[2] In speaking, we are both listening to ourselves as we hear our own words and, reciprocally, hearing other people's words as if they were our own.[3] To engage in genuine conversation is always to strive for an agreed view, a *joint* perspective on the world, rather than a narrowly individual one.[4] We deal with the problem of egocentricity by anticipating how others might correct or in some way shape our perspective, meanwhile helping shape their perspective in turn.[5]

While undoubtedly capable of empathy, our primate relatives are so competitive – so cunning and Machiavellian – that they have little incentive to divulge their real intentions or thoughts.[6] It seems that trickery played a major role in driving primate cognitive evolution. One piece of evidence is that, across primate species, the larger the neocortex, the greater is the rate of deception.[7] Apes find joint attention difficult and in most contexts do not trust each other sufficiently to rely on what others intentionally convey.[8] In fact, as Michael Tomasello points out, apes are so lacking in mutual trust that they do not even point things out for one another in the wild.[9] It is the largely egocentric, competitive and individualistic dispositions of primates of both sexes which block any community-wide, stable sharing of values and goals and, for that reason, make it impossible for language to evolve.[10]

Against this background, it should be clear that language has always been revolutionary. Its emergence was not a random or inexplicably isolated leap, but part of a hugely complex speciation event – the 'human revolution' as it is often called.[11] Although the timescale is open to debate,[12] the concept of such a revolution is today widely accepted, not least by Chomsky himself:

> Nothing much seems to have changed for hundreds of thousands of years, and then, all of a sudden, there was a huge explosion. Around seventy, sixty thousand years ago, maybe as early as a hundred thousand, you start getting symbolic art, notations reflecting astronomical and meteorological events, complex social structures . . . just an outburst of creative energy that somehow takes place in an instant of evolutionary time . . . So it looks as if – given the time involved – there was a sudden 'great leap forward.'[13]

Chomsky presents the transition as a one-step computational leap. When asked whether revolutionary social change might have played a role in giving rise to language, he explicitly rules out that possibility: 'We of course know very little about the sociopolitical conditions that existed at the time, but there's no scenario I can think of that suggests how a sudden change in these conditions could have led to the emergence of language.'[14]

Initially, says Chomsky, the leap to language did not even affect communication. While acknowledging that the transition must eventually have had social consequences, he appears anxious to avoid describing it as a social – let alone an egalitarian – revolution.

Chomsky's insistence on this point has an interesting history. To understand the constraints under which he always operated, it is important to remind ourselves again of the powerful effect of the research into social revolutions conducted by the Pentagon in the immediate post-war period. The US military were investing vast resources in studying how and why revolutions break out. They were asking what exactly a 'communist' revolution is. What are the signs that such a revolution might be about to occur? What are the best techniques for crushing such a revolution? How should the CIA and US army special forces be deployed to subvert a revolution before it even begins?[15] This was not an intellectual atmosphere likely to encourage anyone to condone or celebrate social revolution.

During the 1950s, Charles Hockett was the influential linguistic successor to the mantle of Leonard Bloomfield. Raised as a Quaker and a pacifist, Hockett joined the Communist Party as a young man. When threatened in 1952 with dishonourable discharge from the Reserves for having once been a communist, Hockett managed to avoid this by explaining that his commitment was really to the fundamental rights of human beings.[16]

Despite the intimidating McCarthyite atmosphere, Hockett with his egalitarian politics stood his ground, inspiring a broad coalition of anthropologists – linguists, primatologists, hunter-gatherer specialists and others – to join him in developing a new kind of revolutionary theory. In 1960, he published a landmark paper, 'The Origin of Speech', presenting the transition from primate communication to language as a combined evolutionary and revolutionary event,[17] a theme developed in a subsequent co-authored paper, 'The Human Revolution'.[18] Historian of science Gregory Radick has shown how, by treating linguistics as central to the task of reconnecting physical with cultural anthropology, Hockett helped to establish a promising new framework for the unification of science as a whole.[19]

Among the most eloquent contributors to Hockett's endeavour was a young anthropologist named Marshall Sahlins. Alongside Hockett's 1960 article in *The Scientific American*, he looked at how the transition from a highly competitive, often despotic ape social system to a cooperative and egalitarian human one might have occurred. The establishment of hunter-gatherer egalitarianism was more than an evolutionary step – it was a revolutionary one that established a genuine kind of communism. The article was entitled 'The Origin of Society'.[20] If Sahlins was right, that evolutionary event seemed to be precisely the kind of thing the Pentagon was dedicated to suppressing!

Sahlins pictured the egalitarian ethos of our hunter-gather ancestors – nowadays acknowledged by virtually all anthropologists[21] – as the momentous outcome of a *political* struggle. Anyone who has read Frederick Engels on the origin of the family will recognize how closely Sahlins was working within that classical Marxist tradition.[22] Noting how primate cooperation is recurrently undermined by sexual violence and competition, he wrote:

> It was this side of primate sexuality that forced early culture to curb and repress it. The emerging human primate, in a life-and-death economic struggle with nature, could not afford the luxury of a social struggle. Co-operation, not competition, was essential. Culture thus brought primate sexuality under control. More than that, sex was made subject to regulations, such as the incest tabu, which effectively enlisted it in the service of cooperative kin relations. Among subhuman primates sex had organized society; the customs of hunters and gatherers testify eloquently that now society was to organize sex – in the interests of the economic adaptation of the group.[23]

The author concluded with what sounded like a political rallying cry:

> In selective adaptation to the perils of the Stone Age, human society overcame or subordinated such primate propensities as selfishness, indiscriminate sexuality, dominance and brute competition. It substituted kinship and co-operation for conflict, placed solidarity over sex, morality over might. In its earliest days it accomplished the greatest reform in history, the overthrow of human primate nature, and thereby secured the evolutionary future of the species.[24]

Sahlins' celebration of 'the overthrow of human primate nature' could be taken as not-too-subtle code for the overthrow of capitalism.

He was not alone here. Similar ideas and interests soon led to the convening of a stellar international conference dedicated to exploring how hunter-gatherer egalitarianism first emerged. One of the two young organizers of the 1966 'Man the Hunter' symposium was the Canadian anthropologist Richard Lee, whose fieldwork among the Kalahari Bushmen had led him to celebrate what he termed (following Marx and Engels) 'primitive communism'.[25] His co-organizer was Irven DeVore, known for his recent field studies of baboons. Encouraged by his mentor, the evolutionary anthropologist Sherwood Washburn, DeVore now began turning his attention to human primates, joining Lee on his trips to Botswana.

When DeVore studied hunter-gatherers, he was constantly aware of the contrast between the assertive egalitarianism of these people and the much more despotic arrangements of non-human primates, such as baboons. Like Sahlins, he insisted that to understand human evolution, it was necessary to acknowledge just how stark was the contrast between hunter-gatherer social arrangements and what was then known of the strikingly competitive dominance/subordination dynamics of wild-living apes, such as chimpanzees.

In 1968, while teaching a course in primate behaviour, DeVore wondered aloud why primates would at times kill babies of their own species. Sherwood Washburn, supported by his student Phyllis Dolhinow, explained the behaviour as a form of social pathology resulting from abnormal conditions, such as the presence of humans or overcrowding. Other theorists wondered whether the behaviour might be adaptive in some way. In 1962, the ethologist V.C. Wynne-Edwards had published a massive and highly influential book premised on the idea that resource scarcity is the main problem faced by animals everywhere. Groups or species which reproduce too rapidly during times of plenty, he claimed, risk starvation and population crash later on. The only way for the species as a whole to overcome this problem, he said, is to evolve mechanisms such as infanticide to keep the population within sustainable limits.

Among DeVore's students that year was a young undergraduate called Sarah Hrdy. She was puzzled by reports of infanticide among langur monkeys and was not convinced that it was just a rare pathological occurrence. She resolved to travel to India to see for herself what the causes were. Adult male langurs are sexually competitive, often fighting ferociously to establish a monopoly over a whole harem of females. Hrdy noticed that it was only once a male outsider had attacked and successfully displaced the alpha male that infants in the troop began to be killed. She was struck by two facts: that the newly dominant male only attacked infants born to females he had never mated with; and that his attacks ceased once the infants were over six months old. In this way, the male restricted his murderous activities to the unweaned offspring of other males. Killing these particular infants was the only way in which he could force the members of his newly acquired harem to stop breast feeding and resume fertile cycles, bringing them into receptivity once more. Since his tenure was likely to be brief, he needed to move fast. Were he to allow the troop's females to continue nursing their existing infants, he might in the end father none of his own at all. Group benefits from population control, as proposed by Wynne-Edwards, had nothing to do with it: the male's murderous behaviour made sense as a strategy for perpetuating his own genes.

During her research, Hrdy observed a young infant kidnapped from another troop now being carried by a new female. The current alpha male had previously mated with this female, and he now left that infant alone. This led Hrdy to suspect that one female strategy for protecting future offspring would be to have sex with any male who might one day be in charge of the harem. Hrdy realized that this might also explain why females would simulate ovulation through pseudo-estrous behaviour, mating with outsider males who might later be potential usurpers. Instead of perceiving group members as conforming to long-term species imperatives, Hrdy had the insight that conflicts of interest within and between the sexes were the stuff of political life. She explains:

> Wherever males attempt to constrain female reproductive options, we can expect selection for traits that help females to evade them. What are we to make of such far-flung solicitations and enterprising sexuality as are being documented for creatures as diverse as fireflies, langurs, and chimps? After all, applied to females pejorative-sounding words like *promiscuous* only make sense from the perspective of the males who had been attempting to control them ... From the perspective of the female, however, her behavior is more nearly *assiduously maternal*. For this is a mother doing all she can to secure the survival of her offspring.[26]

Far from assisting with population control, adult female langurs would battle desperately to protect the lives of their infants, forming coalitions with female kin to drive away would-be infanticidal males – often at great cost to themselves. Only when all else had failed – only once their infant had been killed – would they consent to sexual intercourse with the very male who had done the deed. Instead of viewing male infanticide as somehow good for the species or group, Hrdy realized that the two sexes in this instance have radically divergent interests when it comes to perpetuating their genes.

Hrdy soon discovered that the 'selfish gene' theoretical framework being developed at the time by her mentor Robert Trivers fitted her observations much better than Wynne-Edwards' group selection paradigm. Apart from its scientific merits, individual selection seemed to her less ideological, less obviously sexist and much more empowering to any feminist.[27] Here is how Trivers himself sums up the ideological implications of group-selectionist thinking:

Species-advantage reasoning has some important consequences. First, it tends to elevate one individual's self-interest to that of the species, thereby tending to justify that individual's behavior. In our example, the adult male's self-interest has been elevated to that of the species: it is given a new name. What he is concerned with is population regulation, something that is beneficial to all. By contrast, the viewpoint of natural selection should make us suspicious of the notion that one individual's self-interest is the same as that of the species.[28]

Then Trivers notes the unconscious sexism:

Secondly, group-selection reasoning distracts our attention from conflict within social groups and from maneuvers that have evolved to mediate such conflict. Hence, unconsciously such reasoning tends to render other individuals powerless. 'The male has the power and the power is good for the species.'[29]

'Such reasoning', continues Trivers, 'prevents us from predicting female counterstrategies and from seeing the limits to male power. By contrast, an approach based on natural selection demands these counterstrategies. We expect to find them, and we analyze any social interaction from the standpoint of the individuals affected by it.'[30]

The idea here is not that animal behaviour is always selfish, but simply that genes are replicators, and can only replicate themselves. Hrdy observed langur mothers forming coalitions to defend their own and one another's offspring, acting bravely and selflessly precisely because that was the best way to pursue their genetic interests – their Darwinian 'fitness'. Nowadays, this explanation of primate male infanticide – accompanied always by female resistance to it – is almost universally accepted.[31]

So decisive was Hrdy's case study that it quickly became central to Robert Trivers' thinking and, through him, to the new paradigm known as 'sociobiology'. Instead of viewing mothers as united in blissful harmony with their male partners and offspring, scientists came to expect divergent interests in every biological relationship. Much as Marx had revealed class struggle as the engine of history, so cutting-edge biologists now saw conflict as the engine of evolutionary change. Gone at last were the old 'Man the Hunter', 'Man the Toolmaker' and 'Man the Thinker' fables about human origins. In the case of evolving humans as much as any other primate, females were at last recognized as autonomous agents with their own strategies and goals, often in bitter conflict with the priorities of the other sex.

Even though 'a woman needs a man like a fish needs a bicycle',[32] any mother would prefer it if the males in her life could be persuaded to help, rather than hinder, her childcare efforts. Hrdy's contribution was to take this feminist argument and run with it. From her point of view, the whole point of selfish-gene Darwinism is that it is firmly part of natural science, yet focuses relentlessly on social and political issues – on competition and co-operation in the animal world. From its beginnings in the 1970s, this new way of looking at life struck at the root of Chomsky's fundamental assumption that research cannot be scientific if it is social and political. In stark opposition to Chomsky's view that science can solve only highly simplified problems devoid of human interest or relevance,[33] Hrdy saw her own scientific work as a tool for thinking more clearly about the most important political issues of all. Egalitarian hunter-gatherers tend to put childcare first, treating the welfare of each new generation as their overriding social and political priority.[34] The contrast with modern capitalist economic priorities – which make nurseries, parent support and childcare peripheral issues – could hardly be more stark. The welfare of future generations, Hrdy argues, should be the absolute priority for any rationally organized society today.

Far from treating science as value-free and irrelevant to human concerns, and advising activists to look elsewhere for guidance, Hrdy and her colleagues hold that feminists have every right to draw on evolutionary science in support of their cause.[35] Until the 1970s, no one had explained why human sexuality is so radically different from that of our primate relatives. According to the conventional stereotypes of male versus female behaviour, the female of the species puts commitment and maternity before sex, whereas the male is promiscuous, caring little for his young. If these Darwinian stereotypes are accepted, the paradox is that human evolution stood them on their head, to such an extent that the human female became by those standards 'male' in her behaviour, her sexuality assertive and far from monogamous, while the male developed the ability to become committed, caring and in those respects 'female'. Gender stereotypes were in this way revolutionized.

More than anyone else, it was Hrdy who pulled all the evidence together. The stereotypical Darwinian strategy, as exemplified by the langur case, was for the primate male to invest in his own genetic offspring to the exclusion of all others, to the point of violently abusing babies suspected of being fathered by a rival. But the evolving human female, Hrdy realized, *always* had good reason to resist male dominance. This is an illustration of 'reverse dominance', a strategy of communal resistance which, as anthropologist Christopher Boehm has shown, has always been an aspect of primate politics – one which in our case culminated in the overthrow of dominance and its replacement by a revolutionary new moral and cultural order.[36] Just as Boehm deals with power relations in general, Hrdy deals specifically with *sexual* politics. Far from colluding by guaranteeing paternity, the human female from earliest times has had every reason to actively *confuse* paternity, encouraging each of several males to act toward her variously fathered offspring as if each of those offspring has some chance of carrying his genes.[37] In keeping with the core expectation that females and males will pursue divergent reproductive strategies, Hrdy explained how such characteristically human female adaptations as concealed ovulation and proactive sexuality evolved to ensure that the males in our species became increasingly involved with childcare.[38] Developing strategies which permanently curbed alpha male dominance and attempts to monopolize whole harems, females forged permanent alliances with the effect of turning childcare into an increasingly collective enterprise. The most successful alliances, typically matrilineal, were those best at mobilizing not only female help, but also the provisioning energies of as many males as possible.[39]

Not just in her early work with langurs, but subsequently, in a series of ground-breaking books,[40] Hrdy explored precisely the themes which Sahlins had broached in his 1960 article, supplying the missing causal links and evolutionary sequences in a forceful and persuasive way. Instead of following Sahlins in attributing the key changes to 'culture' – something which itself requires explanation – she showed how sexual dynamics gave rise eventually to a way of life guided by moral norms. To a new generation of female primatologists and evolutionary scientists, Hrdy's work struck a long-overdue blow against a wealth of sexist stereotypes and assumptions which had held back progress in the field. Many of these scholars – among them Barbara Smuts, Shirley Strum, Joan Silk, Patricia Gowaty and Meredith Small – derived explicit feminist inspiration from the powerful new paradigm which Hrdy had done so much to pioneer.[41]

Unfortunately, however, just as Hrdy was developing her insights, the new field of socially aware evolutionary biology collided with political passion in a calamitous and damaging way. While it is true that 'selfish gene' theory was not the invention of any one individual, it needs to be understood that the 'founding fathers' of the new theoretical paradigm – George Williams, William Hamilton, E.O. Wilson, John Maynard Smith, Robert Trivers and Richard Dawkins – were neither primatologists nor specialists in human origins. From their standpoint, the new biology was about animals in general, not humans in particular. Yet in a politically disastrous move, the social insect specialist E.O. Wilson blundered into human territory in the last chapter of his 1975 book *Sociobiology*. 'In hunter-gatherer societies', explained Wilson in an accompanying article, 'men hunt and women stay at home.' He then suggested that there might be some genetic basis for this, making it difficult to see how women could ever achieve political equality with men.[42]

Wilson's statement here had in fact no grounding in selfish gene theory or in any new theoretical insights at all. It was a thoughtlessly sexist comment in the spirit of old-fashioned Lorenz-style genetic determinism. But the damage had been done. The embarrassingly inept sentence was seized on by a veritable army of Maoist, Leninist and other 'politically correct' intellectuals across America and much of the West, all denouncing what they called 'sociobiology' as if it were a renovated form of Social Darwinism, its adherents by definition sexist reactionaries.[43] Wilson himself was utterly astonished:

A few observers were surprised that I was surprised. John Maynard Smith, a senior British evolutionary biologist and former Marxist, said that he disliked the last chapter of *Sociobiology* himself and 'it was also absolutely obvious to me – I cannot believe Wilson didn't know – that this was going to provoke great hostility from American Marxists, and Marxists everywhere.' But it was true that I didn't know.[44]

Wilson, then, was at least ready to admit his political illiteracy.

The irony is that Wilson's British critic in the above passage, John Maynard Smith, remained not only a socialist, but also a solid defender of selfish gene Darwinism to the end of his life. Robert Trivers, Hrdy's mentor and selfish gene pioneer, was also an anti-establishment figure. An active member of the Black Panther Party for several years, Trivers was often behind bars and founded an armed group in Jamaica to protect gay men from mob violence.[45] 'Thinking like a revolutionary was something that came naturally to Trivers', writes Sarah Hrdy. She adds:

Thinking back to those days, and to a time when there were anti-sociobiology demonstrations at Trivers' lectures on the grounds that he was seeking biological justification for an oppressive status quo, I can only smile and shake my head. In fact, I doubt that Robert Trivers ever met an institutionalized status quo that he didn't feel driven to destabilize.[46]

In the end, however, the misunderstandings provoked by Wilson's notorious chapter would prove heated, long-lasting and damaging to all sides. In the furore which followed, hardly anyone remembered that, of all sociobiology's pioneers, only DeVore's student Sarah Hrdy had earned the credentials to speak with any authority on the subject of human evolution. Her fresh and original reconstruction of the entire narrative of human origins represented both a landmark contribution to evolutionary theory and a break with the politics of the past. While Wilson had thoughtlessly allowed the politics of his time (compounded by his own sexist prejudices) to influence his pronouncements, Hrdy did the opposite – putting science first in a way that inspired many feminists in particular to champion a radically new politics.

Hrdy is not a linguist: her work touches only tangentially on language. Yet there was logic in her decision to subtitle her 2009 book on cooperative childcare *The evolutionary origins of mutual understanding*. It signalled that language must have begun to emerge in our species for deep-rooted biological reasons. The core psychological disposition underpinning linguistic communication is *intersubjectivity* – my readiness to share what I am thinking with you, while striving to know what you may be thinking of my thoughts.[47] While other great apes are expert at inferring the intentions of others, they seem reluctant and correspondingly ill-equipped to help those others read their minds.[48] Even their dark-on-dark eyes seem designed to prevent one ape from using direction of gaze to infer what another might be thinking.[49] Hrdy argues persuasively that language is unlikely to emerge in a species whose internal conflicts prevent group members from trusting each other or needing to share their feelings or thoughts.

In Hrdy's scenario, intersubjectivity arises as mothers probe whether they can trust one another sufficiently to babysit or help carry their offspring. Precisely when an evolving hominin mother lets someone else take her baby, selection pressures for two-way mind-reading and joint attention are set up. The mother must become socially adept to elicit support and judge how others feel about her infant. Meanwhile, the baby itself, once handed over, must develop the ability to carefully monitor 'where's mum gone?' while assessing the intentions of its new carer. This helper in turn – necessarily a relative in Hrdy's original scenario – adopts a quasi-maternal role. Hrdy describes how, in this way, a whole array of novel dispositions and skills – mutual gazing, babbling, kiss-feeding and so forth – spring up to enable the triad of mum, baby and mother's helper to stay in trusting contact with one another.[50]

Hyper-possessive great ape mothers never needed such elaborate bonding mechanisms. Because female chimpanzees must move out of their natal group on reaching sexual maturity, their access to childcare help is severely constrained, simply by the absence of female relatives living nearby. Even among chimpanzees, however, it happens occasionally that a female does manage to stay close to her own mother after giving birth, a situation likely to prove enormously advantageous to her and her offspring. 'Make no mistake, reproductively, nothing becomes a female more than remaining among kin', remarks Hrdy of primate mothers generally.[51] Across primate species, she notes, two generalizations hold up remarkably well: 'First, females who live among kin are better able to defend their interests than those who leave their natal groups to forage and breed among non-kin. Second, mothers are most prone to share infants when they feel confident that they can readily get them back unharmed.'[52]

As language capacities began developing among our own forebears, continues Hrdy, human groups must have been switching increasingly to female kin-bonding, with mothers and daughters living together throughout life. It is worth noting that Hrdy here finds grounds for reviving a basic tenet of the nineteenth-century evolutionist anthropology of Lewis Henry Morgan and Frederick Engels, who argued that before pastoralism and farming, the most likely pattern would be for a mother to share childcare tasks by continuing to live with her own mother.[53] A growing body of twenty-first-century genetic data has spectacularly confirmed Hrdy's insight that matrilocal residence must indeed have been the original deep-time pattern for our evolving ancestors on the way to becoming behaviourally modern.[54] The consequent new opportunities for cooperative childcare, according to Hrdy, made our pre-linguistic ancestors 'already far more interested in others' intentions and needs than chimpanzees are'.[55]

Of course, none of this yet explains how we evolved the ability to invent and make use of grammar. Hrdy's argument is that, prior to that development, our ancestors must have possessed other ways to communicate complex feelings and thoughts. Instead of imagining a language faculty installed in one step, she argues that the basic emotional and cognitive underpinnings of the new system must have had ancient roots, with decisive changes beginning at least half a million years ago.

With the exception of Chomsky and his colleagues, all those now researching the origins of language agree on at least that point. No evolutionary narrative can appear persuasive if it ignores the bulk of what is today known about the palaeoanthropology of human origins, about how signals evolve in nature, about the lives of human hunter-gatherers as compared with non-human primates and about known processes of cultural evolution in historically documented languages. Against this background, it is astonishing to find Chomsky and his supporters dismissing vast areas of exciting evidence from these and other relevant fields.

In their 2016 book on language and evolution, Berwick and Chomsky mention James Hurford's achievements merely to observe: 'Languages change, but they do not evolve. It is unhelpful to suggest that languages have evolved by biological and nonbiological evolution – James Hurford's term. The latter is not evolution at all.'[56]

According to this doctrine, then, it is 'unhelpful' for language evolution specialists even to speak of cultural evolution. From this comment, no one would realize that Hurford has contributed probably more than any other single scholar to turning evolutionary linguistics – 'language in the light of evolution' – into a serious interdisciplinary field. Hurford has given us two majestic volumes – *The Origins of Meaning* and *The Origins of Grammar* – which have moved the debate forward, piecing together just about all that is known of relevance to this new field.[57] Hurford's strength, inseparable from his personal modesty, has been his ability to inspire students to open up new fields, bringing together scholars from widely differing disciplines to work cooperatively together, in this way placing the modern debate about language origins on a firm scientific footing.

But Hurford is not the only major language origins researcher to be set aside with a dismissive comment. To Berwick and Chomsky, nearly everyone in the field is guilty of the false belief that novel adaptations must have precursors and must evolve incrementally. Rather than work through a list of condemned or omitted research programmes and names, let me draw attention to just one.

In view of the issues at stake, it seems surprising that Berwick and Chomsky make no mention of the artificial intelligence pioneer Luc Steels, who was a student of Chomsky's at MIT in the 1970s. It will be recalled that, early in his career, Chomsky held out the hope of specifying in detail the neural wiring needed for a machine to acquire the grammar of a language. 'There would be no difficulty, in principle', he wrote in 1967, 'in designing an automaton which incorporates the principles of universal grammar and puts them to use to determine which of the possible languages is the one to which it is exposed.'[58] In fact, as we know, Chomsky never accomplished this task or came close to it. One might therefore have expected him to show excitement and interest when a former student of his explained what you would in fact need to do in order to solve this theoretical problem. Steels uses computers that can communicate with one another and learn from each interaction.[59] Unlike Chomsky, he has always assumed that the whole point of language is to render our thoughts communicable to others.[60] Instead of trying to fathom the details of the necessary internal wiring in advance, concluded Steels, the trick is to design machines which can interact socially – and then make them do the difficult work.

Sometimes, Steels' artificial agents are just virtual entities moving around on a screen. In other experiments, they possess arms, legs, ears and eyes, and move about in the real world, encountering objects, picturing scenes and attempting to communicate with neighbouring machines built like themselves. Steels shows how, through processes of self-organization, successive interactions can lead to a repertoire of lexical symbols and grammatical rules. Over time, complexity increases because each agent remembers its successful attempts at communication, while discarding failures.

Two categories of innate cognitive equipment are essential to success. First, the robots must possess sensory, motor, navigational and other capacities which, in addition to their basic functions, can be recruited to serve communicative ends.[61] Second, they must be capable of joint attention and egocentric perspective reversal, which means picturing the world from another agent's point of view.[62] These and other indicators of a cooperative stance are absolutely essential if grammar is to evolve. Steels has been able to show that when his automata compete instead of cooperating – for example if you try the experiment of forcing them to struggle for energy at one another's expense – the resulting payoffs for manipulation and deceit prevent any kind of language from evolving at all.[63]

Although his robots are not gendered, do not have sex and never produce babies, that does not tempt Steels to dispute the theoretical relevance of sex and reproduction to the evolution of language among humans in real life. His robots are incapable of saying one thing while meaning another – a window of freedom central to linguistic creativity and absolutely essential for grammar to evolve.[64] And Steels' otherwise sociable robots never laugh or crack jokes with one another, fail to appreciate sarcasm or irony, and cannot express themselves figuratively at all. Yet these limitations do not prompt Steels to dispute the centrality of metaphor to the evolution of grammar among humans in real life. Interdisciplinary in spirit and respectful of his colleagues, Steels avoids the mistake of so many theorists who allow their beautiful models to outshine and eclipse reality. After all, the whole point of constructing simple models is to deepen our understanding of the world as it really is.

Against this background, it is worth noting that, in designing a robot, getting it to cooperate is the easy part. Indeed, it takes a special effort to design a selfish, devious, competitive machine. Since humans are great apes – by nature highly competitive creatures – Steels writes that it is a 'deep puzzle' how the ultrasociality necessary for language to evolve could have arisen through Darwinian evolution.[65] That is why Steels encourages artificial intelligence specialists to listen carefully to theoretical biologists, primatologists, anthropologists and archaeologists. We need a robust theory – consistent with Darwinian principles but applied in a highly unusual and specific way – to explain how distinctively human sociality emerged. Bearing in mind resource competition, differential reproductive strategies and the many other forces driving conflict between individuals and groups, how did a once-competitive primate evolve to be sufficiently trusting for language to evolve?

Hurford, like Steels, emphasizes the crucial importance of public honesty and trust.[66] He writes:

> Linguistic behaviour is typically trusting behaviour. As a speaker, you trust the hearer not to use to your cost what you tell him. And as a hearer, you believe or do what the speaker says. Trust by one party is an inference of trustworthiness in the other . . . How do trust and trustworthiness arise and what physical phenomena correlate with them?[67]

Some reductionist scientists, continues Hurford, have sought to replace social explanations with neurochemical ones, one idea being that oxytocin may be the hormone behind language. Retorting that there are 'no magic bullets for human language', Hurford cites a comment of mine:

> Tracing it all back to a hormone reminds me of Chomsky, who wants to trace the whole of language including even the semantic component to a box of wires somewhere in the head. The wires are probably there in some sense, but that's not the whole story. An evolutionary account has to explain why placing trust in mere words became an evolutionarily stable strategy in the case of our species. To explain this, we have to explain how and why people had the time and energy to punish free-riders who abused that public trust. In short, we have to explain how and why the rule of law became established among our hunter-gatherer ancestors. That means doing everything within a social and anthropological framework, among other frameworks.[68]

Hurford agrees, going on to point out that, given sufficiently high levels of public honesty and trust, the cultural evolution of communication may accelerate rapidly along biologically unprecedented lines. Hurford's chapter on 'Grammaticalization' in *The Origins of Grammar* explains the relevant processes in a particularly wide-ranging, historically informative and up-to-date way.

Grammaticalization is a process whereby the effects of frequent use – for example abbreviation and routinization – become entrenched as part of the learned structure of a language. The sequences involved tend to be unidirectional, as content words become abbreviated and bleached of their original meaning in being recruited to serve new grammatical functions. These are, of course, cultural rather than biological processes. But unless we follow Chomsky in excluding the very concept of cultural evolution, they offer a compelling way to explain how grammar itself may initially have evolved. Writes Hurford:

> Under the dynamic uniformitarian assumption that early human linguistic behaviour was guided by the same principles as modern behaviour, we can use the revelations of grammaticalization studies to reverse-engineer back from modern syntactic structures through the unidirectional processes that built them up. If protolanguage is words strung together without grammar, and modern mature languages use grammar to put words together, grammaticalization, as I exploit it here, is the process bridging the gap. It is a gradual cultural alternative to any putative 'syntax mutation'.[69]

In a fascinating development, Andrew Smith and Stefan Höfler have been able to show that what used to be considered two separate evolutionary problems – the earliest use of symbols, and the subsequent transition to grammatical structure – actually emerge from one and the same set of underlying cognitive mechanisms. In their model, early humans develop the cooperative capacity to infer intended meanings from context, to recognize common ground, to remember and build upon communicative attempts which proved successful in the past – and on that basis to elicit and invent metaphorical expressions which subsequently become conventionalized.[70] The idea here is that the cognitive mechanisms underlying just one creative principle – metaphor – can provide a single solution to what were previously viewed as two evolutionary sub-problems: the emergence of words and the emergence of grammar. Grammatical structures arise out of metaphor, grammatical markers being in fact metaphorical expressions which have been conventionalized and abbreviated through historical processes which are now well understood. It is significant that these understandings concern the one thing which Chomsky has suggested we are unlikely ever to understand – the creative *use* of language. It is also significant that metaphorical expressions, by definition, are not literally true. Since there is a sense in which they are falsehoods, a great deal of empathy, goodwill and trust in communicative intentions is required.[71]

There are good reasons why the cognitive revolution prompted scholars to ignore sex and gender. After all, if you think the body does not matter because abstract digital information is what we really are, then the particular kind of body you inhabit is hardly relevant. Among nonhuman primates, however, sex is a major source of conflict and mistrust. Against this background, it is impossible to think intelligently about how our ancestors overcame such problems – minimizing them sufficiently for language to evolve – if we restrict ourselves to theoretical models allowing no place for gender or sex. For those of us who follow Hrdy's train of thought, the event described by Sahlins as 'the greatest reform in history, the overthrow of human primate nature' must have been the outcome not only of male fitness-enhancing endeavours, but also coalitionary strategies pursued by child-burdened females in their own genetic interests.[72] It is not difficult to see how far removed is this sophisticated Darwinian approach from Chomsky's fable about a person named Prometheus having his brain rewired for language in a single genetic step.[73]

The 1960s human revolution pioneers were motivated by a shared concern that humanity faced an unprecedented crisis in the nuclear age, that self-interested political agendas were the problem and that a scientific understanding of our species' past, present and possible future had to be part of any solution. The promising start made by Hockett, Sahlins, Lee, DeVore and many others was unfortunately drowned out in part by the cognitive revolution, whose adherents tended to follow Chomsky in picturing science as an assortment of specialist fields of no relevance or interest to political activists. Equally chilling were the mirror-image responses of the anti-sociobiology and sometimes anti-science activists themselves, as they turned their back on what they perceived to be a dehumanized and alien Western ideology justifying the status quo by masquerading as 'science'.

One result of all this was that Sahlins' idea of a human revolution gave way in the 1980s to a newly resurgent gradualist Darwinism, now – thanks to the cognitive revolution – forced to cohabit somewhat uncomfortably with Chomskyan mentalism. Unfortunately, Sahlins' initial interest in monkeys and apes proved to be short lived. A paper he had published in 1959 in the journal *Human Biology* was reprinted in the 1972 edited volume *Primates on Primates*.[74] But that was before sociobiology. Once it became clear that primate field research would be conducted virtually everywhere within the new 'selfish gene' framework, Sahlins backed away, denouncing that entire framework as nothing more than genetic determinism and pro-capitalist myth-making.[75]

To his credit, Chomsky never joined the stampede against sociobiology. He has always considered it perfectly reasonable for socialists to view humans as beings endowed with a genetically determined nature.[76] But while Chomsky refused to join in the baiting of sociobiology, he never fully understood the new Darwinian paradigm or sympathized with its project to adequately explain cooperation in nature. His notion of how genetic factors regulate development remained closer to the simplistic genetic determinism of ethologists such as Konrad Lorenz. The younger Darwinian generation, represented by such figures as Dawkins, Trivers and Hrdy, saw genes as instructions to build proteins involved in bodily form and function; genes, they well understood, cannot specify social arrangements or fix concepts in our heads. But Chomsky's work in shaping the cognitive revolution had the momentous effect of diverting attention away from the body in favour of a relentless new focus on genetic constraints acting on the mind. So momentous was the change in intellectual fashion that Darwinism itself – although of little interest to Chomsky – came under intense pressure to conform.

As a consequence, many self-proclaimed Darwinian psychologists during the 1980s abandoned the traditional idea that selection pressures continuously drive behavioural and anatomical change. Instead, they began arguing that in the human case, the most important selection pressures were those acting on the mind. It was noticeable that these thinkers applied their new mentalist version of Darwinism to just one species – our own. No one tried to explain, for example, the behaviour of primates by invoking innate computational modules inside the skull. But, inspired by Chomsky's language organ idea, the new evolutionary psychologists claimed that human cognition was modular in a special and peculiar way, requiring a special new kind of Darwinism to explain it. The search was now on for a range of additional devices inside the head, their existence decoupled from current or recent selection pressures and instead attributed to conditions prevailing during some unspecified time in our species' past.[77]

Although this mentalist school of evolutionary psychology was confined mostly to the United States, its effects were felt worldwide, as a new generation of self-proclaimed Darwinians began seeing uniquely human computer modules everywhere – one for language, another for morality, another for detecting cheats, another for recognizing faces and so on.[78] The new paradigm was eloquently summed up in these words by Leda Cosmides and John Tooby:

Instead of viewing the world as the force that organizes the mind, researchers now view the mind as imposing (on an infinitely rich and extensive world) its own pre-existing kinds of organization – kinds invented by natural selection during the species' evolutionary history to produce adaptive ends in the species' natural environment. On this view, our cognitive architecture resembles a confederation of hundreds or thousands of functionally dedicated computers (often called modules) designed to solve adaptive problems endemic to our hunter-gatherer ancestors. Each of these devices has its own agendas and imposes its own exotic organization on different fragments of the world.[79]

In addition to Chomsky's Language Acquisition Device – the point of departure for the entire enterprise – this plethora of mini-computers includes a theory of mind module, a device for face recognition, one for construing objects, a special computer for recognizing emotions and yet others to detect animacy, gauge eye direction and detect whether someone might be cheating on a contract.

It has to be said that each evolutionary psychologist imagined his or her own personal assortment of special-purpose computers, the number ranging from Chomsky's three or four (one for grammar, another for science and perhaps another for moral judgement) to the hundreds or even thousands imaginatively envisaged by Tooby and Cosmides. Uncannily mirroring the interdisciplinary fragmentation of modern science, the mind according to Cosmides is a 'Swiss army knife', each separate tool designed for its own special task without reference to its neighbours.

The enthusiasm for what has aptly been termed 'the new phrenology'[80] soon spread to archaeology, rock art research, the study of religion and many other areas. Chomsky had successfully redefined language as a natural component of the human mind; now religious belief was interpreted as an equally natural feature of that same modular mind.[81] The fashion became so irresistible that one ambitious archaeologist, having described how shamanism involves soul flight, soul journey, out-of-body experience and astral projection, tautologically explained Upper Palaeolithic shamanism as the outcome of a dedicated 'soul flight, soul journey, out-of-body experience and astral projection' module located somewhere in the human brain.[82]

Just as Chomsky had striven to detach linguistics from its former ties with the political and social sciences, so the new evolutionary psychology was explicitly aimed at detaching evolutionary science from politics, and especially from Marxism. Its exponents even insisted that the various chambers of the mind could at last be studied with the dispassionate objectivity and detachment of a chemist or physicist. Chomsky took this idea to extremes, picturing himself as a scientist looking down on humanity from a vantage point on Mars.[83] To many intellectuals, there is evidently something alluring about so lofty and detached a standpoint. But the problem is that if you are on Mars, you are all alone. To be cut off to that extent is no guarantee of objectivity at all, since inevitably your isolation will shield you from corrective social feedback. To imagine yourself on Mars is in fact to imagine yourself in the very worst place for a scientist of human nature to be.

Sarah Hrdy's rigorously materialist exploration of human psychology roots the evolving mind in the body and its reproductive, social and other relationships. To explain what is special about human mentality, she turns to an investigation of the social and political strategies pursued by evolving humans in transcending primate dominance dynamics and establishing a cooperative lifestyle. For John Tooby, Leda Cosmides, Steven Pinker and their co-thinkers, on the other hand, the contrasts which matter are located primarily in the head. Since the late 1980s, their ambitious intellectual project has been to repudiate what they term 'standard social science' – especially Marxism – in order to evict and replace it wholesale, installing modular mind psychology throughout its former territories. For these thinkers, the subtle cultural and political understandings accumulated by social science's intellectual giants have turned out to be more or less worthless.[84]

While distancing himself from Darwinian gradualism, Chomsky explains that he feels politically comfortable with the new evolutionary psychology, finding it perfectly compatible with a left-wing perspective on the world: 'Peter Kropotkin was surely on the left. He was one of the founders of what is now called "sociobiology" or "evolutionary psychology" with his book *Mutual Aid*, arguing that human nature had evolved in ways conducive to the communitarian anarchism that he espoused.'[85] Yet, unlike Kropotkin, Chomsky sees no point in recalling or celebrating any kind of communism or egalitarianism associated with humanity's evolutionary past. Asked what lessons he draws from the egalitarian lifestyle of extant hunter-gatherers, Chomsky replied:

Well, suppose it turned out that the Kalahari Bushmen were living in an absolute utopia. That's not true, but suppose it turned out to be true . . . That wouldn't tell us anything about this world. It's a different world. I mean you have to start, if you want to be related to the world in which people live, you have to start with the existence of that world and ask how it can be changed.[86]

For Chomsky, then, activists can understand the world as it is today without needing to worry about history or prehistory.

Chomsky's position is an extreme one, not typical of those American evolutionary psychologists whose modular mind mentalism was initially inspired by his work. But although Steven Pinker and his co-thinkers frequently invoke what they term the 'Environment of Evolutionary Adaptedness', what is striking is their intellectual arrogance – not unlike Chomsky's – with its reluctance to take account of ethnographic details. In place of social anthropology's treasury of painstaking records and in-depth understandings of hunter-gatherer shamanism, kinship networks and rock art, for example, we are regaled with simplistic put-downs. Pinker himself provides such off-the-shelf gems as the following:

Tribal shamans are flim-flam artists who supplement their considerable practical knowledge with stage magic, drug-induced trances, and other cheap tricks . . .[87]

Foraging tribes can't stand one another. They frequently raid neighbouring territories and kill any stranger who blunders into theirs . . .[88]

In foraging cultures, young men make charcoal drawings of breasts and vulvas on rock overhangs, carve them on tree trunks, and scratch them in the sand. Pornography is similar the world over . . .[89]

Of course, the culturally aware anthropologists, historians and others taunted by such reductionism have proved to be less than enthusiastic, many of them mounting a spirited fightback. As so often happens, however, the interdisciplinary animosities stirred up on all sides have led to bitterness and deadlock, in many ways echoing the fruitless disputes of the linguistics wars. Politically correct postmodernists and cultural theorists have tended to hit out wildly, not bothering to discriminate between, say, Sarah Hrdy's subtle explorations of gender in evolutionary perspective and the crass biological reductionism of Pinker and other US-style evolutionary psychologists. Some feminists would too often criticize as 'essentialist' anyone who even dared mention genes.[90]

Rather than perpetuating such exhausting divisions, the final goal of science must surely be to unite what we know about nature with what we know about ourselves. This represents a theoretical challenge, but it is also a practical and sometimes an unavoidably political one. Scientists who make new discoveries have never been content with contemplation. They have always found it necessary to get actively engaged, intervening in the world in order to understand it.

CHAPTER 23

THE HUMAN REVOLUTION

A theme throughout this book has been the Tower of Babel, reminding us that a shared conceptual language sustaining a world without borders or nations will inevitably seem threatening to those in positions of power.

I am not the first to have discerned something perhaps God-like in Chomsky's unique ability to command our scientific understanding of language, to the point where we barely understand one another at all. I have already suggested that Chomsky's linguistic output is not really science but scientism – Science with a capital S. In the words of Bernard Latour, the function of Science is simply to signal 'Keep your mouth shut!' to ordinary people.[1] The Old Testament pictures God as a jealous patriarch, confusing our tongues to keep us ignorant of the heavenly secrets. Rudolf Botha presents us with a similar figure – Chomsky as 'Lord of the Labyrinth' – deliberately constructing a vast intellectual maze to much the same end. Both metaphors nicely capture the suspicion that there is something so empowering about language that its inner workings and secrets are liable to be coveted and monopolized by those in positions of power, condemning the rest of us to speechless bewilderment. We end up with what Pierre Bourdieu terms the 'invisible, silent violence' which elite speakers direct against ordinary people whose inability to express themselves in the fashionable tongue 'leaves them "speechless", "tongue-tied", "at a loss for words", as if they were suddenly dispossessed of their own language.'[2]

In the opening chapter I asked why the edifice which Botha terms Chomsky's 'labyrinth' was ever built. Who financed its construction and why? What secret concealed at the heart of that maze matters so much that none of us should ever be allowed to reach it?

I have shown how the Pentagon played a major role in financing the maze, and suggested a reason for this. I am not suggesting a conscious conspiracy here. My point, rather, is that the state has no interest in facilitating the kind of science which would allow ordinary people to piece together the big picture of what it means to be human. Although Chomsky does sometimes take this line – blaming state propaganda for keeping us ignorant of our creative potential – he can also swing to the opposite extreme, claiming that our limitations in this respect are down to genetics. It is clear that this second line of argument exonerates the state and for that reason may be suspected of being fostered by the state. Exemplifying this reactionary line of argument, James McGilvray writes: 'If reason . . . has the biological basis it seems to have, reason must have limits. Those limits are revealed in an incapacity to make scientific sense of the creative aspect of language.'[3] Chomsky himself says much the same in declaring that 'the Cartesian question of creative [language] use . . . remains as much of a mystery now as it did centuries ago and may turn out to be one of those ultimate secrets that will ever remain in obscurity, impenetrable to human intelligence.'[4] So the very thing which is most important to us – the secret of our nature as humans – is and must always remain incomprehensible even to ourselves.

To be trapped inside Chomsky's labyrinth is to experience just that incomprehension. In our futile attempts to fulfil our human potential, we feel lost in a vastly complex maze, unable to see our way out. Some colossal force seems to be stopping us from taking the final step – putting everything together in order to successfully accomplish the next momentous transition in evolution. We all feel – from climate change to economic crisis – that our world is crashing down around us and nothing can be done. That sense of helplessness seems likely to continue for as long as we feel that things just don't add up, that there is no way of communicating across our multiple specialisms, that it's all so complex that our limited brains cannot cope. If there is a big picture, nobody can see it.

The secret at the heart of the labyrinth is both simple and shocking. Across the world, ordinary people now hold the power. Our lives are intertwined and our brains electronically networked, giving us a capacity for synchronized action with the potential force of a hurricane. But there is a problem: our minds remain trapped somewhere else, keeping us unaware of the power within our grasp. The day we tear off the blindfold – that's when the revolution starts.

Fortunately, science is not what Chomsky says it is – the mechanical output of a 'science-forming capacity' located in each separate skull. In the words of Talmy Givón, 'good science is profoundly communal'.[5] At its best, science is intersubjectivity writ large – a huge, global, collective enterprise through which we creatively participate in *each other's* minds. Although modern science may nowadays seem more than the intersubjectivity which has always characterized our species, the underlying principle remains the same. To be scientific is part of what it means to be human. It is to be rational, which means being open to doubt, to being corrected, to being persuaded by the evidence, even when it comes from someone else.

Far from our minds being restricted by the limits of the physical brain, we humans have arrived at the point where, God-like, we can begin to grasp not only the outlines of our past and present, but also our possible future. Contrary to Chomsky, no relevant discovery is likely to be permanently barred to us owing to the physical limitations of your brain or mine. Yes, deep mysteries remain. But, combined, our brains have already discovered what happened during the first few seconds of the existence of our universe, worked out when and how galaxies formed, reconstructed the origins of our solar system and gained a good idea of the major transitions central to the emergence of life as we know it on Earth. The most recent of these transitions, of special interest to humans, involved the astonishing emergence of our peculiar capacity for articulate speech. And here we are today, empowered with that capacity and with the huge potential for cooperation that it imparts.

In the face of all this, we urgently need to recover our voice, break the silence and lift the confusion of tongues. We need to speak aloud and engage with one another as well-rounded people, our heads joined to our bodies, our activism grounded in our science. Our first requirement is a shared conceptual language. While this can only be the language of science – humanity has no other common tongue – we can no longer afford to treat science as just abstract understanding cut off from laughter, from our emotions, from life in the streets, from joint action to make for ourselves a better world. No individual contribution – not even from a genius – qualifies as science until the rest of us have found that it works. There is nothing special or esoteric about this way of doing things. We consult one another. We check each other's results. 'Do you see what I see?' Humans have been doing this since our species first evolved.

What does it mean to be human? One thing is clear: the more active we become and the more widely we cast the net – the more diverse the voices we hear – the better our chances of finding a voice which empowers us all. If we are to begin talking to one another, as we must, we have to start developing terms and concepts we can all share. The barriers between us have no scientific basis; they are institutionally, *politically* imposed. For progress to be made, we have to break through the terminological, methodological and disciplinary frontiers which have kept us all apart.

This means pursuing science for its own sake, free from worries about what the political repercussions might be. The eminent climate scientist James Hansen makes this point powerfully. Hansen became an environmental activist not in pursuit of a new political career, but to keep faith with his science. Pursuing this course, he soon found himself getting arrested by the police. Unlike Chomsky, however, Hansen regards his own painful transition from scientist to scientifically aware activist as logical and consistent – a matter, as he puts it, of 'joining up the dots'.[6] He is concerned about the future – concerned for his grandchildren. To suspect that we are tipping the planet toward disaster *and to respond by doing nothing* is, from Hansen's standpoint, not a logical option at all.

Together, we need to work out how a biological creature became a speaking and self-aware one, how *Homo sapiens* evolved and is evolving still. In this context, the most damaging artificial border is the one obstructing free movement of ideas between scholars in the humanities and their counterparts in natural science. The best way to dismantle that barrier would be to tunnel through from both ends – from the humanities and social sciences on the one side, the natural sciences on the other – converging on the nature/culture interface. That was the intellectual challenge addressed by Charles Hockett, Marshall Sahlins and other champions of unified science during the optimistic 1960s. Those efforts at unification – given powerful political expression in 1969 in the founding statement of the Union of Concerned Scientists[7] – ebbed away for primarily institutional and political reasons. Today, the problem of our nature and future constitutes an intellectual challenge, but the lingering institutional obstacles make it a political one as well. To solve it, we will need political will.

Chomsky has always displayed an impressive and urgent political will, showing it in his activism more productively than any other prominent academic to date. But in his scientific capacity, he has moved mountains to insist that he has no political will at all. He presents his linguistics as dispassionate, value-free, insulated from social fashion, as well as from his own or anyone else's political agendas or needs. 'The search for theoretical understanding', as he puts it, 'pursues its own paths, leading to a completely different picture of the world, which neither vindicates nor eliminates our ordinary ways of talking and thinking.'[8] The idea that science is somehow value-free is, of course, itself a potent Western myth. Social forces are always active, and those championing such forces have a will of their own. We have seen how, in Chomsky's case, state funding provided the institutional framework needed to ensure the cognitive revolution's success. If there was a political will in operation here, no matter how unconscious, it was indisputably that of the state.

This is not to say that all Chomsky's misgivings about Darwinian gradualism were misplaced. Darwinian theory has certainly struggled to explain
the onset of grammar. Steven Pinker follows Chomsky in picturing language
as a device for combining a limited repertoire of mental atoms (lexical
concepts) in an infinite variety of possible ways. No one imagines that
either the physical brain or its material environment presents itself in digital
format: location, weight, temperature, size and so forth are dimensions
which vary along an unbroken continuum. Summing up the mind–body
contrast, Pinker describes each one of us as a 'digital mind in an analog
world'.[9]

Chomsky differs from Pinker in perceiving a *theoretical* difficulty here.
The computational theory of mind, he has always realized, confronts us
with the baffling problem of 'how a messy system such as the brain could
have developed an infinite digital system in the first place'.[10] Determined at
all costs to marry Chomskyan mentalism with Darwinian gradualism,
Pinker simply bangs the two together. Since the digital mechanism is a
biological adaptation, he states, it *must* have evolved in the usual way,
through Darwinian natural selection. He rests his argument not on any
particular scenario, but on first principles: quite simply, there is no other
way a biological organ *can* evolve. Skating over the difficulties, he argues
that digital infinity emerged so long ago in the evolutionary past that the
details are unknown and, in any event, not required. 'We say virtually
nothing about the precursors and very first forms of language and the
specific sequence leading to its current form', Pinker and his colleagues
explain.[11] Provided no one wants a concrete explanation or detailed evolutionary sequence, no special refinement of Darwin's theory appears necessary. We can rest assured, runs this argument, that the gradualist doctrine of
descent with modification ought to work.[12]

Most American-style evolutionary psychologists side with Pinker here.
They don't have a theory, but accept natural selection as an article of faith.
As against all this, Chomsky maintains that invoking Darwin is no answer
at all. The idea of evolving by incremental steps from a call system to digital
infinity is incoherent in principle. To assume an evolutionary development
from primate vocalizations to grammar is, to quote Chomsky, like 'assuming
an evolutionary development from breathing to walking'.[13] The two are
such different things that evolution in this context makes no sense.

But instead of helping to crack this problem, Chomsky's contribution
has been to render it insoluble. To insist that digital infinity is a real object
in the head is to define it as something which *cannot* evolve. No biological
organ can evolve to the point where it has become partly infinite: anything

less than infinite is not infinite at all. Likewise, no biological organ can be partly digital. By defining language as a digital and infinite organ – by legislating that concept into force – Chomsky erected a categorical barrier separating its existence from all that ever evolved or could possibly have evolved during the story of life on Earth.

In confronting language, Darwinism *does* face a major theoretical challenge. What we need, however, is not an alternative to Darwinism, but a sufficiently sophisticated refinement of this theoretical framework – one that is not simply gradualist, but can accommodate revolution, too. The punctuated equilibrium approach of palaeontologists Niles Eldredge and Steven Jay Gould[14] goes some way toward a solution, but more powerful still is the 'major transitions' paradigm pioneered in the 1990s by John Maynard Smith and Eörs Szathmáry.[15]

Scientists, to repeat, have recently illuminated the basic contours of the story of life on Earth. Operating as always within a 'selfish gene' framework – now accepted by virtually everyone in the field – Maynard Smith and Szathmáry showed this story to be a sequence of eight or so momentous shifts or revolutions, each separated by a seemingly endless period of stability involving only gradual change. The appearance of life itself constitutes the first major transition, the origin of multicellular complexity another and the emergence of language-based human society the most recent in the series. At each stage, the revolution is fundamentally social in character, smaller entities assembling in a radically new way to form a larger cooperative whole. This more complex plane of cooperation is at first blocked by the 'selfish' replicatory strategies of entities still going their own way on their own lower level. In each case, it takes multiple factors accumulating over time to set up the conditions necessary to achieve a sudden social breakthrough. Pursuing a theme reminiscent of Hegel's *Science of Logic*,[16] Engels' *The Dialectics of Nature*[17] and perhaps even Khlebnikov's *Tables of Destiny*,[18] the authors offer an explanation for the emergence of cooperation at all levels of complexity, showing how an understanding of any particular qualitative leap sheds light on all the others in the sequence.

This approach offers fresh and valuable new insights into the origins of language. Whereas for Chomsky installation of the relevant organ is a one-off genetic accident, for Maynard Smith and Szathmáry the emergence of language takes its place as the latest cooperative leap in the history of life on Earth. In a profound sense, each such shift or leap amounts to a social revolution. With that basic assumption in mind, Maynard Smith and Szathmáry address the origin of language only *after* discussing social contract theory and the establishment of society itself.[19] In their co-authored book on the evolution of language, Chomsky and Berwick feel obliged to cite Maynard Smith and Szathmáry in support of the idea of sudden evolutionary change – yet do so while presenting the reader with a peculiarly neutered version of the 'major transitions' paradigm, now stripped entirely of the engine of social conflict, competition and emergent cooperative dynamics.[20]

In reconstructing the origins of society, Maynard Smith and Szathmáry point to the importance of communal rituals of the kind reported among hunter-gatherers to this day. Citing Robert Boyd and Peter Richerson, they suggest that 'between-group selection may have favoured rituals that are particularly effective in binding a group together'.[21] In stressing the importance of ritual, they reinforce an intellectual tradition stretching back to Rousseau and Durkheim and developed in more recent years by such social and cultural anthropologists as Pierre Bourdieu, Roy Rappaport, Camilla Power, Jerome Lewis, Morna Finnegan and myself.[22] Lewis and Finnegan in particular have shown how the playful rituals of the Bayaka and other hunter-gatherers of Central Africa are bound up inseparably with laughter, singing and vocalizations of all kinds, the interdependence of all these capacities offering insight into how the entire complex may have evolved.[23]

Times, then, have changed since Chomsky came up with his simple model of language as an organ installed by a mutation in one gene. While this idea may once have seemed intriguing and exciting, it no longer works now that genetic science has become so very much more ambitious and sophisticated. We will not succeed in explaining the evolutionary emergence of language without embedding our theory in a wider one which encompasses the origins of everything – joint attention, gender-specific reproductive strategies, group-level morality, symbolic ritual, cultural kinship, notions of the sacred and many other things characteristic of a fully human life.

For Chomsky, a mentally represented grammar is part of the physical world, a particular steady state attained by one special component of the physical brain, hence 'a definite real world object, situated in space-time and entering into causal relations', in this respect like any other physical object.[24] If you had to pick out just one foundational error at the root of all Chomsky's other intellectual contradictions and difficulties, this would be it. It is true that the human brain is a biological object with a certain weight and size, situated in space-time and entering into causal relations with other physical objects. But the mind, being intersubjective, cannot be pinned down in this way. Minds reflect back on each other, interpenetrate one another and so transcend the confines of the skull. In the previous chapter, I chose to focus on Sarah Hrdy's work because she, more successfully than anyone, has set out to explain just how, when and why the peculiar phenomenon of intersubjectivity began to evolve in our species and no other.

Unlike Chomsky, Hrdy takes for granted that a serious evolutionary theory to explain a biological adaptation must assume the existence of precursors at an earlier stage. Her approach is a social one. For language to begin to evolve, trust-based coalitions of the kind she describes had to be scaled up from mother–daughter beginnings to become more and more inclusive – to the point where intersubjectivity through mutual understanding might in principle be extended across entire communities.[25] This idea is consistent with the work of Michael Tomasello, who points out that whereas primate-style dominance impedes intersubjectivity, egalitarianism fosters it. 'It is in social interaction and discourse with others who are equal in terms of knowledge and power,' Tomasello explains, 'that children are led to go beyond rule-following and to engage with other moral agents who have thoughts and feelings like their own.'[26] Endorsing this line of thought, evolutionary psychologist Andrew Whiten points out that while mind-reading exposes minds to public appraisal, thereby encouraging counter-dominance and egalitarianism, the causal arrow works equally the other way: '*Egalitarianism* encourages a certain kind of cooperative *mind-reading*' without which language would be impossible.[27] Happily, the notion of an egalitarian political revolution is also consistent with Maynard Smith and Szathmáry, giving additional anthropological substance to their intuition of a relatively sudden leap.[28]

Hrdy herself refrains from speculating about the culminating stages of human evolution, when art and other signs of symbolic culture first make their appearance in the archaeological record. In this sense, her work leaves many of the big questions unanswered. How *exactly* did our hominin ancestors manage to overcome their primate instincts sufficiently to establish a social contract and live by its rules?

What we can say is that humans feel healthiest and happiest when relaxing among equals, taking orders from no one, free to play, laugh and sing together as egalitarian hunter-gatherers still do to this day.[29] To be linguistically and psychologically human, as Friedrich Engels once pointedly observed, is to have *something to say to one another*.[30] Being on speaking terms means *not* following instructions like a computer. It means rejoicing in metaphor – the creative evolutionary source of both words and grammatical rules.[31] This in turn means allowing ourselves a certain kind of freedom – freedom to joke, to laugh, to make fun of one another, to express our thoughts in ways that are not literally true, to allow a gap between what we say and what we intend our words to mean.[32] Without such freedom, no part of language could ever have evolved.

Primate social systems, to repeat an obvious point, are neither just nor fair. Since one can make exactly the same criticism of human social and political arrangements today, many people find it tempting to conclude that a peaceful and united planet is an unrealistic goal – unrealistic because it takes no account of primate and human nature. Revolutions have come and gone, but ultimately each one failed. If human nature is the problem, we just have to accept that a genuine socialist or communist society is impossible. After all, not even a revolution can change human nature.

Chomsky's views are subtly different from this. Although he treats human nature as fixed, he argues that it is precisely our species' innate creativity – our felt need to exercise control over all aspects of our lives – that allows activists to hope for success in resisting authoritarian social structures. But, while expressing solidarity with national liberation movements in the so-called Third World, Chomsky denies that his concept of human nature justifies any hope for revolution in countries like the US. If you had mass support for revolutionary change in the US, he argues, questions about revolutionary action might at least arise. 'But we are so remote from that point that I don't even see any point speculating about it and we may never get there.'[33] Note that Chomsky's scepticism does not just reflect recent disappointments, but has been a life-long characteristic. Even during the strikes and riots in Paris in May 1968, when students and workers came close to provoking a revolutionary crisis, Chomsky still could not bring himself to express any support. In his own words, 'I paid virtually no attention to what was going on in Paris as you can see from what I wrote – rightly, I think.'[34]

Chomsky does advocate the transformation of Western capitalism, sometimes associating this with revolution. But he insists that such change ought to be gradual:

> In the case of workers taking control of the workplace . . . I think what we should do is try [the changes] piecemeal. In fact I have a rather conservative attitude toward social change: since we're dealing with complex systems which nobody understands very much, the sensible move I think is to make changes and then see what happens and if they work, make further changes. That's across the board actually.[35]

This approach sounds very much like that of Zellig Harris in his only political book, *The Transformation of Capitalist Society*, where he advocates 'non-capitalist production growing inside capitalist society' and argues that there must first be significant 'employee-owning and cooperative forms before any large post-capitalist political changes are reached'.[36] Like Harris,

Chomsky has little sympathy for calls for 'revolution', which he condemns as 'insidious . . . at a time when not even the germs of new institutions exist, let alone the moral and political consciousness that could lead to a basic modification in social life'.[37]

Chomsky here expresses the idea that mass political consciousness must change first, any revolutionary action being kept under wraps until mental states have been sufficiently transformed. Applied to politics in this way, Chomskyan mentalism stands the materialist tradition of revolutionary activism on its head.

Despite decades of often fruitless disputes between anarchists and Marxists, activists in both camps have usually been able to unite around the insight that practice comes first. To postpone action until you have managed to raise consciousness to an ideal level is a recipe for paralysis stretching into the indefinite future.

The fundamental Marxist insight is this: it is not consciousness which determines conditions, but the other way round. Experience counts for more than abstract ideas. Under capitalism, people tend to feel competitive and isolated. This leads to deep feelings of fragmentation and helplessness – which are a logical response under the circumstances. Newspaper and other mass media proprietors will then find a ready market for individualistic, racist, sexist and other divisive ideas. If that is true, it cannot be propaganda that is the root cause of the low level of consciousness – as Chomsky argues in his influential article, 'The Manufacture of Consent'.[38] Rather, it is the lack of community, solidarity and activism which gives rise to a profitable market in reactionary ideas.

Anyone who has stood on a picket line or been involved in an occupation will know how radically consciousness can change, once the possibility of collective action has opened up. Think how often we are inclined to say of our activist selves that 'we did not really know what we felt or wanted until we acted'.[39] Once people feel empowered and connected, they are likely to experience a thirst for newly relevant ideas, perhaps utterly revolutionary ones. The insight that *practice is primary* is expressed beautifully in the words of the young Marx:

> Both for the production on a mass scale of this communist conscious-
> ness, and for the success of the cause itself, the alteration of men on a
> mass scale is necessary, an alteration which can only take place in a prac-
> tical movement, a revolution; this revolution is necessary, therefore, not
> only because the ruling class cannot be overthrown in any other way, but
> also because the class overthrowing it can only in a revolution succeed

in ridding itself of all the muck of ages and become fitted to found society anew.[40]

For Chomsky, the urgent priority is not revolutionary social transformation but survival. In the face of nuclear proliferation, climate catastrophe and related threats, dreams of revolution must be put on hold while we struggle simply to survive.[41] But what if science-inspired revolution has become a fundamental *condition* of our survival?

Chomsky comes close to revolutionary politics, yet always steps back at the last moment. By now, the reader will be familiar with this pattern. Just as he has felt torn between institutional dependence on the US military-industrial complex and political hostility to that complex, so he appears torn between the anarchist dream of overthrowing the state and a variety of all-too-familiar proposals for reforming it. As noted in my discussion of mindless activism and tongue-tied science, Chomsky has himself conceded that his insistence on separating science from activism may have limited his own political effectiveness. As he puts it, his doubts about the very possibility of doing social science 'probably lead me to underestimate the potentialities of activism' and 'restrict, no doubt improperly, my own personal involvement'.[42]

I think this is exactly right. Irrespective of the doctrines of liberal Western academics, ordinary people faced with the need to explain poverty and injustice will continue to blame human nature. It is widely held – often on supposedly scientific grounds – that poverty, sexism, inequality and war will always be with us, just as humans will continue to be born with five digits on each hand.

Without going back to our beginnings, there is no way to counter such deeply ingrained ideas. The best riposte is Hugh Brody's hymn to the hunter-gatherer lifestyle, *The Other Side of Eden*:

> What makes us who we are? Things we inherit, be they aspects of body or the hard-wiring of the mind. But language means that much of who we are does not lie within us as individuals so much as between us . . . And this is where we can see a particular importance of hunter-gatherer societies: they have established and relied upon respect for children, other adults and the resources on which people depend. If these relationships are not respectful, then everything will go wrong. The sickness of particular individuals, the failure of the hunt, the weather itself – these are all expressed in terms of relationship. The egalitarianism of hunter-gatherer societies, arguably their greatest achievement and their most compelling lesson for other peoples, relies on many kinds of respect.[43]

So how did such respectful egalitarianism come to be established and shape human nature for more than a hundred thousand years?

Compared with current theories that locate the origins of language in an egalitarian revolution, Chomsky's obsessive focus on a random mutation is no help at all. The chief value of the study of human origins is that it teaches us that everything *distinctively human* about our nature is the result of a remarkable evolutionary process which culminated in *revolutionary* social change. We can be confident that this human revolution succeeded because here we all are with those linguistic and other capacities which make us unique. Our ability to see ourselves as others see us, to establish moral principles and commit to them, to trust one another sufficiently to use language to share our dreams – all these abilities were once utterly revolutionary. Human nature is a complex mix, part ancient and shared with other primates, part uniquely human and without evolutionary precedent. The most remarkable components of human genetic nature were fine-tuned and given scope for development thanks to the revolution which worked. To end on an optimistic note in these bleak times – when revolution has been written off even by the left – the conclusion must surely be that having won the revolution once, we can do it again.

GLOSSARY

Behaviourist psychology – The Ivan Pavlov/B.F. Skinner school of psychology and language-acquisition theory, now discredited. According to this view, 'mind' is an unscientific concept, all that matters being behaviour, since this is the only thing that can be accurately observed and measured. During cultural conditioning or 'learning', the human brain is held to resemble a blank slate on which anything can be written. The founders of behaviourist psychology conducted learning experiments on rats and other laboratory animals, claiming that their findings apply equally to human beings.

Blank slate psychology – The theory that individuals are born without any built-in mental content, all knowledge deriving from experience. In terms of the nature versus nurture debate, blank slate theorists downplay genetics and emphasize experience.

Cartesian – Relating to the French philosopher René Descartes (1596–1650). Since Descartes is often regarded as the founder of modern science, the term Cartesian is sometimes used to imply 'scientific'. Since Chomsky began describing his linguistics as Cartesian, however, it has come to mean scientific in a rather special sense. To explain his own use of the term, Chomsky resurrected Descartes's idea that certain truths can be discovered by reason alone, without requiring experiments or evidence.

Cognitive psychology – The theory that the mind is best understood as an information-processing device on the model of a digital computer. Because the theory downplays computations that machines are unable to perform, cognitive psychology tends to marginalize complex emotions, sexuality and interpersonal dynamics.

Cognitive revolution – An intellectual movement, beginning in the 1950s, that rejected behaviourist and 'blank slate' approaches to psychology in favour of the idea that the mind is a computational device whose innate wiring determines how it works.

Cybernetics – The scientific study of control and communication in the animal and the machine. From the Greek *kybernetike*, meaning 'governance'.

Deep structure – The underlying syntactic structure of a sentence, determining what it means. During the 1970s, 'deep structure' was popularly imagined to be 'deep' in the sense of profound and meaningful. Although he invented the idea, deep structure has no place in Chomsky's current thinking.

Digital infinity – A technical term in theoretical linguistics. Alternative formulations are 'discrete infinity' and 'the infinite use of finite means'. The idea is that all human languages follow a simple logical principle, according to which a limited set of digits – irreducible atomic elements – are combined to produce an infinite range of potentially meaningful expressions.

Distinctive features theory – The theory that each of the world's culturally variable linguistic sound systems represents a particular selection from a pan-human alphabet of natural oppositions, such as the distinction between 'voiced' and 'unvoiced', or between 'lips open' and 'lips closed'.

Evolutionary psychology – Neo-Darwinian psychology. The basic argument is that, in understanding the human mind, it helps to know what specific functions it was designed to perform in the course of natural selection. In its most recent and influential American form, evolutionary psychology views the mind as an assemblage of specialized modules or mini-computers.

Generative grammar – A set of rules or principles that formally defines each and every one of the set of well-formed expressions of a natural language as distinct from a computer language.

Generative semantics – An unsuccessful attempt by some American scholars (ex-followers of Noam Chomsky) to extend Chomskyan syntactic theory into the domain of semantics. The basic idea is that it is social pressure to convey meaning which determines syntactic structure, rather than hard wiring in the individual brain.

Intersubjectivity – A term used by philosophers and psychologists, describing how human minds are not cut off from one another but *interpenetrate*. Each person is shown their own mind thanks to the mirror provided by others. The concept has no place in Chomsky's scientific thinking.

Lexicon – A dictionary, especially the mental dictionary consisting of a person's intuitive knowledge of words and their meanings. From the Greek *lexikon (biblion)*, '(book) of words'.

Merge – To combine into a single entity. In Chomsky's Minimalist Program, Merge means combining words to form phrases and then combining these in turn. An example would be taking the words *the* and *cat* and putting them together, giving you *the cat*. Now, with this result and others like it, you can assemble such combinations to build up whole sentences – for example *the cat ate the canary*.

Minimalism – Chomsky's current approach to theoretical linguistics. Dating from the early 1990s, Minimalism appeals to the idea that the genetically installed language faculty is not complex but simple. The core mechanism consists of putting words together to form combinations and putting these together in turn. The capacity to do this is said to be not only unique to humans and unique to language but perfect in design.

Phoneme – The smallest sound unit in a language that is capable of conveying a distinction in meaning, as in the *m* of *mat* and the *b* of *bat*. Phonemes are not genetically fixed items but culturally variable ones, with those permissible in French, for example, comprising a range quite different from those characteristic of English.

Phonetics – The study of the sounds made by the human voice, including the production, perception and analysis of speech sounds from both an acoustic and a physiological point of view. Phonetics is the study of speech sounds independently of linguistic considerations.

Phonology – The system of contrastive relationships – e.g. the distinction between *b* and *p* – among the speech sounds of a language. Phonologists study speech sounds with reference to their linguistic distribution and patterning and to tacit rules governing pronunciation.

Pragmatics – The branch of linguistics dealing with how and in what contexts people put their linguistic abilities to use.

Rationalism – The philosophical doctrine according to which reason takes precedence over other ways of acquiring knowledge. Rationalists hold that the criterion of truth is not sensory, but intellectual and deductive. Reflecting their confidence in reason, rationalists argue that their theories must be accepted regardless of empirical proof or evidence.

Recursion – The process of combining objects and combining the combinations, in principle without limit. Technically, it means treating the output of one operation as the input of the next. As defined within Chomsky's Minimalist Program, recursion is the core mechanism of the innately installed language faculty, equivalent to Merge, Universal Grammar and digital infinity.

Rewrite rule – A mechanical procedure for breaking down a sentence into its constituent parts.

Semantics – The branch of linguistics concerned with meaning.

Syntax – Coordination or arrangement. The rules or principles specifying how words are ordered to form sentences.

Transformation – In Chomskyan linguistics, a transformation is an operation performed upon one sentence – perhaps a simple one such as *John read the book* – to arrive at a subtly different equivalent, such as *The book was read by John*.

Transformational grammar – The use of transformations to produce new sentences from existing ones. In Chomsky's early work, the technique was conceptualized as a way of mapping between the surface structure of a sentence (close to its phonology) and its deep structure (viewed as closer to its meaning).

Universal Grammar – Whatever it is about human nature that enables members of our own species, but not apes or other animals, to internalize the grammar of a natural language.

NOTES

Preface to the paperback edition

1. Chomsky 2016c. See also Chomsky 2016f.
2. The Pentagon funded machine translation in order to translate Soviet bloc documents, to aid communication between the Western allies and to develop new methods of weapons 'command and control'. However, as Jerome Wiesner says, 'We did not know enough about language to attempt computer translation, so we decided to take a detour and study linguistics.' Many decades later, the Pentagon continues to look to MIT for advanced ideas about 'command and control'. One comment by Chomsky is particularly interesting:

   > MIT has projects right now on efforts to try to control the motions of animals by computers, and maybe even to pick up signals from human brains and translate them into commands to control what other organisms do. This is presented – and maybe people believe it – as the great new frontier in fighting wars. You will be able to get a commander to tell a pilot, just by thinking, 'Anything that flies on anything that moves,' or something like that.

   Gordin 2016, 208–223; Nielsen 2010, 39–42, 194, 338; Wiesner 2003, 215–217, 493; Chomsky 2004b.
3. Garfinkel n.d.
4. Chomsky 2009b.
5. Chomsky 2009b.
6. Wiesner 2003, 524–534.
7. Wiesner 1986, 2 mins.
8. *Naval Research Reviews*, July 1966, 4. Between 1949 and 1976, all three RLE directors had involvement with missiles and/or nuclear weaponry. Snead 1999, 57–59, 72; Wiesner 2003, 569; Zimmermann 1991.
9. 'Tri-Services honor MIT achievements in military electronics research and development', *Army Research and Development News Magazine* 12(4), July–August 1971, 68. HQ Department of the Army, Washington.
10. Jerome Wiesner was also involved with 'political warfare' and 'psychological warfare' in ideological opposition to Soviet Marxism. In 1954, he complained: 'We don't have the overt ideological philosophy that we can tell to the natives . . . We can say such things as "freedom" and "economic development," but we do not have a positive goal . . . in the same sense that the Russians do.' By 1958, he was suggesting that Marxism could be countered by presenting the US as a 'classless society', one which had already accomplished what 'the communists have claimed as their goals'. However, Wiesner probably achieved more in this project to undermine Marxism when, in 1953, his representative, Roman Jakobson, organized seminars in Paris whose contributions to philosophy were later to evolve into postmodernism. *New Yorker*, 19 January 1963, 40; UCLA Historical Journal, No. 10, 1990, 18; Engerman 2003, 84–85; Rydell 1993, 198; Geoghegan 2011, 111–113, 118–119, 124–126.
11. *Chicago Tribune*, 29 June 1969, 24. See also Wiesner 1958.
12. Wiesner 1961. Wiesner also influenced British nuclear policy, advising President Kennedy 'to sell Polaris missiles, excluding warheads, at a cost of about $1 million each to the UK'. Priest 2006, 42–44.
13. Brennan 1969, 33; Snead 1999, 118.

14. Wiesner 2003, 103; *New York Times*, 2 July 1971, 12; Finkbeiner 2007, 65–66, 75–76; Feldman 2008; Bridger 2015, ch. 5. In view of Wiesner's deep military involvement, it may seem surprising that Chomsky described him as 'the best choice' for MIT president. *Time*, 15 March 1971, 43.

15. *Science*, 3924, 13 March 1970, 1475; *Technology Review*, June 1970, 82.

16. UPI Archives 1989.

17. *The Tech*, 24 February 1989, 5 and 28 February 2006, 13; Chu 2015; Chu 2012; *MIT News* 2013; US Department of Defense 2017; *Technology Review*, 20 March 2002; Ippolito 1990.

18. This claim comes from the student newspaper, *The Thistle*, which also claims that John Deutch was a strong supporter of 'using chemical and biological weapons together in order to increase their killing efficiency'. *The Thistle*, (9)7, available at: http://web.mit.edu/activities/thistle/v9/9.07/tv9.07. html (accessed 2 April 2017); *The Tech*, 7 March 1989, 2 and 27 May 1988, 2, 11; *Science for the People*, March–April 1988, 6.
As an influential member of several Pentagon panels, Deutch not only called for the deployment of US chemical weapons, he also succeeded in getting the MX missile deployed and the Midgetman missile developed. Later, when Deutch became No. 2 at the Pentagon and, in 1995, director of the CIA, student radicals demanded that MIT cut all ties with him. Chomsky, however, disagreed, telling the *New York Times* that Deutch 'has more honesty and integrity than anyone I've ever met in academic life, or any other life … If somebody's got to be running the CIA, I'm glad it's him.' *Chemical and Engineering News*, 60(1), February 1982, 24–25; Scowcroft 1983, frontispiece, 20–21; *Washington Post*, 26 December 1986, 23; Deutch 1989, 1447–1449; UPI Archives 1989; *New York Times*, 2 February 1986, 1 and 10 December 1995.

19. Chomsky 2017, 2 hours 18 mins.

20. Chomsky 2016c. In recent years, Chomsky has minimized the significance of war research at MIT in the following talks and interviews: Chomsky 2010d; Chomsky 2011c, 8–12 mins; Chomsky 2011d; Chomsky 2014b, 59 mins; Chomsky 2014c, 12–14 mins; Chomsky 2016d, 1 hour 7–11 mins; Chomsky 2016e, 33 mins; Chomsky 2016f; Chomsky 2017, 2 hours 18 mins.

21. Chomsky 2016f.

22. *Science*, 3880, 9 May 1969, 653; *Boston Globe*, 23 February 1969, 47; Chomsky 1969, 59–69.

23. Chomsky 2009b; Chomsky 2008–9, 530, 534; Bridger 2015, 159–162, 178.

24. Chomsky 1988g [1977], 247–248. In December 1969, Chomsky told anti-militarist students that 'you ought to have the Department of Chemical and Biological Warfare right in the center of the campus so you can see who is coming and going'. In his talk, one of Chomsky's justifications for relying on moral persuasion to stop war research was that, otherwise, you would have to use force, since 'only coercion could eliminate the freedom to undertake such work'. *Boston Globe*, 30 December 1969, 29; *Los Angeles Times*, 30 December 1969, 12; Chomsky and Otero 2003, 189–190, 288–290. See also *Columbia Daily Spectator*, 13 May 1968, 1, 4; Barsky 1997, 140–141; Rai 1995, 129–131.

25. Chomsky 1967d.

26. Chomsky 2016c.

27. Greenberg 1999, 151.

28. Killian 1977, 59.

29. *Naval Research Reviews*, March 1968, 1, 9–10.

30. *London Review of Books*, 15 June 2017, 4.

31. Liberman 2016; Hutchins 2000a, 78, 301; Barsky 1997, 54.

32. *MIT President's Report, 1957*, 104.

33. Lees 1957, 406.

34. Bar-Hillel 1959, Appendix 2, 1, 6. See also Nielsen 2010, 340–342 and Harris 1993, 250.

35. Chomsky 2016f.

36. Renehan 2007, 247. Professor King had good reason to want to expose MIT's military involvement after his own Biology Department was pressured to apply for Pentagon funding under threat of having its other funding cut. *Science for the People*, January–February 1988, 17–20; *The Tech*, 27 May 1988, 2, 27.

37. Chomsky 2016d, 1 hour 7–11 mins.

38. Glantz and Albers 1974, 706, 710; *Science*, 4189, 16 May 1975, 678–683; *Stanford Daily*, 14 February 1973, 2. These anti-militarist academics specifically criticized Pentagon-funded researchers who claim that their research 'is not dictated by any military problem' and who thereby 'ignore the fact that the DoD funds their research *because it contributes explicitly* to solving a military problem'. This quote, from their 1971 report, is consistent with MIT President James Killian's 1952 statement that his institution received funding from the Pentagon as 'an integral part of programmed research committed to specific military aims'. As an Air Force official put it in 1971: 'We don't support broad research programs … which have little direct and apparent mission applicability to the Air Force.' Glantz et al. 1971, 7; *Historical Studies in the Physical and Biological Sciences*, 18(1), 1987, 201; Glantz and Albers 1974, 711. Of course, any university that restricted its research to military technology would soon run out of ideas. Despite the Pentagon's self-justifying claims, it remains true that, in practice, as Chomsky

points out, military funding necessarily found its way also to music, various arts programmes and much purely theoretical work including, of course, theoretical linguistics. In terms of student recruitment and public relations, MIT benefited hugely from its non-military research, as was made clear by the lecturer who criticized an art symposium at the university as 'brilliant camouflage for the "military-industrial complex"'. Thompson 1971, 62, 67.

39. Chomsky 2002b, 10; Chomsky and Otero 2003, 290. Chomsky has repeatedly defended the academic freedom of 'war criminals' to do military research, even if that research was being used to 'murder and destroy'. In these statements, he is clearly including fellow academics at MIT. *Quadrant*, April 1982, 7; *Dissent*, Spring 1982, 220; *Daily Camera*, September 1985.

40. Nielsen 2010, 260.

Chapter 1: The revolutionary

1. Harris and Harris 1974.
2. Chomsky; quoted in Chepesiuk 1995.
3. Neil Smith; quoted in Jaggi 2001.
4. Harris 1998.
5. Dean 2003, viii.
6. Barsky 2007, ix; *Guardian*, 18 October 2005.
7. MIT News 1992; quoted in Achbar 1994, 17.
8. Strazny 2013, 207.
9. Harman 1974, vii.
10. Chomsky 1989.
11. Barsky 1997, 3.
12. Golumbia 2009, 31.
13. Smith 1999, 1.
14. Albert 2006, 63.
15. Leiber 1975, 19.
16. Cogswell 1996, 7.
17. Milne 2009.
18. Cartlidge 2014, 8.
19. Edgley 2000, 1.
20. Chomsky 1988a, 2. Asked about the connection between his academic work and his political activity, Chomsky replied that it was 'almost non-existent … There is a kind of loose, abstract connection in the background. But if you look for practical connections, they're non-existent. I'd do the same things if I was an algebraic pathologist and somebody could have the same linguistic views as I do and be a fascist or a Stalinist. There'd be no contradiction.' *Irish Times*, 21 January 2006.
21. Chomsky 1988f, 98–99.
22. Chomsky 1996a, 15.
23. Chomsky 2000b, 115.
24. Chomsky 1988h, 16.
25. Mailer 1968.
26. Chomsky 1967b.
27. Chomsky 1992, 86–87.
28. Chomsky 1996a, 128.
29. Chomsky 1988e, 225–226.
30. Chomsky 1988e, 225–226.
31. Chomsky 1986, xxvii.
32. Chomsky 1986, xxix.
33. Chomsky 1980a, 66.
34. Letter dated 15 December 1992; in Barsky 1997, 208.
35. Chomsky 1988a, 189–190.
36. Chomsky 1986, xxvii ff.
37. Chomsky 1986, xxv.
38. Chomsky 1991a, 15.
39. Chomsky 1991a, 15.
40. Article on 25 February 1979; quoted in Achbar 1994, 19.
41. Scruton 2016, 118–119.
42. Chomsky 1988k, 744.
43. Chomsky 1998a, 17–18.
44. Chomsky 1988j, 697.
45. Barsky 1997, 95.
46. Williamson 2004, 234.

47. Chomsky 2007a.
48. Berwick and Chomsky 2016.
49. Chomsky 1988a, 36–37.
50. Chomsky, personal communication to A. Edgley, 21 February 1995; Edgley 2000, 154.
51. Chomsky 2000b, 2.
52. Lakatos 1970.
53. Chomsky 1986, 40.
54. Chomsky 1986, 15, 24.
55. Chomsky 1986, 15, 24.
56. Chomsky 2000b, 7.
57. Chomsky 1988a, 174.
58. Chomsky 2016b, 1 hour 17 seconds.
59. Chomsky 2000b, 65–66.
60. Chomsky 2006a, 59.
61. Chomsky 1996b, 29–30.
62. Chomsky 2012, 88.
63. Harris 1993.
64. Berwick and Chomsky 2011; cf. Chomsky 2007b, 20.
65. Baker and Hacker 1984; Seuren 2004.
66. Baker and Hacker 1984, back cover.
67. Searle 2003, 55.
68. Rappaport 1999.
69. Atran 2002, 264.
70. Wolpert 2006.
71. Alcorta and Sosis 2005; Atran and Norenzayan 2004.
72. Rappaport 1999.
73. Baker and Hacker 1984, back cover.
74. Review comment on back cover of paperback edition of Seuren 2004.
75. Trask 1999, 109.
76. Postal 1995, 140.
77. Botha 1989.

Chapter 2: The language machine

1. Maclay 1971, 163.
2. Lightfoot 2002.
3. Boeckx and Hornstein 2010, 116.
4. Gardner 1987, 195.
5. Fromkin 1991, 79.
6. Koerner 1994, 3–17.
7. Newmeyer 1996, 23–24.
8. Koerner 1994, 3–17.
9. Newmeyer 1986b, 43.
10. Chomsky 1975a, 40.
11. Chomsky 1957, 1.
12. Chomsky 1965, iv.
13. Chomsky 1988h, 15–16.
14. Newmeyer 1986a, 85–86.
15. Golumbia 2009, 60.
16. Golumbia 2009, 60.
17. Chomsky 1988h, 17.
18. Chomsky 1957, 18.
19. Chomsky 1965.
20. Harris 1993, 179–180.
21. Bach 1974, 158.
22. Chomsky 1981.
23. Chomsky 2000b, 8.
24. Chomsky 1995.
25. Newmeyer 2003.
26. Chomsky 1979a, 57.
27. Chomsky 1965, 3.
28. Chomsky 1976a, 15.
29. Chomsky 1998b.

30. Chomsky 2000b, 64–66.
31. Chomsky 1976a, 57–69. See Chomsky 2002a, 148.
32. Hauser, Chomsky and Fitch 2002, 1569.
33. Chomsky 1976a, 186, 123.
34. Chomsky 1988a, 167.
35. Chomsky 1991c, 50.
36. Chomsky 1998b, 17.
37. Hauser, Chomsky and Fitch 2002.
38. Chomsky 2000b, 106–133.
39. Chomsky 1996b, 30.

Chapter 3: A man of his time

1. George Miller interviewed in Baars 1986, 203; quoted in Edwards 1996, 223.
2. Skinner 1960.
3. Skinner 1957, 3.
4. Harris 1993, 55.
5. Chomsky 1988i [1984], 131.
6. Radick 2016.
7. See for example Vygotsky 1986, Piaget 1929. In 1975, Chomsky did turn his attention to Piaget; see Piatelli-Palmarini 1980.
8. Bloomfield 1970, 227.
9. McDavid 1954.
10. Chomsky 1988a, 174.
11. Chomsky 2002a, 53.
12. Chomsky 2006a, 59.
13. Chomsky 1988a, 137–138.
14. Chomsky 1988a, 137–138.
15. Gleitman and Newport 1995.
16. Chomsky 1959, 57.

Chapter 4: The most hideous institution on this earth

1. Edwards 1996, 51.
2. Chomsky 1988h, 14.
3. Chomsky 1988h, 9.
4. Hughes 2006, 86–87.
5. Chomsky 2003; Jaggi 2001.
6. Chomsky 2009b, 44 mins.
7. Chomsky 2013a, 1 hour 31 mins.
8. Barsky 1997, 82–83; Sperlich 2006, 22.
9. White 2000, 445.
10. Edwards 1996, 47.
11. Forman 1987, 156–157; quoted in Edwards 1996, 47.
12. Adams 1982; quoted in Edwards 1996, 47.
13. Chomsky 1988h, 55.
14. Chomsky 1997c, 144.
15. Chomsky 1967b.
16. Chomsky 1988g, 247.
17. Mehta 1974, 152.
18. Chomsky 1988g, 248.
19. Chomsky 1988g, 248.
20. Leslie 1993, 181.
21. Albert 2006, 97–99.
22. Chomsky 1988g, 247.
23. Chomsky 2015, 43–50 mins.
24. Wallerstein and Starr 1972, 240–241.
25. 'Why smash MIT?' in Wallerstein and Starr 1972, 240–241.
26. Chomsky 2015, 43–50 mins.
27. Chomsky; quoted in Mehta 1974, 153.
28. *The Tech*, 29 April 1969, 2 May 1969. It is noteworthy that Jon Kabat went on to develop mindfulness meditation techniques which are now used not only by health services but also by the US military. Kabat-Zinn 2014, 556–559.
29. Johnson 2001, 174, 191.

30. *New York Times Magazine,* 18 May 1969.
31. Chomsky 1969, 37–38.
32. Letter to Barsky dated March 1995; Barsky 1997, 140.
33. Shalom 1997.
34. Katsiaficas 1969, 92.
35. Albert 2006, 98.
36. *The Tech,* 21 November 1969.
37. Barsky 1997, 122.
38. Skolnikoff 2011, 1 hour 40 mins. This appears to be confirmed by Nelkin 1972, 81–82. See also Chomsky 1969, 17, 31.
39. Segel 2009, 206–207. See Chomsky 2011c on the occupation tactic: 'I wasn't in favor of it myself, and didn't like those tactics.'
40. *The Tech,* 22 May 1970. The two jailed students were George Katsiaficas and Peter Bohmer. Another, Stephen Krasner, was later sentenced to a year in prison for constructing the metal ram used to break into Johnson's office. George Katsiaficas's mother was also imprisoned for contempt of court. In all, seven students were expelled from MIT, although three were later readmitted. *The Tech,* 5 October and 14 December 1971; Johnson 2001, 201.
41. Segel 2009, 216; Isadore Singer video interview, MIT 150 Infinite History Project, available at: http://mit150.mit.edu/infinite-history/isadore-singer (accessed 2 April 2016).
42. *The Tech,* 16 January 1970; Johnson 2001, 202–203.
43. Barsky 1997b, 122; Chomsky 1971b. Chomsky also said that 'the student movement has focused too much on preventing people from doing this or that, and not enough on creating alternatives. So while I share a lot the indignation and outrage of the kids, I think they're misled.' *New York Times,* 27 October 1968.
44. Chomsky 1971b.
45. *The Tech,* 5 May 1970.
46. *Technology Review,* 72 (December 1969), 96.
47. *The Tech,* 25 and 28 April 1972. In 1972, three students were sentenced to 30 days' imprisonment for staging a 21-hour occupation of MIT's military officer training department; *The Tech,* 4 August 1972. This punishment was clearly acceptable to MIT, but when Stephen Krasner was sentenced to a whole year in prison, Jerome Wiesner did make a failed attempt to prevent such a harsh punishment. Wiesner 2003, 532.
48. Chomsky 1967d.
49. Chomsky 1967e.
50. Chomsky and Otero 2003, 311.
51. *Time,* 21 November 1969, 68 and 15 March 1971, 43.
52. Chomsky 1996a, 137.
53. Chomsky interviewed in Chepesiuk 1995, 145.
54. *Technology Review,* 72 (October 1969), 93.
55. Chomsky 1980c.
56. Milne 2008, 7, 60, 71, 93–95, 174–175, 255–257.
57. Johnson 2001, 189–190; Wiesner 2003, 582.
58. Barsky 1997, 140–141.
59. *The Tech,* 11 April 1969.
60. Personal communication to the author, 19 May 2016.
61. Eun-jung 2015, 78.
62. Albert 2006, 99.
63. Albert 2006, 9. Interviewed in 1975 (*The Tech,* 95(3), 9–10), Albert had plenty to say about the distorting effect of MIT on the psychology of even the most morally sensitive researcher:

> I still think of MIT the same way I did when I was there. It seems to me that it's a masquerade as an institution of higher learning and objective science. That's a byproduct. What it really is is a place where scholars who are really mandarins get together and promulgate a lot of ideas and theories which uphold the status quo and which aid people who are trying to further American interests. That's on the intellectual side. On the technological side, it's a place which . . . creates a breed of scientists who don't question the reasons why they're doing things but just do them. MIT creates people who are willing to do scientific research as if it's value-free, as if it isn't plugging into a system that's very value-laden. Every so often . . . it gives somebody a good education who then becomes a critical activist. I think it's a shithole. If you were trying to create somebody who was going to do scientific work without considering the value implications, you'd want someone who in fact had partially lost touch with his own feelings and with the reality of people around him. MIT does exactly that . . . I just think that's barbaric. You can try to take advantage of it, but it's a very risky proposition. It can warp you just as much as you get some good out of it. I don't even know with respect to myself the extent to which it or I won.

64. MIT tried hard to make Albert's expulsion more damaging for him by calling the Draft Board and 'urging reclassification' to get him sent to Vietnam. Albert 2007; 2006, 107.
65. Chomsky 1988g, 247.
66. Price 2011.
67. Chomsky 1996b, 31.
68. Joseph and Taylor, 1990, 2.

Chapter 5: The cognitive revolution

1. Minsky; quoted in Hayles 1999, 244–245.
2. Hayles 1999, 18.
3. Wiener 1948, 132; quoted in Hayles 1999, 14.
4. Hayles 1999, 12–13.
5. Hayles 1999, 13.
6. Shannon 1948.
7. Hayles 1999, 19.
8. Hayles 1999, 18.
9. Chomsky 1991a, 3–25.
10. Bruner 1990, 2–3.
11. Turing 1950.
12. von Neumann 1958.
13. Putnam 1960.
14. Putnam 1988, 73.
15. Putnam 1960.
16. Pinker 1997b, 24.
17. Chomsky 1996b, 30.
18. Edwards 1996, 180.
19. Edwards 1996, 210.
20. Although Neil Smith has pointed out to me that during the war against Japan, the US military ingeniously used Navajo to encrypt their messages.
21. Quoted in Edwards 1996, 199.
22. Stroud 1949; quoted in Hayles 1999, 68.
23. Stroud 1949; quoted in Hayles 1999, 68.
24. Hayles 1999, 68, paraphrasing Stroud 1949.
25. Bray 1946; quoted in Edwards 1996, 206.
26. Gross and Lentin 1970, 111.
27. Newmeyer 1986a, 85–86.
28. Miller, Wiener and Stevens 1946; quoted in Edwards 1996, 212.
29. Capshew 1986; quoted in Edwards 1996, 212.
30. Edwards 1996, 211.
31. Bruner, Jolly and Sylva 1976.
32. Edwards 1996, 235.
33. Edwards 1996, 214.
34. Edwards 1996, 217.

Chapter 6: The Tower of Babel

1. Chomsky 1975a, 40.
2. King James Bible, Genesis 11: 1–9.
3. King James Bible, Genesis 11: 9.
4. Weaver 1955a, vii.
5. Kahn 1960.
6. Booth and Locke 1955, 2.
7. Weaver 1955b.

Chapter 7: The Pentagon's 'New Tower'

1. Quoted in Mehta 1974, 148.
2. Dostert 1955, 133–134.
3. N. Glazer; quoted in Barsky 2011, 32.
4. Newmeyer 1986b, 43.
5. Harris 2010, 44.
6. Postal 2004.

7. Postal 2009.
8. Sampson 2001 [1979], 153–156.
9. Sampson 1975, 9.
10. Searle 1972, 17.
11. Newmeyer 1986b, 41.
12. Miller 1979; quoted in Gardner 1987, 28.
13. Mehta 1974, 167.
14. Edwards 1996, 235.
15. Barsky 1997, 54.
16. Hutchins 2000b, 304.
17. Locke and Booth 1955, 232.
18. Booth and Locke 1955, 5.
19. Chomsky 1956, 124.
20. Chomsky 2006a, 405.
21. Chomsky 2006a, 405.
22. Chomsky 1975a, 40.
23. Harris 2010, 254.
24. Chomsky 1965, 25.
25. Quoted in Barsky 1997, 86.
26. Bar-Hillel 1961 [1958]; quoted in Hutchins 2000b, 305.
27. Hutchins 1986, 89.
28. Delavenay 1960; quoted in Hutchins 1986, 151.
29. Taube 1961; quoted in Hutchins 1986, 162.
30. Hutchins 1986, 89–90.

Chapter 8: Machine translation: the great folly

1. Barsky 1997, 86.
2. Otero 1988b, 64–65.
3. Otero 1988b, 64–65.
4. Yngve 1956, 44–45.
5. Harris 2010, 254.
6. Chomsky 1956, 124.
7. Beaugrande 1991, 181.
8. Chomsky 1975a, 40.
9. Gigerenza and Goldstein 1996, 135.
10. Chomsky 1996b, 30.
11. Bar-Hillel 1960; quoted in Boden 2006.
12. Taube 1961; quoted in Boden 2006.
13. Bar-Hillel 1966 [1962], 23.
14. Yngve 1964, 279.
15. Chomsky 1957, 106.
16. Chomsky 1967a, 415.
17. Siodmak 1971 [1942].
18. Hutchins 2000a, 10.
19. Hutchins 2000a, 11.

Chapter 9: A universal alphabet of sounds

1. Watson and Crick 1953, 737.
2. Chomsky 1956.
3. Chomsky 1956, 124.
4. Harris 2010, 253–254.
5. Hauser, Chomsky and Fitch 2002, 1569.
6. Lorenz 1996, 201.
7. Lorenz 1996, 242.
8. Chomsky 2012, 21.
9. Lorenz 1937.
10. Tinbergen 1951.
11. Putnam 1988, 73.
12. Postal 2009.
13. Dosse 1997, 52.
14. Mehta 1974, 177.

15. Quoted in Mehta 1974, 182.
16. Quoted in Mehta 1974, 181–182.
17. See Geoghegan 2011 for Jakobson's wartime technological enthusiasms.
18. Jakobson et al., 1951, 40.
19. Chomsky 2012, 22.
20. Chomsky 1983, 81–82.
21. Chomsky 1983, 81–82.
22. Chomsky 2006a, 65.
23. Newmeyer 1986b, 32.
24. Leach 1983, 10–16.
25. Lévi-Strauss 1991, 41.
26. Lévi-Strauss 1963 [1956], 233.
27. Atran 2009.
28. Lévi-Strauss 1963 [1956], 228.
29. Lévi-Strauss 1973, 249.
30. Sperber 1985, 65–66.
31. Mehta 1974, 218.
32. Mehta 1974, 173.
33. Quoted in Mehta 1974, 183.
34. Jakobson and Waugh 2002, 88.
35. Jakobson et al. 1951, 40.
36. Chomsky 1983, 81–82.
37. Chomsky 1983, 81–82.
38. Chomsky 2012, 22.
39. Jakobson and Pomorska 1983, 55.
40. Chomsky 1957, 17.
41. Chomsky 1957, 93.
42. Chomsky 1964c [1962], 936.
43. Chomsky 2006a, 107. See also Chomsky 1965, 148–163.
44. Chomsky 1967a, 402–403.
45. Chomsky 1967a, 402–403.
46. Matthews 1993, 223.
47. Chomsky 1979a, 141.
48. Chomsky 1964c [1962], 936.
49. Chomsky 1979a, 142.
50. Bromberger and Halle 1991 [1989], 72.

Chapter 10: Russian formalist roots

1. Joseph 2002, 167.
2. Jakobson 1997, 153–155.
3. Jakobson 1997, 331, note 5.
4. Jakobson 1997, 81.
5. Jakobson 1997, 82.
6. Newmeyer 1986b, 32.
7. Kruchenykh and Khlebnikov 1912.
8. Cooke 1987, 50.
9. Markov 2006 [1968], 147–148.
10. Markov 2006 [1968], 193.
11. Sapir 1929.
12. Jakobson and Waugh 2002, 181–234.
13. Jakobson and Pomorska 1983, 137.
14. Khlebnikov 1987e [1919], 147; see also Jakobson and Pomorska 1983, 137.
15. Khlebnikov 1987e [1919], 146–149.
16. Jakobson and Pomorska 1983, 138.
17. Shelley 1840 [1839].

Chapter 11: Incantation by laughter

1. Chomsky 1967a, 402–403.
2. Fromkin 1991, 79.
3. Khlebnikov 1990 [1910], 20.

4. Khlebnikov 1986; extract translated in Gasparov 1997, 109–110.
5. Quoted in Vroon 1997, 1.
6. Schmidt 1989, xi.
7. Quoted in Gasparov 1997, 109–110.
8. Khlebnikov 1968, 27.
9. Gasparov 1997, 105.
10. Khlebnikov, V. 1987b [1913], 294–295.
11. Khlebnikov, V. 1987f [1921], 400.
12. Khlebnikov, V. 1987e [1919], 147.
13. Cooke 1987, 68.
14. Markov 2006 [1968], 146–147.
15. Borchardt-Hume 2014, 24–25.
16. Markov 2006 [1968], 144.
17. Khlebnikov, 1987i [1914], 262.
18. Markov 2006 [1968], 306.
19. Cooke 1987, 104.
20. Douglas 1985, 155.
21. Cooke 1987, 104.
22. Cooke 1987, 104.
23. Cooke 1987, 104. *The Lay of Igor's Tale* is an anonymous epic poem written in the twelfth century in the Old East Slavic language.
24. Khlebnikov 1987a [1912], 284.
25. Khlebnikov 1989 [1918], 116.
26. Khlebnikov 1989 [1918], 121.
27. Khlebnikov 1989 [1918], 122.
28. Volume I of the 1987 English-language version of Khlebnikov's *Collected Works* contains an excellent introduction summarizing much of his life; I acknowledge editor Charlotte Douglas for many of the details touched on here. Among other secondary sources is my own 1977 MPhil thesis, held in the library of the University of Sussex and researched when almost nothing of Khlebnikov had yet been translated into English.
29. Khlebnikov 1987j, 130.
30. Khlebnikov 1987g [1921], 392.
31. Jakobson 1997, 212.
32. Jakobson 1997, 217–218.
33. Jakobson 1997, 245.
34. Quoted in Mehta 1974, 181.
35. Ivanov 1983, 50.

Chapter 12: Tatlin's tower

1. Khlebnikov 1987b [1913], 288–289.
2. Khlebnikov 1987d [1914], 89–90.
3. Jakobson 1985, 101–114.
4. Cooke 1987, 104.
5. Daniil Danin; quoted in Milner 1983, 203.
6. Milner-Gulland 2000, 209.
7. Quoted in Milner 1983, 139.
8. Quoted in Milner 1983, 139.
9. Milner 1983, 143.
10. My main source in what follows is John Milner's insightful 1983 study, *Vladimir Tatlin and the Russian Avant-garde.*
11. Andersen 1968; quoted in Milner 1983, 163.
12. Milner 1983, 163.
13. Milner 1983, 161.
14. Milner 1983, 154.
15. Shklovsky 1988 [1921], 343
16. Punin 1920, 96
17. Milner 1983, 169
18. Shklovsky 1988 [1921], 343.
19. Khlebnikov 1987g [1921], 392.
20. Dirac 1963.
21. Smith 1999, 86.

Chapter 13: An instinct for freedom

1. Ivanov 1983, 49.
2. Rudy 1997, x.
3. Quoted in Mehta 1974, 181.
4. Joseph 2002, 167.
5. Barsky 1997, 87.
6. Foucault and Chomsky 1997, 143.
7. Foucault and Chomsky 1997, 144.
8. Chomsky 1985, 252.
9. Rai 1995, 115.
10. *New York Times*, 27 October 1968. See also *Boston Globe*, 27 February 1968, 15.
11. Branfman 2012.
12. Chomsky 1976a, 7.
13. Chomsky 1976a, 133.
14. Barsky 1997, 208.
15. Chomsky 1988l [1984], 597.
16. Chomsky 1988l [1984], 594.
17. Chomsky in Barsamian 1992, 355.
18. Chomsky 1976a, 134.
19. Chomsky 1988h, 21.
20. Tonkin 1989; quoted in Rai 1995, 138.
21. Chomsky 1975b, 219; quoted in Rai 1995, 138.
22. Rai 1995, 138.
23. Kuhn 1970; Latour and Woolgar 1979.
24. Haraway 1989; Nader 1996.
25. Chomsky 1988h, 16.
26. Chomsky 1988a, 36–37.
27. Chomsky 1996b, 31.
28. Chomsky 1988a, 36.
29. Saussure 1974 [1915], 14.
30. Saussure 1983 [1912], 77.
31. Chomsky 1988a, 36–37.
32. Chomsky 1988a, 173.
33. Lorenz 1996, 201.
34. Chomsky 1976a, 9.
35. Chomsky 1988a, 174.

Chapter 14: The linguistics wars

1. Bloomfield 1933, 139; quoted in Newmeyer 1986b, 5.
2. Chomsky 1957, 15.
3. Chomsky 1957, 108.
4. Newmeyer 1986b, 162.
5. Chomsky 1967a, 433.
6. Halle 1959.
7. For Chomsky's ability to legislate, see Beaugrande 1998.
8. Chomsky 1964a [1963]; Chomsky 1966a [1964].
9. Chomsky 1965, 160.
10. Chomsky 1965, 152.
11. Chomsky 1965, 107–111, 148–192.
12. Katz and Postal 1964.
13. Chomsky 1966a [1964], 16–18.
14. Chomsky 1965, 16.
15. Chomsky 1966b, 35.
16. Chomsky 1964c [1962], 914–977.
17. Chomsky 1966b, 39.
18. Joseph 2002, 191.
19. Joseph 2002, 191.
20. Mey 1993, 7.
21. Langacker 1988, 16–17.
22. Bolinger 1965.
23. Lakoff 1977, 172; quoted in Newmeyer 1986b, 228.
24. Leech 1983, 2; Mey 1993, 21–22.

25. Searle 1969; Searle 1971b; Searle 1996.
26. Ross 1995, 125.
27. Postal 1995, 141.

Chapter 15: Between colliding tectonic plates

1. Chomsky 1988h, 13.
2. Chomsky 1991b; quoted in Rai 1995, 7.
3. Chomsky 1988h, 13.
4. Chomsky 1996a, 93.
5. Chomsky 1996a, 93.
6. Chomsky 1988h, 5.
7. Chomsky 1988h, 6.
8. Chomsky 1990; quoted in Rai 1995, 8.
9. Chomsky 1988h, 13.
10. Barsky 1997, 16.
11. Quoted in Otero 1981, 31.
12. Quoted in Rai 1995, 103.
13. Quoted in Rai 1995, 102.
14. Rai 1995, 102.
15. Chomsky 1996a, 101.
16. Chomsky 1988c, 649.
17. Chomsky 1996a, 37–38.
18. *The Thistle*, 9(7), available at: http://web.mit.edu/thistle/www/v9/9.07/3vest.html (accessed 2 April 2017); *Science for the People*, 20(2), 6.
19. Chomsky 1988d, 318.
20. Chomsky 1988a, 156.
21. Interview with David Barsamian; in Chomsky 1996a, 15.
22. Fodor 1985, 3.
23. Otero 1988a, 98–99.
24. Chomsky 1988c, 653–654.
25. Chomsky 2002b, 228–229, 230. Chomsky is referring to Goering's alleged statement 'When I hear the word culture, I reach for my revolver.'
26. Personal communication to the author, 18 October 2015.
27. Hegel 1929 [1812–1816].

Chapter 16: The escapologist

1. Chomsky 1998c, 128; quoted in Otero 1981, 35–36.
2. Chomsky 1988g [1977], 248.
3. Postal 1995, 141.
4. Barsky 1997, 99–100.
5. Barsky 1997, 152.
6. Chomsky 1988l [1984], 598.
7. Barsky 1997, 140.
8. Barsky 1997, 141.
9. Chomsky 1988b [1983], 419.
10. Chomsky 1988b [1983], 419.
11. Barsky 1997, 141.
12. Barsky 1997, 124.
13. Chomsky 1975a, 4.
14. Nevin 2009, 473 note 210.
15. Chomsky 1957, 6.
16. Harris 1957, 283–284 note 1.
17. Barsky 1997, 53.
18. Barsky 1997, 54.
19. Goldsmith 2005, 270.
20. Harris, forthcoming. Cited in Harris, personal communication, 4 March 2016.
21. Descartes 1985 [1649], 341.
22. Descartes 1985 [1649], 345.
23. Descartes 1991 [1633], 40–41.
24. Letter dated 31 March 1995; in Barsky 1997, 106.
25. Chomsky 1997c, 112–113.
26. Chomsky 1997c, 113–114.

27. Chomsky 2002a, 53.
28. Chomsky, 1997d; quoted in Barsky 1997, 108.
29. Chomsky 1997c, 114.
30. Chomsky 1976a, 23.
31. Chomsky 2012, 12.
32. Chomsky 2012, 27.
33. Chomsky 2012, 27.
34. Chomsky 2012, 28.
35. Chomsky 2011a, 178.
36. Chomsky 1996a, 102.
37. Chomsky 2012, 37.
38. Chomsky 1996b, 30.

Chapter 17: The soul mutation

1. Chomsky 1991c, 50.
2. Chomsky 1976b, 57.
3. Chomsky 1976b, 57.
4. Chomsky 1988a, 183.
5. Hauser, Chomsky and Fitch 2002.
6. Chomsky 2000a, 4.
7. Chomsky 2005b, 11–12.
8. Chomsky 2005b, 12.
9. Cartlidge 2014.
10. Chomsky 2000a, 18.
11. Hauser, Chomsky and Fitch 2002.
12. Hauser, Chomsky and Fitch 2002, 1571.
13. Bickerton 2014, 84.
14. Bickerton 2010, 131.
15. Chomsky 1966b, 30.
16. Bouchard 2013, 41.
17. Chomsky 2006a, 59.
18. Chomsky 2000b, 162.
19. Chomsky 2012, 49.
20. Chomsky 2000a, 4.
21. Chomsky 2002a, 146.
22. Chomsky 2010a, 59.
23. Chomsky 2016a, 40 minutes 3–17 seconds.
24. Chomsky 2012, 44.
25. Chomsky 2002a, 76.
26. Chomsky 2002a, 148.
27. Chomsky 1996b, 30.
28. Berwick and Chomsky 2011; cf. Chomsky 2007b, 20.
29. Chomsky 2002a, 143.
30. Chomsky 1991a, 23.
31. Descartes 1984 [1641], 114.
32. Chomsky 2002a, 109.
33. Chomsky 2002a, 118–119.
34. Chomsky 2002a, 119.
35. Chomsky 2006a, 398.
36. Chomsky 2006a, 405.

Chapter 18: *Carburettor* and other innate concepts

1. Bouchard 2013, 5.
2. Bouchard 2013, 5.
3. Seuren 2004, 74.
4. Chomsky 2010a.
5. Bickerton 2014, 91.
6. Berwick and Chomsky 2011, 20.
7. Berwick and Chomsky 2011, 21.
8. Chomsky 2000b, 65.

9. Fodor 1975.
10. Olson and Faigley 1991, 1–36.
11. Chomsky 2000a, 75.
12. Chomsky 2000b, 66.
13. Chomsky 2000a, 75.
14. Putnam 1988, 15. Putnam is criticizing the 'strong innateness hypothesis' as presented by Chomsky's colleague Jerry Fodor (1975). Fodor's position is equally Chomsky's.
15. Chomsky 2000b, 61.
16. Chomsky 2000b, 64–66.
17. Dennett 1991, 192–193, note 8.
18. Bourdieu 1991.
19. Bloch 1975.
20. Durkheim 1976 [1915].
21. Chomsky 2009a, 386.
22. Chomsky 2009a, 386.
23. Chomsky 2000a, 4.
24. Chomsky 2008, 22 (abbreviated slightly).
25. Chomsky 1988a, 183.
26. Chomsky 2012, 51.
27. Chomsky 2011b
28. Berwick and Chomsky 2016.
29. Berwick 1998.
30. Chomsky 2000a, 4.
31. Berwick and Chomsky 2016, 87.
32. Dawkins 2015, 384.
33. Maynard Smith and Harper 2003, 113.

Chapter 19: A scientific revolution?

1. Otero 1988a, 406.
2. Harman 2001, 265.
3. Chomsky 1988a, 91–92.
4. Gray 1976; quoted in Newmeyer 1996, 24.
5. Antilla 1975; quoted in Newmeyer 1996, 24.
6. Murray 1980, 81; quoted in Newmeyer 1996, 23.
7. Koerner 1994, 10.
8. Lees 1957.
9. Koerner 1994, 10.
10. Chomsky 1975a, 3.
11. Newmeyer 1986b, 30.
12. Mehta 1974, 165.
13. Chomsky 2002a, 124.
14. Chomsky, 2002a, 98.
15. Chomsky 2002a, 102.
16. Chomsky 2009a, 36.
17. Seuren 2004, 29.
18. Mehta 1974, 165.
19. Mehta 1974, 165.
20. Chomsky 1976a, 15.
21. Lakoff 1995, 115.
22. Ross 1995, 125.
23. Lakoff 1995, 115.
24. Chomsky 1988a, 157–158.
25. Chomsky 1988a, 157–158.
26. Chomsky 2002a, 151.
27. Chomsky 1982a, 58.
28. Chomsky 1982a, 40.
29. Chomsky 1982a, 41.
30. Chomsky 1988b [1983], 418.
31. Chomsky 1957, 13–17; Chomsky 1962, 127; Chomsky 1976a, 80.
32. Chomsky 1986, 82–83; Chomsky 1995, 25.
33. Chomsky 1957, 45; Chomsky 1962, 136.

34. Chomsky and Lasnik 1977, 41.
35. Chomsky 2000c, 130.
36. Chomsky 1957, 13–14; Chomsky 1966a [1964], 32; Chomsky 1972b, 64.
37. Chomsky 1995, 194.
38. Chomsky 1995, 213n.
39. Chomsky 1967a, 433.
40. Chomsky 1976a, 76.
41. Chomsky 1993, 346n.
42. Chomsky 1966a [1964], 91.
43. Chomsky 1995, 189.
44. Hauser et al. 2014, 2.
45. Chomsky 1979a, 81.
46. Chomsky; quoted in Piatelli-Palmarini 1980, 182.
47. Chomsky; quoted in Piatelli-Palmarini 1980, 76.
48. Hauser, Chomsky and Fitch 2002.
49. Chomsky 2009a, 387.
50. Hauser, Chomsky and Fitch 2002.
51. Chomsky 1986, 3.
52. Chomsky 1988a, 410–411.
53. Chomsky 2012, 41.
54. Chomsky 1965, 8.
55. Newmeyer 1996, 25.
56. Chomsky 1980a, 76.
57. Chomsky 1964b, 50–51.
58. Chomsky 1972a, 101.
59. Chomsky 2012, 41.
60. Smith 1999, 86.
61. Matthews 1993.
62. Chomsky 1995, 233.
63. Pinker; quoted in Kenneally 2007, 271.
64. Fiengo 2006, 471.
65. Gardner 1987, 185.
66. Newmeyer 1996, 30.
67. Newmeyer 2003, 596.
68. Chomsky 1982a, 42–43.
69. Chomsky 2002a, 151.
70. Chomsky 2012, 84.
71. Chomsky 1980b, 48
72. Chomsky 1980b, 48.
73. Evans and Levinson 2009.
74. Behme 2014, 30.
75. Evans and Levinson 2009.
76. Botha 1989, 84–87.
77. Botha 1989, 207.
78. Chomsky 1979c; quoted in Botha 1989, 86.
79. Chomsky 1979c; quoted in Botha 1989, 87.
80. Chomsky 1982a, 110; quoted in Botha 1989, 87.
81. Botha 1989, 87.
82. Chomsky 1982b, 76.
83. Mehta 1974, 139.
84. Achbar and Wintonick 1992; quoted in Beaugrande 1998, 765.
85. Postal 1995.
86. Chomsky 1996a, 102.
87. Cartlidge 2014.
88. Cartlidge 2014.
89. Cartlidge 2014.
90. Cartlidge 2014.
91. Chomsky 2014a.
92. Hauser et al. 2014.
93. Hauser et al. 2014, 10.
94. Cartlidge 2014.
95. Chomsky 2014a.

Chapter 20: Mindless activism, tongue-tied science

1. Chomsky 1971b.
2. Rohde 2012, 137-138.
3. Rohde 2012, 140.
4. Rohde 2012, 142.
5. Lybrand 1962, 11.
6. Lybrand 1962, 11.
7. Lybrand 1962, 14-15.
8. Lybrand 1962, 16.
9. Lybrand 1962, 17.
10. Eckstein 1962, 262.
11. Eckstein 1962, 252-253.
12. Eckstein 1964.
13. Chomsky 1979a, 5.
14. Chomsky 1979a, 3.
15. Edgley 2000.
16. Chomsky 1971a, 71.
17. Marx and Engels 2000 [1845], 192.
18. Marx and Engels 2000 [1845], 180-181.
19. Kuhn 1970.
20. Doyle, Mieder and Shapiro 2012, 51.
21. Golumbia 2009, 32.
22. Golumbia 2009, 32.
23. Marx 2000a [1845], 173.
24. Marx 2000a [1845], 173.
25. Marx 1961a [1844], 85.
26. Chomsky 1988l [1984], 592.
27. Chomsky 2000b, 115.
28. Chomsky 2000b, 115.
29. Chomsky 1988d, 318
30. Chomsky 2008, 23.
31. Chomsky 2008, 23.
32. Chomsky 2008, 23.
33. Marx 1961a [1844], 88.
34. Engels 1957 [1886], 266.
35. Khlebnikov 1987f [1921], 400.
36. Marx 1961a [1844], 87.
37. Marx 1961c [1843-1844], 190.
38. Engels 1957 [1886], 266.
39. Golumbia 2009, 31-32.
40. Golumbia 2009, 32.
41. Golumbia 2009, 31.
42. Nelkin 1972, 63.
43. Golumbia 2009, 32.
44. Chomsky 1988l [1984], 599.
45. Chomsky 1988i [1984], 136.
46. Chomsky 1967c; quoted in McGilvray 1999, 28
47. *The New Yorker*, 8 May 1971.
48. Barsky 1997, 58.
49. Otero 1988a, 595. On another occasion, however, Chomsky remarks that Lenin was absolutely right to say that 'the victory of socialism requires the joint efforts of workers in a number of advanced countries.' Chomsky 1971a, 64.
50. Quoted in Otero 1988a, 595.
51. While Chomsky has denounced the Russian Bolsheviks of 1917, he has been less hostile towards the so-called communist regimes which later took power in Asia, asserting during a 1969 TV debate that, despite 'a great deal of authoritarianism', peasants in China 'control the institutions of their lives. They control the organization of work.' A decade later, he also seemed reluctant to acknowledge the full horror of the 'communist' regime in Cambodia. The explanation I favour is that it pained Chomsky's conscience to denounce people anywhere who were being threatened by the very war machine that was funding his research. Buckley and Chomsky 1969; Chomsky and Herman 1979, ch. 6; Otero 1988a, 153-155.

Chapter 21: Chomsky's tower

1. Chomsky 1988l [1984], 589.
2. Chomsky 1988l [1984], 590.
3. Chomsky 1988l [1984], 589.
4. Chomsky 1988l [1984], 591.
5. Chomsky 1988l [1984], 589.
6. Chomsky 1988a, 157.
7. Chomsky (1988a, 157) concedes that people who lack the necessary special abilities may still be able to understand and evaluate a proposed scientific theory, despite being unable to actively produce science themselves.
8. Chomsky 1988h, 16.
9. Chomsky 2006b.
10. Einstein 1954, 292.
11. Chomsky 1996a, 128.
12. Letter dated 18 February 1993 in Barsky 1997, 192.
13. Chomsky 1988i [1984].
14. Trotsky 1940 [1925].
15. Amadae 2003, 2.
16. Amadae 2003, 1, quoting Gary Hart.
17. McCumber 2001, 34.
18. Quoted in McCumber 2001, 39.
19. McCumber 2001, 39.
20. Quine 1959, xi.
21. Searle 1971a, 132.
22. McCumber 2001, 99.
23. Barsky 1997, 21.
24. Balari and Lorenzo 2010, 125.
25. Lakoff 1971, ii; quoted in Harris 1993, 208.
26. Chomsky 1996a, 128.

Chapter 22: Before language

1. Tomasello 2000.
2. Steels 2015.
3. Hurford 1989.
4. Tomasello and Farrar 1986; Tomasello et al. 2005.
5. Whiten 1999.
6. Byrne and Whiten 1988.
7. Byrne and Corp 2004.
8. Tomasello 2006.
9. Tomasello 2006.
10. Tomasello and Rakoczy 2003; Knight and Lewis 2014.
11. Mellars and Stringer 1989.
12. Mellars et al. 2007; McBrearty 2007.
13. Chomsky 2012, 13.
14. Chomsky 2008, 21.
15. Lybrand 1962, 17; Rohde 2012, 140.
16. Radick 2016, 69.
17. Hockett 1960.
18. Hockett and Ascher 1964.
19. Radick 2016.
20. Sahlins 1964 [1960].
21. For a concise account of the egalitarianism typical of hunter-gatherers, see Gowdy 2005, 391–398.
22. Engels 1972 [1884].
23. Sahlins 1964 [1960], 60–61.
24. Sahlins 1964 [1960], 65.
25. Lee 1968; Lee 1988; Lee 1992.
26. Hrdy 2000, 87–88.
27. On this point, see Haraway 1989.
28. Trivers 1985, 77–78.
29. Trivers 1985, 78.
30. Trivers 1985, 78.

31. Van Schaik and Janson 2000; Rees 2009.
32. Attributed to Australian journalist Irina Dunn (1970).
33. Chomsky 2000b, 115.
34. Hewlett and Lamb 2005.
35. Knight and Power 2005.
36. Boehm 1999.
37. Hrdy 1981. For ethnographic case studies, see Beckerman and Valentine 2002; for further discussion, see Knight 2008c.
38. Hrdy 1981.
39. Hrdy 2009, 233–272.
40. Hrdy 1981; Hrdy 2000; Hrdy 2009.
41. Smuts, 1985; Strum 1987; Silk 1993; Small 1993; Gowaty 1997; Power and Aiello 1997.
42. Wilson 1975.
43. Segerstråle 2000.
44. Wilson 1995.
45. Trivers 2015.
46. Hrdy 2000, 427.
47. Tomasello et al. 2005.
48. Call 2009.
49. Kobayashi and Kohshima 2001; Tomasello et al. 2007.
50. Hrdy 2009.
51. Hrdy 2000, 50–51.
52. Hrdy 2009, 239.
53. Morgan 1877; Engels 1972 [1884]; see Knight 2008c.
54. Destro-Bisol et al. 2004; Wood et al. 2005; Verdu et al. 2013; Schlebusch 2010.
55. Hrdy 2009, 38.
56. Berwick and Chomsky 2016, 92.
57. Hurford 2007; Hurford 2012.
58. Chomsky 1967a, 415.
59. Steels 1998; Steels 2012.
60. Steels 2014.
61. Steels 2007.
62. Steels and Kaplan 2001; Steels 2009, 52–54.
63. Steels 2009.
64. Deutscher 2005; Smith and Höfler 2014; Smith and Höfler 2016.
65. Steels 2009, 57.
66. Hurford 2007, 201; citing Knight 2002, 148.
67. Hurford 2007, 325.
68. Knight; quoted in Hurford 2007, 328–329. On the 'rule of law', see Knight 2008a.
69. Hurford 2012, 646.
70. Smith and Höfler 2014; Smith and Höfler 2016.
71. Knight 1998; Knight 1999; Knight 2008b; Knight 2014; Knight and Lewis 2014.
72. Power and Aiello 1997; Power 2009.
73. Chomsky 2010a, 59.
74. Sahlins 1972 [1959].
75. Sahlins 1977.
76. Chomsky 2008, 19.
77. Bowlby 1969.
78. Barkow, Cosmides and Tooby 1992; Buss 2008.
79. Tooby and Cosmides 1995, xiii–xiv.
80. Uttal 2001.
81. Boyer 2001.
82. Winkelman 2002.
83. Hauser, Chomsky and Fitch 2002.
84. Tooby and Cosmides 1992; Pinker 2002.
85. Chomsky 2008, 19.
86. Chomsky 1991e, 27–29.
87. Pinker 1997b, 305.
88. Pinker 1997b, 375–376.
89. Pinker 1997b, 472.
90. Fine 2010.

Chapter 23: The human revolution

1. Latour 1999, 59.
2. Bourdieu 1991, 52.
3. McGilvray 2009, 34.
4. Chomsky 2010c, 29.
5. Givón 1995, 22.
6. Hansen 2009, 112–113.
7. Gottfried 1969. Here is a key passage from this historic document:

> Our community has the ... responsibility ... to evaluate the long-term social conse-
> quences of its endeavor, and to provide guidance in the formation of relevant public
> policy. This is a role it has largely failed to fulfill and it can only do so if it enters the
> political arena.
>   Only the scientific community can provide a comprehensive and searching evaluation
> of the capabilities and implications of advanced military technologies. Only the scientific
> community can estimate the long-term global impact of an industrialized society on our
> environment. Only the scientific community can attempt to forecast the technology that
> will surely emerge from the current revolution in the fundamentals of biology.
>   The scientific community must ... engage effectively in planning for the future of
> mankind, a future free of deprivation and fear.

8. Chomsky 2000b, 115.
9. Pinker 1999, 269–287.
10. Chomsky 1991c, 50.
11. Pinker and Bloom 1990, 765.
12. Pinker and Bloom 1990, 765–766; Pinker and Jackendoff 2005; Pinker 1994; Pinker 1997a; Pinker 1999.
13. Chomsky 2006a, 59.
14. Eldredge and Gould 1972.
15. Maynard Smith and Szathmáry 1995.
16. Hegel, 1929 [1812–1816]
17. Engels 1964 [1876]
18. Khlebnikov 1987k.
19. Maynard Smith and Szathmáry 1995, 257–278.
20. Berwick and Chomsky 2016, 27–28.
21. Maynard Smith and Szathmáry 1995, 272, citing Boyd and Richerson 1985.
22. Bourdieu 1991; Power 2014; Lewis 2009; Knight 1998; Knight and Lewis 2014; Finnegan 2014; Rappaport 1999.
23. Lewis 2009; Knight and Lewis 2014; Finnegan 2014.
24. Chomsky 1980b, 156–157.
25. Knight 1996; Knight 1998; Knight 1999; Knight 2000; Knight 2002; Knight 2008a; Knight 2008b; Knight 2009; Knight 2014; Lewis 2014; Knight and Lewis 2014.
26. Tomasello 1999, 180.
27. Whiten 1999, 190.
28. Maynard Smith and Szathmáry 1995, 271–278.
29. Woodburn 1982; Solway 2006; Lewis 2008; Finnegan 2008; Finnegan 2009; Finnegan 2014.
30. Engels 1964 [1876], 175.
31. Lakoff and Johnson 1980; Deutscher 2005; Smith and Höfler 2014.
32. Sperber and Wilson 1986; Lewis 2009; Knight and Lewis 2014.
33. Chomsky 2013b.
34. Barsky 1997, 131.
35. Chomsky 2013b.
36. Harris 1997, 4–5.
37. Chomsky 2002c, 17–18.
38. Chomsky 1988i [1984].
39. Taylor 1975, 16.
40. Marx and Engels 2000 [1845], 195.
41. Chomsky 1971a, 110.
42. Chomsky 1971.
43. Brody 2001, 308.

BIBLIOGRAPHY

Achbar, M. 1994. *Manufacturing Consent: Noam Chomsky and the media*. Montreal: Black Rose Books.

Achbar, M. and P. Wintonick. 1992. *Manufacturing Consent: Noam Chomsky and the media*. Documentary film. Toronto: Necessary Illusions.

Adams, G. 1982. *The Politics of Defense Contracting: The iron triangle*. New Brunswick, NJ: Transaction Books.

Albert, M. 2006. *Remembering Tomorrow: From the politics of opposition to what we are for*. New York: Seven Stories Press.

Albert, M. 2007. 'From SDS to life after capitalism', Michael Albert interview, 17 April, available at: www. democracynow.org/2007/4/17/from_sds_to_life_after_capitalism (accessed 1 April 2016).

Alcorta, C.S. and R. Sosis. 2005. 'Ritual, emotion, and sacred symbols: The evolution of religion as an adaptive complex', *Human Nature*, 16(4): 323–359.

Amadae, S.M. 2003. *Rationalizing Capitalist Democracy: The Cold War origins of rational choice liberalism*. Chicago and London: University of Chicago Press.

Andersen, T. 1968. *Vladimir Tatlin*. Stockholm: Moderna Museet.

Antilla, R. 1975. 'Comments on K.L. Pike and W.P. Lehmann's papers', in R. Austerlitz (ed.), *The Scope of American Linguistics*. Lisse: Peter de Ridder.

Atran, S. 2002. *In Gods We Trust: The evolutionary landscape of religion*. Oxford: Oxford University Press.

Atran, S. 2009. A memory of Claude Lévi-Strauss, available at: www.huffingtonpost.com/ scott-atran/a-memory-of-claude-lvi-st_b_349597.html (accessed 9 April 2016).

Atran, S. and A. Norenzayan. 2004. 'Religion's evolutionary landscape', *Behavioural and Brain Sciences*, 27: 713–730.

Baars, B.J. 1986. *The Cognitive Revolution in Psychology*. New York: Guilford Press.

Bach, E. 1974. 'Explanatory inadequacy', in D. Cohen (ed.), *Explaining Linguistic Phenomena*. New York: Wiley.

Baker, G.P. and P.M.S. Hacker. 1984. *Language, Sense and Nonsense*. Oxford: Blackwell.

Balari, S. and G. Lorenzo. 2010. 'Specters of Marx: A review of "Adam's Tongue", by Derek Bickerton', *Biolinguistics*, 4(1): 116–127.

Bar-Hillel, Y. 1959. *Report on the State of Machine Translation in the United States and Great Britain*. Jerusalem: US Office of Naval Research.

Bar-Hillel, Y. 1960. 'The present status of automatic translation of languages', *Advances in Cognition*, 1: 91–163.

Bar-Hillel, Y. 1961 [1958]. 'Some linguistic obstacles to machine translation', *Proceedings of the Second International Congress on Cybernetics*. Namur.

Bar-Hillel, Y. 1966 [1962]. 'Four lectures on algebraic and machine translation', in *Automatic Translation of Languages: Papers presented at NATO Summer School held in Venice, July 1962*. Oxford: Pergamon Press.

Barkow, J. H., L. Cosmides and J. Tooby (eds). 1992. *The Adapted Mind: Evolutionary psychology and the generation of culture*. New York: Oxford University Press.

Barsamian, D., 1992. *Noam Chomsky: Chronicles of dissent*. Stirling: AK Press.

Barsky, R.F. 1997. *Noam Chomsky: A life of dissent*. Cambridge, MA: MIT Press.

Barsky, R.F. 2007. *The Chomsky Effect: A radical works beyond the ivory tower*. Cambridge, MA: MIT Press.

Barsky, R.F. 2011. *Zellig Harris: From American linguistics to socialist Zionism*. Cambridge, MA: MIT Press.

Beaugrande, R. de. 1991. *Linguistic Theory: The discourse of fundamental works*. London and New York: Longman.

Beaugrande, R. de. 1998. 'Performative speech acts in linguistic theory: The rationality of Noam Chomsky', *Journal of Pragmatics*, 29: 765–803.

Beckerman, S. and P. Valentine (eds). 2002. *Cultures of Multiple Fathers: The theory and practice of partible paternity in Lowland South America*. Gainesville, FL: University Press of Florida.

Behme, C. 2014. 'A potpourri of Chomskyan science', *Philosophy in Science*, available at: http://ling.auf. net/lingbuzz/001592 (accessed 10 May 2014).

Berwick, R.C. 1998. 'Language evolution and the Minimalist Program: The origins of syntax', in J. R. Hurford, M. Studdert-Kennedy and C. Knight (eds), *Approaches to the Evolution of Language: Social and cognitive bases*. Cambridge: Cambridge University Press.

Berwick, R. and A.N. Chomsky. 2011. 'The biolinguistic programme: The current state of its development', in A.M. di Sculio and C. Boeckx (eds), *The Biolinguistic Enterprise: New perspectives on the evolution and nature of the human language faculty*. Oxford: Oxford University Press.

Berwick, R.C. and N. Chomsky. 2016. *Why Only Us: Language and evolution*. Cambridge, MA: MIT Press.

Bickerton, D. 2010. 'Response to Balari and Lorenzo', *Biolinguistics*, 4(1): 128–132.

Bickerton, D. 2014. 'Some problems for biolinguistics', *Biolinguistics*, 8: 73–96.

Bloch, M. 1975. 'Introduction', in M. Bloch (ed.), *Political Language and Oratory in Traditional Society*. London: Academic Press.

Bloomfield, L. 1933. *Language*. New York: Holt, Rinehart and Winston.

Bloomfield, L. 1970. *A Leonard Bloomfield Anthology*, ed. C. F. Hockett. Bloomington, IN: Indiana University Press.

Boden, M. 2006. *Mind as Machine: A history of cognitive science*. Oxford: Oxford University Press.

Boeckx, C. and N. Hornstein. 2010. 'The varying aims of linguistic theory', in J. Bricmont and J. Franck (eds), *Chomsky Notebook*. New York: Columbia University Press.

Boehm, C. 1999. *Hierarchy in the Forest*. Cambridge, MA: Harvard University Press.

Bolinger, D.L. 1965. 'The atomisation of meaning', *Language*, 41: 553–573.

Booth, D.A. and W.H. Locke. 1955. 'Historical introduction', in W.H. Locke and A.D. Booth (eds), *Machine Translation of Languages*. Cambridge, MA: MIT Press.

Borchardt-Hume, A. (ed.). 2014. *Malevich*. London: Tate Publishing.

Botha, R. 1989. *Challenging Chomsky: The generative garden game*. London: Blackwell.

Bouchard, D. 2013. *The Nature and Origin of Language*. Oxford: Oxford University Press.

Bourdieu, P. 1991. *Language and Symbolic Power*. Oxford: Blackwell.

Bowlby, J. 1969. *Attachment*. New York: Basic Books.

Boyd, R. and P. Richerson. 1985. *Culture and the Evolutionary Process*. Chicago: University of Chicago Press.

Boyer, P. 2001. *Religion Explained: The evolutionary origins of religious thought*. New York: Basic Books.

Branfman, F. 2012. 'When Chomsky wept', available at: www.salon.com/2012/06/17/when_chomsky_wept/ (accessed 20 March 2016).

Bray, C.W. (ed.). 1946. *Human Factors in Military Efficiency: Summary Technical Report of the Applied Psychology Panel, NDRC*, Vol. 1. Washington, DC: US Government Printing Office.

Brennan, D.G. 1969. *ABM, Yes or No?* Center occasional papers, 2(2). Santa Barbara, CA: Center for the Study of Democratic Institutions.

Bridger, S. 2015. *Scientists at War: The ethics of Cold War weapons research*. Cambridge, MA: Harvard University Press.

Brody, H. 2001. *The Other Side of Eden: Hunter-gatherers, farmers and the shaping of the world*. London: Faber & Faber.

Bromberger, S. and M. Halle 1991. 'Why phonology is different', in A. Kasher (ed.), *The Chomskyan Turn*. Oxford: Blackwell, 56–77.

Bruner, J. 1990. *Acts of Meaning*. Cambridge, MA: Harvard University Press.

Bruner, J.S., A. Jolly and K. Sylva (eds). 1976. *Play: Its role in development and evolution*. New York: Basic Books.

Buckley, W. and N. Chomsky. 1969. 'Buckley–Chomsky debate transcript part 5', available at: http:// buckley-chomsky.weebly.com/debate-part-5.html (accessed 19 March 2016).

Buss, D.M. 2008. *Evolutionary Psychology: The new science of the mind*. Boston, MA: Omegatype Typography, Inc.

Byrne, R. and N. Corp. 2004. 'Neocortex size predicts deception rate in primates', *Proceedings of the Royal Society of London*, B 271: 1693–1699.

Byrne, R. and A. Whiten (eds). 1988. *Machiavellian Intelligence: Social expertise and the evolution of intellect in monkeys, apes, and humans.* Oxford: Clarendon Press.

Call, J. 2009. 'Contrasting the social cognition of humans and non-human apes: the shared intentionality hypothesis', *Topics in Cognitive Science*, 1: 368–379.

Capshew, J.H. 1986. 'Psychology on the march'. Unpublished PhD dissertation, University of Pennsylvania.

Cartlidge, E. 2014. 'Faith and science', *Tablet*, 1 February.

Chepesiuk, R. 1995. *Sixties Radicals, Then and Now: Candid conversations with those who shaped the era.* Jefferson, NC: McFarland.

Chomsky, N. 1956. 'Three models for the description of language', *Institute of Radio Engineers Transactions on Information Theory*, 2: 113–124.

Chomsky, N. 1957. *Syntactic Structures.* The Hague: Mouton.

Chomsky, N. 1959. 'Review of B.F. Skinner's *Verbal Behavior*', *Language*, 35(1): 26–58.

Chomsky, N. 1962. 'A transformational approach to syntax', in Archibald Hill (ed.), *Third Texas Conference on Problems of Linguistic Analysis in English.* Austin, TX: University of Texas.

Chomsky, N. 1964a [1963]. *Current Issues in Linguistic Theory.* The Hague: Mouton.

Chomsky, N. 1964b. 'Current issues in linguistic theory', in J.A. Fodor and J.J. Katz (eds), *The Structure of Language.* Englewood Cliffs, NJ: Prentice Hall.

Chomsky, N. 1964c [1962]. 'The logical basis of linguistic theory', in H.G. Lunt (ed.), *The Proceedings of the Ninth International Congress of Linguists.* The Hague: Mouton.

Chomsky, N. 1965. *Aspects of the Theory of Syntax.* Cambridge, MA: MIT Press.

Chomsky, N. 1966a [1964]. *Topics in the Theory of Generative Grammar.* The Hague: Mouton.

Chomsky, N. 1966b. *Cartesian Linguistics: A chapter in the history of rationalist thought.* New York: Harper and Row.

Chomsky, N. 1967a. 'Appendix A: The formal nature of language', in E.H. Lenneberg, *Biological Foundations of Language.* Malabar, FL: Krieger.

Chomsky, N. 1967b. 'On resistance', *New York Review of Books*, 7 December.

Chomsky, N. 1967c [1959]. 'Review of B.F. Skinner, *Verbal Behavior*', in L.A. Jakobovits and M.S. Miron (eds), *Readings in the Psychology of Language.* Englewood Cliffs, NJ: Prentice Hall.

Chomsky, N. 1967d. 'Letter', *New York Review of Books*, 23 March.

Chomsky, N. 1967e. 'Letter', *New York Review of Books*, 20 April.

Chomsky, N. 1969. *Statement by Noam A. Chomsky to the Massachusetts Institute of Technology Review Panel on Special Laboratories, Final Report.* MIT Libraries Retrospective Collection.

Chomsky, N. 1971a. *Problems of Knowledge and Freedom.* London and New York: The New Press.

Chomsky, N. 1971b. 'In defence of the student movement', available at: https://chomsky.info/1971_03/ (accessed 5 March 2016).

Chomsky, N. 1972a. *Language and Mind* (expanded edition). New York: Harcourt Brace Jovanovich.

Chomsky, N. 1972b. 'Some empirical issues in the theory of transformational grammar', in S. Peters (ed.), *Goals of Linguistic Theory.* Englewood Cliffs, NJ: Prentice Hall.

Chomsky, N. 1975a. *The Logical Structure of Linguistic Theory.* Chicago: Chicago University Press.

Chomsky, N. 1975b. 'Towards a humanistic conception of education', in W. Feinberg and H. Rosemount (eds), *Work, Technology, and Education: Dissenting essays in the intellectual foundations of American education.* Chicago and London: University of Illinois Press.

Chomsky, N. 1976a. *Reflections on Language.* London: Fontana.

Chomsky, N. 1976b. 'On the nature of language', in S.R. Harnard, H.D. Steklis and J. Lancaster (eds), *Origins and Evolution of Language and Speech.* Annals of the New York Academy of Sciences, Vol. 280. New York: New York Academy of Sciences.

Chomsky, N. 1979a. *Language and Responsibility: Based on conversations with Mitsou Ronat*, trans. John Viertel. New York: Pantheon.

Chomsky, N. 1979b [1951]. *Morphophonemics of Modern Hebrew.* Reprint of unpublished MA thesis, University of Pennsylvania. New York: Garland Publications.

Chomsky, N. 1979c. 'Markedness and core grammar'. Mimeo. Subsequently published in A. Belletti, L. Brandi and L. Rizzi (eds). 1981. *Theory of Markedness in Generative Grammar.* Pisa: Scuola Normale Superiore.

Chomsky, N. 1980a. *Rules and Representations.* New York: Columbia University Press.

Chomsky, N. 1980b. 'On cognitive structures and their development: A reply to Piaget. As well as other contributions to the Abbaye de Royaumont debate (October 1975)', in M. Piatelli-Palmarini (ed.), *Language and Learning: The debate between Jean Piaget and Noam Chomsky.* London: Routledge and Kegan Paul.

Chomsky, N. 1980c. 'Some elementary comments on the rights of freedom of expression', available at: https://chomsky.info/19801011 (accessed 12 March 2016).

Chomsky, N. 1981. *Lectures on Government and Binding*, first edition. Dordrecht: Foris.

Chomsky, N. 1982a. *The Generative Enterprise: A discussion with Riny Huybregts and Henk van Riemsdijk.* Dordrecht: Foris.

Chomsky, N. 1982b. *Some Concepts and Consequences of the Theory of Government and Binding.* Cambridge, MA: MIT Press.

Chomsky, N. 1983. Contribution in *A Tribute to Roman Jakobson 1896–1982.* Berlin: de Gruyter.

Chomsky, N. 1985. *Turning the Tide: US intervention in Central America and the struggle for peace.* Boston: South End.

Chomsky, N. 1986. *Knowledge of Language: Its nature, origin, and use.* Westport, CT: Praeger.

Chomsky, N. 1988a. *Language and Problems of Knowledge: The Managua lectures.* Cambridge, MA: MIT Press.

Chomsky, N. 1988b [1983]. 'Things no amount of learning can teach', in C.P. Otero (ed.), *Noam Chomsky: Language and politics.* Montreal: Black Rose, pp. 407–419.

Chomsky, N. 1988c. 'The "right turn" in US policy (22 October 1986)', in C.P. Otero (ed.), *Noam Chomsky: Language and politics.* Montreal: Black Rose, pp. 648–660.

Chomsky, N. 1988d. 'The treachery of the intelligentsia: A French travesty. Interview dated 26 October 1981', in C.P. Otero (ed.), *Noam Chomsky: Language and politics.* Montreal: Black Rose, pp. 312–323.

Chomsky, N. 1988e. 'Address given at the Community Church of Boston, December 9 1984', reprinted as 'Afghanistan and South Vietnam', in J. Peck (ed.), *The Chomsky Reader.* London: Serpent's Tail, pp. 223–226.

Chomsky, N. 1988f [1968]. 'The intellectual as prophet', in C.P. Otero (ed.), *Noam Chomsky: Language and politics.* Montreal: Black Rose, pp. 85–99.

Chomsky, N. 1988g [1977]. 'Language theory and the theory of justice', in C.P. Otero (ed.), *Noam Chomsky: Language and politics.* Montreal: Black Rose, pp. 233–250.

Chomsky, N. 1988h. 'Interview', in J. Peck (ed.), *The Chomsky Reader.* London: Serpent's Tail.

Chomsky, N. 1988i [1984]. 'The manufacture of consent', in J. Peck (ed.), *The Chomsky Reader.* London: Serpent's Tail, pp. 1–55.

Chomsky, N. 1988j [1986]. 'Political discourse and the propaganda system', in C.P. Otero (ed.), *Noam Chomsky: Language and politics.* Montreal: Black Rose, pp. 662–697.

Chomsky, N. 1988k. 'The cognitive revolution II', in C.P. Otero (ed.), *Noam Chomsky: Language and politics.* Montreal: Black Rose, pp. 744–759.

Chomsky, N. 1988l [1984]. 'Knowledge of language, human nature, and the role of intellectuals. Interview with Hannu Reime', in C.P. Otero (ed.), *Noam Chomsky: Language and politics.* Montreal: Black Rose, pp. 586–603.

Chomsky, N. 1989. 'Noam Chomsky: An interview', *Radical Philosophy*, 53.

Chomsky, N. 1990. 'Transcript, interview by David Barsamian, 2 February 1990'. MIT, Cambridge, MA.

Chomsky, N. 1991a. 'Linguistics and adjacent fields: a personal view', in A. Kasher (ed.), *The Chomskyan Turn.* Oxford: Blackwell.

Chomsky, N. 1991b. Transcript of 'Reflections on the Gulf War', *Alternative Radio.* Recorded by D. Barsamian in Cambridge, MA on 21 May.

Chomsky, N. 1991c. 'Linguistics and cognitive science: Problems and mysteries', in A. Kasher (ed.), *The Chomskyan Turn.* Oxford: Blackwell.

Chomsky, N. 1991d. 'Language, politics and composition', in G. Olsen and I. Gales (eds), *Interviews: Cross-disciplinary perspectives on rhetoric and literacy.* Carbondale, IL: Southern Illinois University Press.

Chomsky, N. 1991e. 'A brief interview with Noam Chomsky on anarchy, civilization & technology', *Anarchy: A Journal of Desire Armed*, 29: 27–29.

Chomsky, N. 1992. *What Uncle Sam Really Wants.* Tucson, AZ: Odonian Press.

Chomsky, N. 1993. *Lectures on Government and Binding*, seventh edition. Berlin and New York: Mouton de Gruyter.

Chomsky, N. 1995. *The Minimalist Program.* Cambridge, MA: MIT Press.

Chomsky, N. 1996a. *Class Warfare: Interviews with David Barsamian.* London: Pluto Press.

Chomsky, N. 1996b. *Powers and Prospects. Reflections on human nature and the social order.* London: Pluto Press.

Chomsky, N. 1997a. Letter to R.F. Barsky, dated 15 December 1992, quoted in R.F. Barsky, *Noam Chomsky: A life of dissent.* Cambridge, MA: MIT Press.

Chomsky, N. 1997b. Letter, dated 3 March 1995, quoted in R.F. Barsky, *Noam Chomsky: A life of dissent.* Cambridge, MA: MIT Press.

Chomsky, N. 1997c. 'Noam Chomsky and Michel Foucault. Human Nature: Justice versus power', in A.I. Davidson (ed.), *Foucault and his Interlocutors.* Chicago and London: Chicago University Press.

Chomsky, N. 1997d. 'Creation and culture', *Alternative Radio.* Recorded 25 November 1992.

Chomsky, N. 1997e. Letter to R.F. Barsky, dated 31 March 1995, quoted in R.F. Barsky, *Noam Chomsky: A life of dissent.* Cambridge, MA: MIT Press.

Chomsky, N. 1998a. *The Common Good.* Chicago: Common Courage Press.

Chomsky, N. 1998b. 'Language and mind: Current thoughts on ancient problems. Part I and Part II', lectures presented at Universidad de Brasilia, published in *Pesquisa Linguistica*, 3(4).

Chomsky, N. 1998c [1977]. *Language and Responsibility*, Part I of *On Language: Chomsky's classic works – language and responsibility and reflections on language in one volume*. New York: New Press.

Chomsky, N. 2000a. *The Architecture of Language*. Oxford: Oxford University Press.

Chomsky, N. 2000b. *New Horizons in the Study of Language and Mind*. Cambridge: Cambridge University Press.

Chomsky, N. 2000c. 'Minimalist inquiries: The framework', in R. Martin, D. Michaels and J. Uriagerika (eds), *Step by step: Essays on Minimalist syntax in honor of Howard Lasnik*. Cambridge, MA: MIT Press.

Chomsky, N. 2002a. *On Nature and Language*. Cambridge: Cambridge University Press.

Chomsky, N. 2002b. *Understanding Power*. New York: The New Press.

Chomsky, N. 2002c [1969]. *American Power and the New Mandarins*. New York: Pantheon Books.

Chomsky, N. 2003. 'Anti-Semitism, Zionism, and the Palestinians', *Variant*, 16 (Winter).

Chomsky, N. 2004a. 'Beyond explanatory adequacy', in A. Belletti (ed.), *Structures and Beyond: The cartography of syntactic structures*, Vol. 3. Oxford: Oxford University Press.

Chomsky, N. 2004b. 'War crimes and imperial fantasies', *International Socialist Review*, Issue 37.

Chomsky, N. 2005a. 'Language and freedom', in N. Chomsky, *Chomsky on Anarchism*, ed. B. Pateman. Edinburgh and Oakland, CA: AK Press.

Chomsky, N. 2005b. 'Three factors in language design', *Linguistic Inquiry*, 36(1): 1–22.

Chomsky, N. 2005c. *Chomsky on Anarchism*, ed. B. Pateman. Edinburgh and Oakland, CA: AK Press.

Chomsky, N. 2006a. *Language and Mind*, third edition. Cambridge: Cambridge University Press.

Chomsky, N. 2006b. 'Science in the Dock. Interview with L. Krauss and S.M. Carroll', *Science and Technology News*, March.

Chomsky, N. 2007a. 'Approaching UG from below', in U. Sauerland and H.M. Gartner (eds), *Interfaces + Recursion = Language?* Berlin: Mouton.

Chomsky, N. 2007b. 'Of minds and language', *Biolinguistics*, 1: 9–27.

Chomsky, N. 2008. 'Interview with Noam Chomsky', *Radical Anthropology*, 2: 19–23.

Chomsky, N. 2008–9. 'Interview with Noam Chomsky', *Works and Days*, No. 51–4, Vol. 26/27: 527–537.

Chomsky, N. 2009a. *Of Minds and Language: A dialogue with Noam Chomsky in the Basque Country*, ed. M. Piatelli-Palmarini, J. Uriagereka and P. Salaburu. Oxford: Oxford University Press.

Chomsky, N. 2009b. Video interview for MIT 150 Infinite History Project, available at: https://archive.org/details/NoamChomsky-InfiniteHistoryProject-2009/ (accessed May 2017).

Chomsky, N. 2010a. 'Some simple Evo-devo theses: How true might they be for language?', in R. Larson, V. Déprez and H. Yamakido (eds), *The Evolution of Human Language*. Cambridge: Cambridge University Press.

Chomsky, N. 2010b. 'Poverty of stimulus: Unfinished business', transcript of oral presentation, Johannes-Gutenberg University Mainz, 24 March.

Chomsky, N. 2010c. 'The mysteries of nature how deeply hidden?', in J. Bricmont and J. Franck (eds), *Chomsky Notebook*. New York: Columbia University Press.

Chomsky, N. 2010d. 'Government involvement with science and art', *Z Magazine*, 3 May 2010.

Chomsky, N. 2011a. 'Appendix: Interview with Noam Chomsky', in P. Ludlow, *The Philosophy of Generative Linguistics*. Oxford: Oxford University Press.

Chomsky, N. 2011b. 'On the poverty of the stimulus', talk given to UCL division of psychology and language sciences, available at: www.youtube.com/watch?v=068Id3Grjp0 (accessed 5 April 2016).

Chomsky, N. 2011c. 'Democracy and the public university', 8 April 2011, available at: www.youtube.com/watch?v=QiNqz38YI8M (accessed December 2016).

Chomsky, N. 2011d. 'Academic freedom and the corporatization of universities', available at: https://chomsky.info/20110406 (accessed April 2017).

Chomsky, N. 2012. *The Science of Language: Interviews with James McGilvray*. Cambridge: Cambridge University Press.

Chomsky, N. 2013a. 'After 60+ years of generative grammar', available at: www.youtube.com/watch?v=Rgd8BnZ2-iw (accessed 9 April 2016).

Chomsky, N. 2013b. 'On revolutionary violence, communism and the American left. Noam Chomsky interviewed by Christopher Helali', *Pax Marxista*, 12 March, available at: https://chomsky.info/20130312 (accessed 13 March 2016).

Chomsky, N. 2014a. 'Science, mind, and limits of understanding', Science, Theology and the Ontological Quest Foundation, Vatican, available at: https://chomsky.info/201401__/ (accessed 9 April 2016).

Chomsky, N. 2014b. 'Surviving the 21st century', available at: www.youtube.com/watch?v=wJtfWZGxnGI (accessed April 2017).

Chomsky, N. 2014c. 'Driving forces in US policy/diskussion', available at: www.youtube.com/watch?v=Av8uFvDTvw4 (accessed April 2017).

Chomsky, N. 2015. 'Noam Chomsky and Subrata Ghoshroy: From the Cold War to the Climate Crisis', MIT Video, available at: www.youtube.com/watch?v=7tVG3sRcuU4 (accessed 9 April 2016).

Chomsky, N. 2016a. 'On the evolution of language', UNAM Skype talk, 4 March, available at: http://132.248.212.20/videos/video/1338/ (accessed 1 June 2016).

Chomsky, N. 2016b. 'Language, creativity and the limits of understanding', lecture at the University of Rochester, New York, 21 April, available at: www.youtube.com/watch?v=XNSxj0TVeJs (accessed 1 June 2016).

Chomsky, N. 2016c. 'Chomsky's carburetor', Cited Podcast No. 23, 21 September 2016, available at: http://citedpodcast.com/23-chomskys-carburetor/ (accessed September 2016).

Chomsky, N. 2016d. 'A conversation with Noam Chomsky and Howard Gardner', 10 October 2016, Harvard Graduate School of Education's Gutman Library, available at: www.youtube.com/watch?v=IWGhJ63OXxM (accessed December 2016).

Chomsky, N. 2016e. 'Ralph Nader radio hour', Episode 137, available at: https://ralphnaderradiohour.com/wp-content/uploads/2016/10/Ralph-Nader-Radio-Hour-Episode-137.pdf (accessed December 2016).

Chomsky, N. 2016f. 'On being Noam Chomsky', interview with Sam Tanenhaus, New York Times, 31 October 2016.

Chomsky, N. 2017. ' "The Responsibility of Intellectuals" – 50 years on, Part 2', available at: www.youtube.com/watch?v=B9fE6XFvU0o (accessed April 2017).

Chomsky, N. and E. Herman. 1979. After the Cataclysm: Postwar Indochina and the reconstruction of imperial ideology. Cambridge, MA: South End Press.

Chomsky, N. and H. Lasnik. 1977. 'Filters and control', Linguistic Inquiry, 8(3): 425–504.

Chomsky, N. and C.P. Otero. 2003. Chomsky on Democracy and Education. New York: RoutledgeFalmer.

Chu, J. 2012. 'Driving drones can be a drag', MIT News, available at: http://news.mit.edu (accessed February 2017).

Chu, J. 2015. 'MIT cheetah robot lands the running jump', MIT News, available at: http://news.mit.edu (accessed February 2017).

Cogswell, D. 1996. Chomsky for Beginners. London and New York: Writers and Readers.

Cooke, R. 1987. Velimir Khlebnikov: A critical study. Cambridge: Cambridge University Press.

Dawkins, R. 2015. Brief Candle in the Dark: My life in science. London: Bantam Press.

Dean, M. 2003. Chomsky: A Beginner's Guide. London: Hodder and Stoughton.

Delavenay, E. 1960. An Introduction to Machine Translation. London: Thames and Hudson.

Dennett, D. 1991. Consciousness Explained. London: Penguin.

Descartes, R. 1984 [1641]. 'Objections and replies', in J. Cottingham, R. Stoothoff and D. Murdoch (trans.), The Philosophical Writings of Descartes, Vol. II. Cambridge: Cambridge University Press.

Descartes, R. 1985 [1649]. 'The passions of the soul', in J. Cottingham, R. Stoothoff and D. Murdoch (trans.), The Philosophical Writings of Descartes (two-volume edition), Vol. I. Cambridge: Cambridge University Press.

Descartes, R. 1991 [1633]. Letter to Mersenne, in J. Cottingham, R. Stoothoff, D. Murdoch and A. Kenny (trans.), The Philosophical Writings of Descartes, Vol. III. Cambridge: Cambridge University Press.

Destro-Bisol, G., F. Donati, V. Coia, I. Boschi, F. Verginelli, A. Caglia, S. Tofanelli, G. Spedini and C. Capelli. 2004. 'Variation of female and male lineages in sub-Saharan populations: The importance of sociocultural factors', Molecular Biology and Evolution, 21(9): 1673–1682.

Deutch, J.M. 1989. 'The decision to modernize US intercontinental ballistic missiles', Science, 244 (4911): 1445–1450.

Deutscher, G. 2005. The Unfolding of Language: The evolution of mankind's greatest invention. London: Random House.

Dirac, P. 1963. 'The evolution of the physicist's picture of nature', Scientific American, 208(5): 45–53.

Dosse, F. 1997. History of Structuralism, Vol. 1. Minneapolis, MN: University of Minnesota Press.

Dostert, L.E. 1955. 'The Georgetown–IBM Experiment', in W.N. Locke and A.D. Booth (eds), Machine Translation of Languages. Cambridge, MA: MIT Press.

Douglas, C. (ed.). 1985. The King of Time: Selected writings of the Russian futurian, trans. P. Schmidt. Cambridge, MA: Harvard University Press.

Doyle, C.D., W. Mieder and F.R. Shapiro (eds). 2012. The Dictionary of Modern Proverbs. London and New Haven, CT: Yale University Press.

Durkheim, É. 1976 [1915]. The Elementary Forms of the Religious Life, trans. J.W. Swain. London: Allen and Unwin.

Eckstein, H. 1962. 'Internal wars', in W.A. Lybrand (ed.), Proceedings of the Symposium 'The US Army's Limited-War Mission and Social Science Research'. Washington, DC: American University, Special Operations Research Unit.

Eckstein, H. 1964. Internal War. New York: Free Press of Glencoe.

Edgley, A. 2000. The Social and Political Thought of Noam Chomsky. London and New York: Routledge.

Edwards, P. 1996. The Closed World: Computers and the politics of discourse in cold war America. Cambridge, MA: MIT Press.

Einstein, A. 1954. 'Physics and reality', in A. Einstein, *Ideas and Opinions*, trans. Sonja Bargmann. New York: Bonanza.

Eldredge, N. and S.J. Gould. 1972. 'Punctuated equilibrium: An alternative to phyletic gradualism', in T.J.M. Schopf (ed.), *Models in Paleobiology*. San Francisco, CA: Freeman.

Engels, F. 1957 [1886]. 'Ludwig Feuerbach and the end of classical German philosophy', in K. Marx and F. Engels, *On Religion*. Moscow: Foreign Languages Publishing House.

Engels, F. 1964 [1876]. *The Dialectics of Nature*. Moscow: Progress Publishers.

Engels, F. 1972 [1884]. *The Origin of the Family, Private Property and the State*. New York: Pathfinder Press.

Engerman, D. 2003. *Staging Growth: Modernization, development, and the global Cold War*. Cambridge, MA: MIT Press.

Eun-jung, S. 2015. *Verita$: Harvard's Hidden History*. Oakland, CA: PM Press.

Evans, N. and S.C. Levinson. 2009. 'The myth of language universals: Language diversity and its importance for cognitive science', *Behavioral and Brain Sciences*, 32(5): 429–492.

Feldman, B. 2008. 'Columbia University's IDA Jason Project 1960s work – Part 9', available at: http:// bfeldman68.blogspot.co.uk/2008/04/columbia-universitys-ida-jason-project.html (accessed December 2016).

Fiengo, R. 2006. 'Review of "Chomsky's Minimalism"', *Mind*, 115(458): 469–472.

Fine, C. 2010. *Delusions of Gender: How our minds, society and neurosexism create difference*. New York: Norton.

Finkbeiner, A. 2007. *The Jasons: The secret history of science's post-war elite*. New York: Penguin.

Finnegan, M. 2008. 'The personal is political: Eros, ritual dialogue, and the speaking body in Central African hunter-gatherer society'. Unpublished PhD thesis, Edinburgh University.

Finnegan, M. 2009. 'Political bodies: Some thoughts on women's power among Central African hunter-gatherers', *Radical Anthropology*, 3: 31–37.

Finnegan, M. 2014. 'The politics of Eros: Ritual dialogue and egalitarianism in three Central African hunter-gatherer societies', *Journal of the Royal Anthropological Institute*, NS 19: 697–715.

Fodor, J.A. 1975. *The Language of Thought*. Scranton, PA: Crowell.

Fodor, J.A. 1985. 'Precis of *The Modularity of Mind*', *Behavioral and Brain Sciences*, 8: 1–42.

Forman, P. 1987. 'Behind quantum electronics: National security as basis for physical research 1940–1960', *Historical Studies in the Physical and Biological Sciences*, 18(1): 156–157.

Foucault, M. and N. Chomsky. 1997. 'Human nature: Justice versus power', in A.I. Davidson (ed.), *Foucault and his Interlocutors*. Chicago, IL, and London: Chicago University Press.

Fromkin, V.A. 1991. 'Language and brain: Redefining the goals and methodology of linguistics', in A. Kasher (ed.), *The Chomskyan Turn*. Oxford: Blackwell.

Gardner, H. 1987. *The Mind's New Science: A history of the cognitive revolution*. New York: Basic Books.

Garfinkel, S. No date. 'Building 20, a survey', available at: http://ic.media.mit.edu/projects/JBW/ARTICLES/SIMSONG.HTM (accessed January 2017).

Gasparov, B. 1997. 'Futurism and phonology: Futurist roots of Jakobson's approach to language', *Cahiers de l'ILSL*, 9: 105–124.

Geoghegan, B. 2011. 'From information theory to French theory: Jakobson, Lévi-Strauss, and the cybernetic apparatus', *Critical Inquiry*, 38: 96–126, available at: http://criticalinquiry.uchicago.edu/uploads/pdf/Geoghegan,_Theory.pdf (accessed October 2016).

Gigerenza, G. and D.G. Goldstein. 1996. 'Mind as computer: Birth of a metaphor', *Creativity Research Journal*, 9(2–3): 131–144.

Givón, T. 1995. *Functionalism and Grammar*. Amsterdam and Philadelphia: John Benjamins.

Glantz, S.A. et al. 1971. *DOD Sponsored Research at Stanford: Its impact on the university*, Vol. 1, *The Perceptions: The investigator's and the sponsor's*. Stanford, CA: SWOPSI.

Glantz, S.A. and N.V. Albers. 1974. 'Department of Defense R&D in the university', *Science*, 186 (4165): 706–711.

Gleitman, L.R. and E.L. Newport. 1995. 'The invention of language by children: Environmental and biological influences on the acquisition of language', in D.N. Osherson (series ed.), *An Invitation to Cognitive Science*, Vol. 1, *Language*, ed. L.R. Gleitman and M. Liberman. Cambridge, MA: MIT Press.

Goldsmith, J.A. 2005. 'Review article on Nevin (ed.) 2002', *Language*, 81(3): 719–773.

Golumbia, D. 2009. *The Cultural Logic of Computation*. Harvard, MA: Harvard University Press.

Gordin, M.D. 2016. 'The Dostoevsky machine in Georgetown: Scientific translation in the Cold War', *Annals of Science*, 73(2): 208–223.

Gottfried, K. 1969. 'Beyond March 4', founding document of the Union of Concerned Scientists, MIT, available at: www.ucsusa.org/about/founding-document-beyond.html#.V1a-gWZH1pk (accessed 7 June 2016).

Gowaty, P.A. 1997. 'Sexual dialectics, sexual selection, and variation in mating behavior', in P.A. Gowaty (ed.), *Feminism and Evolutionary Biology: Boundaries, intersections, and frontiers*. New York: Chapman and Hall.

Gowdy, J. 2005. 'Hunter-gatherers and the mythology of the market', in R.B. Lee and R. Daly (eds), *The Cambridge Encyclopedia of Hunters and Gatherers*. Cambridge: Cambridge University Press.

Gray, B. 1976. 'Counter-revolution in the hierarchy', *Forum Linguisticum*, 1: 38–50.

Greenberg, D.S. 1999. *The Politics of Pure Science*. Chicago: University of Chicago Press, p. 151.

Gross, M. and A. Lentin. 1970. *Formal Grammars*. London: Springer Verlag.

Halle, M. 1959. *The Sound Pattern of Russian*. The Hague: Mouton.

Hansen, J. 2009. *Storms of my Grandchildren*. London: Bloomsbury.

Haraway, D. 1989. *Primate Visions: Gender, race and nature in the world of modern science*. New York and London: Routledge.

Harman, G. (ed.). 1974. *On Noam Chomsky: Critical essays*. Modern Studies in Philosophy. Garden City, NY: Anchor Press.

Harman, G. 2001. 'Review of Noam Chomsky, *New Horizons in the Study of Language and Mind*', *Journal of Philosophy*, 98(5): 265–269.

Harris, F. and J. Harris. 1974. 'The development of the linguistics program at the Massachusetts Institute of Technology', 50 years of Linguistics at MIT: A scientific reunion, 9–11 December 2011, MIT, available at: http://ling50.mit.edu/harris-development (accessed 9 April 2016).

Harris, R.A. 1993. *The Linguistics Wars*. New York and Oxford: Oxford University Press.

Harris, R.A. 1998. 'The warlike Chomsky. Review of R.F. Barsky, 1997. *Noam Chomsky: A Life of Dissent*. Cambridge, MA: MIT Press', *Books in Canada*, March.

Harris, R.A. 2010. 'Chomsky's other revolution', in D.A. Kibbee (ed.), *Chomskyan Revolutions*. Amsterdam: John Benjamins.

Harris, R.A. Forthcoming. *Linguistic Wars*, second edition. Oxford: Oxford University Press.

Harris, Z. 1957. 'Co-occurrence and transformation in linguistic structure', *Language*, 33(3): 289–340.

Harris, Z. 1997. *The Transformation of Capitalist Society*. Lanham, MD: Rowman and Littlefield.

Hauser, M.D., N. Chomsky and W.T. Fitch. 2002. 'The faculty of language: What is it, who has it, and how did it evolve?', *Science*, 298(5598): 1569–1579.

Hauser, M.D., C. Yang, R.C. Berwick, I. Tattersall, M.J. Ryan, J. Watumull, N. Chomsky and R.C. Lewontin. 2014. 'The mystery of language evolution', *Frontiers in Psychology*, 5(1).

Hayles, K. 1999. *How We Became Posthuman: Virtual bodies in cybernetics, literature, and informatics*. Chicago, IL: University of Chicago Press.

Hegel, G.W.F. 1929 [1812–1816]. *Science of Logic*, trans. W.H. Johnston and L.G. Struthers. London: Allen & Unwin.

Hewlett, B.S. and M.E. Lamb (eds) 2005. *Hunter-Gatherer Childhoods: Evolutionary, developmental and cultural perspectives*. New Brunswick and London: Aldine.

Hockett, C. 1960. 'The origin of speech', *Scientific American*, 203: 89–96.

Hockett, C.F. and R. Ascher. 1964. 'The human revolution', *Current Anthropology*, 5(3): 135–168.

Hrdy, S.B. 1981. *The Woman that Never Evolved*. Cambridge, MA: Harvard University Press.

Hrdy, S.B. 2000. *Mother Nature*. London: Vintage.

Hrdy, S.B. 2009. *Mothers and Others: The evolutionary origins of mutual understanding*. London and Cambridge, MA: Belknap Press of Harvard University Press.

Hughes, S. 2006. 'Interview with Noam Chomsky, 21 April', in *Penn in Ink: Pathfinders, Swashbucklers, Scribblers & Sages: Portraits from the Pennsylvania Gazette*. Bloomington, IN: Xlibris Corporation.

Hurford, J. 1989. 'Biological evolution of the Saussurean sign as a component of the language acquisition device', *Lingua*, 77(2): 187–222.

Hurford, J.R. 2007. *The Origins of Meaning: Language in the light of evolution*, Vol. 1. Oxford: Oxford University Press.

Hurford, J.R. 2012. *The Origins of Grammar: Language in the light of evolution*, Vol. 2. Oxford: Oxford University Press.

Hutchins, W.J. 1986. *Machine Translation: Past, present, future*. Chichester: Ellis Horwood.

Hutchins, W.J. (ed.). 2000a. *Early Years in Machine Translation*. Amsterdam: John Benjamins.

Hutchins, W.J. 2000b. 'Yehoshua Bar-Hillel: A philosopher's contribution to machine translation', in W.J. Hutchins (ed.), *Early Years in Machine Translation*. Amsterdam: John Benjamins.

Ippolito, T. 1990. 'Effects of variation of uranium enrichment on nuclear submarine reactor design'. Unpublished MSc thesis, MIT.

Ivanov, V.V. 1983. 'Roman Jakobson: The future', in *A Tribute to Roman Jakobson, 1896–1982*. New York and Amsterdam: Mouton.

Jaggi, M. 2001. 'Conscience of a nation', *Guardian*, 20 January.

Jakobson, R. 1985. 'The Byzantine mission to the Slavs', in *Roman Jakobson: Selected Writings*, VI *Early Slavic Paths and Crossroads*, ed. Stephen Rudy: Part One: Comparative Slavic Studies: The Cyrillo-Methodian Tradition. Amsterdam: Mouton.

Jakobson, R. 1997. *My Futurist Years*, ed. B. Jangfeldt, trans. Stephen Rudy. New York: Marsilio.

Jakobson, R. and K. Pomorska. 1983. *Dialogues*. Cambridge: Cambridge University Press.

Jakobson, R. and L.R. Waugh. 2002. *The Sound Shape of Language*, third edition. Berlin and New York: Mouton de Gruyter.

Jakobson, R., C. Gunnar, M. Fant and M. Halle. 1951. *Preliminaries to Speech Analysis: The distinctive features and their correlates*. Cambridge, MA: MIT Press.

Johnson, H. 2001. *Holding the Center: Memoirs of a life in higher education*. Cambridge, MA: MIT Press.

Joseph, J.E. 2002. *From Whitney to Chomsky: Essays in the history of American linguistics*. Amsterdam: Benjamins.

Joseph, J.E. and T.J. Taylor. 1990. 'Introduction', in J.E. Joseph and T.J. Taylor (eds), *Ideologies of Language*. London and New York: Routledge.

Kabat-Zinn, J. 2014. *Coming to Our Senses*. London: Piatkus.

Kahn, H. 1960. *On Thermonuclear War*. Princeton, NJ: Princeton University Press.

Katsiaficas, G. 1969. *A Personal Statement to the Massachusetts Institute of Technology Review Panel on Special Laboratories, Final Report*. MIT Libraries Retrospective Collection.

Katz, J.J. and P. Postal. 1964. *An Integrated Theory of Linguistic Description*. Cambridge, MA: MIT Press.

Kenneally, C. 2007. *The First Word: The search for the origins of language*. London: Viking Penguin.

Khlebnikov, V. 1968. *Sobranie proizvedenij Velimira Xlebnikova*, ed. Ju. Tynjanova and I.N. Stepanova, Vol. III, *Stixotvorenija*, reprint ed. Dmitri Tschizewskij. München: Wilhelm Fink.

Khlebnikov, V. 1986, 'Pust' na mogil'noj plite pročtut [Let it be written on his gravestone]', in B. Gasparov. 1997. 'Futurism and phonology: Futurist roots of Jakobson's approach to language', *Cahiers de l'ILSL*, 9: 105–124.

Khlebnikov, V. 1987a [1912]. 'Teacher and student', in C. Douglas (ed.), *Collected Works of Velimir Khlebnikov*, Vol. I, *Letters and Theoretical Writings*, trans. P. Schmidt. Cambridge, MA: Harvard University Press.

Khlebnikov, V. 1987b [1913]. 'Two individuals: A conversation', in C. Douglas (ed.), *Collected Works of Velimir Khlebnikov*, Vol. I, *Letters and Theoretical Writings*, trans. P. Schmidt. Cambridge, MA: Harvard University Press.

Khlebnikov, V. 1987c [1913]. 'The warrior of the kingdom', in C. Douglas (ed.), *Collected Works of Velimir Khlebnikov*, Vol. I, *Letters and Theoretical Writings*, trans. P. Schmidt. Cambridge, MA: Harvard University Press.

Khlebnikov, V. 1987d [1914]. Letter to Vasily Kamensky, in C. Douglas (ed.), *Collected Works of Velimir Khlebnikov*, Vol. I, *Letters and Theoretical Writings*, trans. P. Schmidt. Cambridge, MA: Harvard University Press.

Khlebnikov, V. 1987e [1919]. 'Self-statement', in C. Douglas (ed.), *Collected Works of Velimir Khlebnikov*, Vol. I, *Letters and Theoretical Writings*, trans. P. Schmidt. Cambridge, MA: Harvard University Press.

Khlebnikov, V. 1987f [1921]. 'Tasks for the President of Planet Earth', in C. Douglas (ed.), *Collected Works of Velimir Khlebnikov*, Vol. I, *Letters and Theoretical Writings*, trans. P. Schmidt. Cambridge, MA: Harvard University Press.

Khlebnikov, V. 1987g [1921]. 'The radio of the future', in C. Douglas (ed.), *Collected Works of Velimir Khlebnikov*, Vol. I, *Letters and Theoretical Writings*, trans. P. Schmidt. Cambridge, MA: Harvard University Press.

Khlebnikov, V. 1987h [1919]. 'Our fundamentals', in C. Douglas (ed.), *Collected Works of Velimir Khlebnikov*, Vol. I, *Letters and Theoretical Writings*, trans. P. Schmidt. Cambridge, MA: Harvard University Press.

Khlebnikov, V. 1987i [1914]. '! Futurian', in C. Douglas (ed.), *Collected Works of Velimir Khlebnikov*, Vol. I, *Letters and Theoretical Writings*, trans. P. Schmidt. Cambridge, MA: Harvard University Press.

Khlebnikov, V. 1987j [1921]. Letter to his sister Vera, in C. Douglas (ed.), *Collected Works of Velimir Khlebnikov*, Vol. I, *Letters and Theoretical Writings*, trans. P. Schmidt. Cambridge, MA: Harvard University Press.

Khlebnikov, V. 1987k [1922]. 'Excerpt from The Tables of Destiny', in C. Douglas (ed.), *Collected Works of Velimir Khlebnikov*, Vol. I, *Letters and Theoretical Writings*, trans. P. Schmidt. Cambridge, MA: Harvard University Press.

Khlebnikov, V. 1989 [1918]. 'October on the Neva', in R. Vroon (ed.), *Collected Works of Velimir Khlebnikov*, Vol. II, *Prose, Plays and Supersagas*, trans. P. Schmidt. Cambridge, MA: Harvard University Press.

Khlebnikov, V. 1990 [1910]. 'Incantation by Laughter', in C. Douglas (ed.), *Velimir Khlebnikov: The king of time*, trans. P. Schmidt. Cambridge, MA: Harvard University Press.

Killian, J.R. 1977. *Sputnik, Scientists, and Eisenhower: A memoir of the first special assistant to the president on science and technology*. Cambridge, MA: MIT Press.

Knight, C. 1996. 'Darwinism and collective representations', in J. Steele and S. Shennan (eds), *The Archaeology of Human Ancestry: Power, sex and tradition*. London and New York: Routledge.

Knight, C. 1998. 'Ritual/speech coevolution: a solution to the problem of deception', in J.R. Hurford, M. Studdert-Kennedy and C. Knight (eds), *Approaches to the Evolution of Language: Social and cognitive bases*. Cambridge: Cambridge University Press.

Knight, C. 1999. 'Sex and language as pretend-play', in R. Dunbar, C. Knight and C. Power (eds), *The Evolution of Culture*. Edinburgh: Edinburgh University Press.

Knight, C. 2000. 'Play as precursor of phonology and syntax', in C. Knight, M. Studdert-Kennedy and J.R. Hurford (eds), *The Evolutionary Emergence of Language: Social function and the origins of linguistic form*. Cambridge: Cambridge University Press.

Knight, C. 2002. 'Language and revolutionary consciousness', in A. Wray (ed.), *The Transition to Language*. Oxford: Oxford University Press.

Knight, C. 2008a. 'Language co-evolved with the rule of law', *Mind and Society: Cognitive Studies in Economics and Social Sciences*, 7(1): 109–128.

Knight, C. 2008b. ' "Honest fakes" and language origins', *Journal of Consciousness Studies*, 15(10–11): 236–248.

Knight, C. 2008c. 'Early human kinship was matrilineal', in N.J. Allen, H. Callan, R. Dunbar and W. James (eds), *Early Human Kinship*. Oxford: Blackwell.

Knight, C. 2009. 'Language, ochre and the rule of law', in R. Botha and C. Knight (eds), *The Cradle of Language*. Oxford: Oxford University Press.

Knight, C. 2014. 'Language and symbolic culture: An outcome of hunter-gatherer reverse dominance', in D. Dor, C. Knight and J. Lewis (eds), *The Social Origins of Language*. Oxford: Oxford University Press.

Knight, C. and J. Lewis. 2014. 'Vocal deception, laughter, and the linguistic significance of reverse dominance', in D. Dor, C. Knight and J. Lewis (eds), *The Social Origins of Language*. Oxford: Oxford University Press.

Knight, C. and C. Power. 2005. 'Grandmothers, politics, and getting back to science', in E. Voland, A. Chasiotis and W. Schiefenhövel (eds), *Grandmotherhood: The evolutionary significance of the second half of female life*. New Brunswick, NJ, and London: Rutgers University Press.

Kobayashi, H. and S. Kohshima. 2001. 'Unique morphology of the human eye and its adaptive meaning: Comparative studies on external morphology of the primate eye', *Journal of Human Evolution*, 40(5): 419–435.

Koerner, E.F.K. 1994. 'The anatomy of a revolution in the social sciences: Chomsky in 1962', *Dhumbadji!*, 1(4): 3–17.

Kruchenykh, A. and V. Khlebnikov. 1912. *Mirskontsa* [Worldbackwards]. Moscow: G.L. Kuzmin and S.D. Dolinskyi.

Kuhn, T. 1970. 'The structure of scientific revolutions', in *International Encyclopedia of Unified Science*, Vol. 2, second edition. Chicago, IL: University of Chicago Press.

Lakatos, I. 1970. 'Falsification and the methodology of scientific research programmes', in I. Lakatos and A. Musgrave (eds), *Criticism and the Growth of Knowledge*. Cambridge: Cambridge University Press.

Lakoff, G. 1971. 'Foreword', in S. Andres, A. Borkin and D. Peterson (eds), *Where the Rules Fail: A student's guide: An unauthorized appendix to M.K. Burt's 'From Deep to Surface Structure'*. Bloomington, IN: Indiana University Linguistics Club.

Lakoff, G. 1977. 'Interview', in H. Parret (ed.), *Discussing Language*. The Hague: Mouton.

Lakoff, G. 1995. 'In conversation with John Goldsmith', in J.H. Huck and J.A. Goldsmith (eds), *Ideology and Linguistic Theory: Noam Chomsky and the deep structure debates*. London: Routledge.

Lakoff, G. and M. Johnson. 1980. *Metaphors We Live By*. Chicago, IL: University of Chicago Press.

Langacker, R.W. 1988. 'An overview of cognitive grammar', in B. Rudzka-Ostyn (ed.), *Topics in Cognitive Linguistics*. Amsterdam: John Benjamins.

Latour, B. 1999. *Pandora's Hope: Essays on the reality of science studies*. Cambridge, MA: Harvard University Press.

Latour, B. and S. Woolgar. 1979. *Laboratory Life: The social construction of scientific facts*. London: Sage.

Leach, E. 1983. 'Roman Jakobson and social anthropology', in *A Tribute to Roman Jakobson 1896–1982*. New York and Amsterdam: Mouton.

Lee, R.B. 1968. 'What hunters do for a living, or, how to make out on scarce resources', in R. Lee and I. DeVore (eds), *Man the Hunter*. Chicago, IL: Aldine.

Lee, R.B. 1988. 'Reflections on primitive communism', in T. Ingold, D. Riches and J. Woodburn (eds), *Hunters and Gatherers*, Vol. 1, *History, Evolution and Social Change*. Chicago, IL: Aldine.

Lee, R.B. 1992. 'Demystifying primitive communism', in C.W. Gailey (ed.), *Dialectical Anthropology: Essays in honor of Stanley Diamond*, Vol. 1, *Civilization in Crisis: Anthropological perspectives*. Gainesville, FL: University of Florida Press.

Leech, G. 1983. *Principles of Pragmatics*. London: Longman Linguistics Library.

Lees, R.B. 1957. 'Review of Noam Chomsky, *Syntactic Structures*', *Language*, 33(3): 375–408.

Leiber, J. 1975. *Noam Chomsky: A philosophic overview*. New York: St Martin's Press.

Leslie, S.W. 1993. *The Cold War and American Science: The military-industrial complex at MIT and Stanford*. New York: Columbia University Press.

Lévi-Strauss, C. 1963 [1956]. 'Structure and dialectics', in C. Lévi-Strauss, *Structural Anthropology*, Vol. 1, trans. C. Jacobson and B.G. Schoepf. Harmondsworth: Penguin.

Lévi-Strauss, C. 1973. 'From honey to ashes', *Introduction to a Science of Mythology*, Vol. 2. London: Cape.

Lévi-Strauss, C. 1991. *Conversations with Claude Lévi-Strauss*, trans. Paula Wissing. Chicago, IL: University of Chicago Press.

Lewis, J. 2008. 'Ekila: Blood, bodies and egalitarian societies', *Journal of the Royal Anthropological Institute*, NS 14: 297–315.

Lewis, J. 2009. 'As well as words: Congo Pygmy hunting, mimicry, and play', in R. Botha and C. Knight (eds), *The Cradle of Language*. Oxford: Oxford University Press.

Lewis, J. 2014. 'BaYaka Pygmy multi-modal and mimetic communication traditions', in D. Dor, C. Knight and J. Lewis (eds), *The Social Origins of Language*. Oxford: Oxford University Press.

Liberman, M. 2016. 'Morris Halle: An appreciation', *Annual Review of Linguistics*, 2 (January): 1–9.

Lightfoot, D. 2002. 'Introduction', in Noam Chomsky, *Syntactic Structures*, second edition. Berlin: Mouton.

Locke, W.N. and A.D. Booth (eds). 1955. *Machine Translation of Languages*. Cambridge, MA: MIT Press.

Lorenz, K. 1937. 'On the formation of the concept of instinct', *Natural Sciences*, 25(19): 289–300.

Lorenz, K. 1996. *The Natural Science of the Human Species: The 'Russian Manuscript'*. Cambridge, MA: MIT Press.

Lybrand, W.A. 1962. *Proceedings of the Symposium 'The US Army's Limited-War Mission and Social Science Research'*. Washington, DC: American University, Special Operations Research Unit.

McBrearty, S. 2007. 'Down with the revolution', in P. Mellars, K. Boyle, O. Bar-Yosef and C. Stringer (eds), *Rethinking the Human Revolution: New behavioural and biological perspectives on the origin and dispersal of modern humans*. Cambridge: McDonald Institute for Archaeological Research.

McCumber, J. 2001. *Time in the Ditch: American philosophy and the McCarthy era*. Evanston, IL: Northwestern University Press.

McDavid, R.I. 1954. 'Review of Warfel, 1952', *Studies in Linguistics*, 12: 27–32.

McGilvray, J. 1999. *Language, Mind, and Politics*. Cambridge: Polity Press.

McGilvray, J. 2009. 'Introduction to the third edition', in N. Chomsky, *Cartesian Linguistics: A chapter in the history of rationalist thought*. Cambridge: Cambridge University Press.

Maclay, H. 1971. 'Linguistics: Overview', in D. Steinberg and L. Jakobovits (eds), *Semantics*. Cambridge: Cambridge University Press.

Mailer, N. 1968. *The Armies of the Night*. New York: New American Library.

Markov, V. 2006 [1968]. *Russian Futurism: A history*. Washington, DC: New Academia Publishing.

Marx, K. 1961a [1844]. 'Economic and philosophic manuscripts', in T.B. Bottomore and M. Rubel (eds), *Karl Marx: Selected writings in sociology and social philosophy*. Harmondsworth: Penguin.

Marx, K. 1961b [1845]. 'Theses on Feuerbach', in T.B. Bottomore and M. Rubel (eds), *Karl Marx: Selected writings in sociology and social philosophy*. Harmondsworth: Penguin.

Marx, K. 1961c [1843–44]. 'Towards a critique of Hegel's philosophy of right', in T.B. Bottomore and M. Rubel (eds), *Karl Marx: Selected writings in sociology and social philosophy*. Harmondsworth: Penguin.

Marx, K. 2000 [1845]. 'Theses on Feuerbach', in D. McLellan (ed.), *Karl Marx: Selected writings*, second edition. Oxford: Oxford University Press.

Marx, K. and F. Engels. 1961 [1845]. 'The German ideology', in T.B. Bottomore and M. Rubel (eds), *Karl Marx: Selected writings in sociology and social philosophy*. Harmondsworth: Penguin.

Marx, K. and F. Engels. 2000 [1845]. 'The German ideology', in D. McLellan (ed.), *Karl Marx: Selected writings*, second edition. Oxford: Oxford University Press.

Matthews, P.H. 1993. *Grammatical Theory in the United States from Bloomfield to Chomsky*. Cambridge: Cambridge University Press.

Maynard Smith, J. and D. Harper. 2003. *Animal Signals*. Oxford: Oxford University Press.

Maynard Smith, J. and E. Szathmáry. 1995. *The Major Transitions in Evolution*. Oxford: W.H. Freeman.

Mehta, V. 1974. *John is Easy to Please*. Harmondsworth: Penguin.

Mellars, P. and C. Stringer (eds). 1989. *The Human Revolution: Behavioural and biological perspectives in the origins of modern humans*. Edinburgh: Edinburgh University Press.

Mellars, P., K. Boyle, O. Bar-Yosef and C. Stringer (eds). 2007. *Rethinking the Human Revolution: New behavioural and biological perspectives on the origin and dispersal of modern humans*. Cambridge: McDonald Institute for Archaeological Research.

Mey, J. 1993. *Pragmatics: An introduction*. Oxford: Blackwell.

Miller, G.A. 1979. 'A very personal history: Talk to Cognitive Science Workshop', MIT, June.

Miller, G.A., F.M. Wiener and S.S. Stevens. 1946. *Transmission and Reception of Sounds under Combat Conditions*, Summary Technical Report of Division 17, Section 3, National Defense Research Council (Washington, DC, NDRC), 2.

Milne, D. 2008. *America's Rasputin: Walt Rostow and the Vietnam War*. New York: Hill and Wang.

Milne, S. 2009. 'Noam Chomsky: US foreign policy is straight out of the mafia', *Guardian*, 7 November.

Milner, J. 1983. *Vladimir Tatlin and the Russian Avant-garde*. New Haven, CT, and London: Yale University Press.

Milner-Gulland, R. 2000. 'Khlebnikov's eye', in C. Kelly and S. Lovell (eds), *Russian Literature, Modernism and the Visual Arts*. Cambridge: Cambridge University Press.

MIT News. 1992. http://newsoffice.mit.edu/1992/citation-0415 (accessed 9 April 2016).

MIT News. 2013. 'Ocean engineering students set stage for a smarter fleet', 14 August, available at: http://news.mit.edu (accessed February 2017).

Morgan, L.H. 1877. *Ancient Society*. London: MacMillan.

Murray, S.O. 1980. 'Gatekeepers and the "Chomskyan revolution" ', *Journal of the History of the Behavioral Sciences*, 16: 73–88.

Nader, L. (ed.). 1996. *Naked Science: Anthropological inquiry into boundaries, power, and knowledge*. London and New York: Routledge.

Nelkin, D. 1972. *The University and Military Research: Moral politics at MIT (science, technology and society)*. New York: Cornell University Press.

Nevin, B.E. 2009. 'More concerning the roots of transformational generative grammar', *Historiographia Linguistica*, 36(2/3): 459–479.

Newmeyer, F.J. 1986a. *The Politics of Linguistics*. Chicago, IL, and London: University of Chicago Press.

Newmeyer, F.J. 1986b. *Linguistic Theory in America*, second edition. New York and London: Academic Press.

Newmeyer, F.J. 1996. *Generative Linguistics*. London: Routledge.

Newmeyer, F.J. 2003. 'Review article', *Language*, 79(3): 583–599.

Nielsen, J. 2010. 'Private knowledge, public tensions: Theory commitment in post-war American linguistics'. Unpublished PhD thesis, University of Toronto.

Olson, G. and L, Faigley. 1991. 'Politics and composition: a conversation with Noam Chomsky', *Journal of Advanced Composition*, 11(1): 1–36.

Otero, C.P. (ed.) 1981. *Radical Priorities*. Oakland, CA: AK Press.

Otero, C.P. (ed.). 1988a. *Noam Chomsky: Language and politics*. Montreal: Black Rose Books.

Otero, C.P. 1988b. 'Introduction: The third emancipatory phase of history', in C.P. Otero (ed.), *Noam Chomsky: Language and politics*. Montreal: Black Rose Books.

Parret, H. (ed.). 1977. *Discussing Language*. The Hague: Mouton.

Piaget, J. 1929. *The Child's Conception of the World*. London: Kegan Paul, Trench, Trubner & Co.

Piatelli-Palmarini, M. (ed.). 1980. *Language and Learning: The debate between Jean Piaget and Noam Chomsky*. London: Routledge and Kegan Paul.

Pinker, S. 1994. *The Language Instinct*. London: Penguin.

Pinker, S. 1997a. 'Language as a psychological adaptation', in G.R. Bock and G. Cardew (eds), *Characterizing Human Psychological Adaptations*. Chichester: Wiley.

Pinker, S. 1997b. *How the Mind Works*. London: Penguin.

Pinker, S. 1999. *Words and Rules*. London: Weidenfeld and Nicolson.

Pinker, S. 2002. *The Blank Slate*. New York: Penguin.

Pinker, S. and P. Bloom. 1990. 'Natural language and natural selection', *Behavioral and Brain Sciences*, 13: 707–784.

Pinker, S. and R. Jackendoff. 2005. 'The faculty of language: What's special about it?', *Cognition* 95(2): 201–236.

Postal, P. 1995. 'In conversation with John Goldsmith and Geoffrey Huck', in J.H. Huck and J.A. Goldsmith (eds), *Ideology and Linguistic Theory: Noam Chomsky and the deep structure debates*. London: Routledge.

Postal, P. 2004. *Skeptical Linguistic Essays*. Oxford: Oxford University Press.

Postal, P. 2009. 'The incoherence of Chomsky's "Biolinguistic" ontology', *Biolinguistics*, 3(1): 104–123.

Power, C. 2009. 'Sexual selection models for the emergence of symbolic communication: Why they should be reversed', in R. Botha and C. Knight (eds), *The Cradle of Language*. Oxford: Oxford University Press.

Power, C. 2014. 'The evolution of ritual as a process of sexual selection', in D. Dor, C. Knight and J. Lewis (eds), *The Social Origins of Language*. Oxford: Oxford University Press.

Power, C. and L.C. Aiello. 1997. 'Female proto-symbolic strategies', in L.D. Hager (ed.), *Women in Human Evolution*. New York and London: Routledge.

Price, D.H. 2011. *Weaponizing Anthropology: Social science in service of the militarized state*. Oakland, CA: AK Press.

Priest, A. 2006. *Kennedy, Johnson and NATO: Britain, America and the dynamics of alliance, 1962–68*. London and New York: Routledge.

Punin, N. 1920. *Pamyatnik III Internatsionala*. Petrograd: Izdanie Otdela Izobrazitelnykh Iskusstv.

Putnam, H. 1960. 'Minds and machines', in S. Hook (ed.), *Dimensions of Mind*. New York: New York University Press.

Putnam, H. 1988. *Representation and Reality*. Cambridge, MA: MIT Press.

Quine, W.V. 1959. *Methods of Logic*, second edition. Cambridge, MA: Harvard University Press.

Radick, G. 2016. 'The unmaking of a modern synthesis: Noam Chomsky, Charles Hockett and the politics of behaviorism, 1955–1965', *Isis*, 107(1): 49–73.

Rai, M. 1995. *Chomsky's Politics*. London and New York: Verso.

Rappaport, R.A. 1999. *Ritual and Religion in the Making of Humanity*. Cambridge: Cambridge University Press.

Rees, A. 2009. *The Infanticide Controversy: Primatology and the art of field science*. Chicago, IL, and London: University of Chicago Press.

Renehan, C. 2007. 'Peace activism at the Massachusetts Institute of Technology from 1975 to 2001: A case study'. Unpublished PhD thesis, Boston College.

Rohde, J. 2012. 'From expert democracy to beltway banditry: How the antiwar movement expanded the military-academic-industrial complex', in M. Solovey and H. Cravens (eds), *Cold War Social Science: Knowledge, production, liberal democracy, and human nature*. New York: Palgrave Macmillan.

Ross, J.R. 1995. 'In conversation with John Goldsmith and Geoffrey Huck', in J.H. Huck and J.A. Goldsmith (eds), *Ideology and Linguistic Theory: Noam Chomsky and the deep structure debates*. London: Routledge.

Rudy, S. 1997. 'Introduction', in Roman Jakobson, *My Futurist Years*. New York: Marsilio.

Rydell, R. 1993. *World of Fairs: The century-of-progress expositions*. Chicago and London: University of Chicago Press.

Sahlins, M. 1964 [1960]. 'The origin of society', in P.B. Hammond (ed.), *Physical Anthropology and Archaeology*. New York and London: Macmillan (originally published in *Scientific American*).

Sahlins, M. 1972 [1959]. 'The social life of monkeys, apes and primitive man', in D.D. Quiatt (ed.), *Primates on Primates*. Minneapolis, MN: Burgess (originally published in *Human Biology*).

Sahlins, M.D. 1977. *The Use and Abuse of Biology: An anthropological critique of sociobiology*. London: Tavistock.

Sampson, G. 1975. *The Form of Language*. London: Weidenfeld & Nicolson.

Sampson, G. 2001 [1979]. 'What was transformational grammar?', in G. Sampson, *Empirical Linguistics*. London: Continuum (originally published in *Lingua*).

Sapir, E. 1929. 'A study in phonetic symbolism', *Journal of Experimental Psychology*, 12: 225–239.

Saussure, F. de. 1974 [1915]. *Course in General Linguistics*, trans. W. Baskin. London: Fontana/Collins.

Saussure, F. de. 1983 [1912]. *Course in General Linguistics*, trans. R. Harris. London: Duckworth.

Schlebusch, C.M. 2010. 'Genetic variation in Khoisan-speaking populations from southern Africa'. Dissertation, University of Witwatersrand.

Schmidt, P. 1989. 'Translator's introduction', in R. Vroon (ed.), *Collected Works of Velimir Khlebnikov*, Vol. II, *Prose, Plays and Supersagas*. Cambridge, MA: Harvard University Press.

Scott, F.D. 2016. *Outlaw Territories: Environments of insecurity/architectures of counterinsurgency*. Cambridge, MA: MIT Press.

Scowcroft, B. 1983. *Report of the President's Commission on Strategic Forces*. Washington, DC: The White House.

Scruton, R. and M. Dooley. 2016. *Conversations with Roger Scruton*. London: Bloomsbury.

Searle, J. 1969. *Speech Acts: An essay in the philosophy of language*. Cambridge: Cambridge University Press.

Searle, J. 1971a. *The Campus War: A sympathetic look at the university in agony*. New York: World Publishing.

Searle, J. 1971b. *The Philosophy of Language*. Oxford: Oxford University Press.

Searle, J. 1972. 'Chomsky's revolution in linguistics', *New York Review of Books*, 29 June.

Searle, J. 1996. *The Construction of Social Reality*. London: Penguin.

Searle, J. 2003. *Conversations with John Searle*, ed. Gustavo Faigenbaum. Buenos Aires: LibrosEnRed.

Segel, J. (ed.). 2009. *Recountings: Conversations with MIT mathematicians*. Natick, MA: A.K. Peters.

Segerstråle, U. 2000. *Defenders of the Truth: The battle for science in the sociobiology debate and beyond*. Oxford: Oxford University Press.

Seuren, P.A.M. 2004. *Chomsky's Minimalism*. Oxford: Oxford University Press.

Shalom, S.R. 1997. 'Review of *Noam Chomsky: A Life of Dissent*, by Robert F. Barsky', *New Politics*, NS 6(3), available at: http://nova.wpunj.edu/newpolitics/issue23/shalom23.htm (accessed 9 April 2016).

Shannon, C. 1948. 'A mathematical theory of communication', *Bell System Technical Journal*, 27: 379–423; 623–656.

Shelley, P.B. 1840 [1839]. *Essays, Letters from Abroad, Translations and Fragments*, ed. Mary Shelley. London: Edward Moxon.

Shklovsky, V. 1988 [1921]. 'The monument to the Third International', in L.A. Zhadova, *Tatlin*. New York: Rizzoli.

Silk, J.B. 1993. 'Primatological perspectives on gender hierarchies', in D. Miller (ed), *Sex and Gender Hierarchies*. Cambridge: Cambridge University Press.

Siodmak, C. 1971 [1942]. *Donovan's Brain*. London: Barrie and Jenkins.

Skinner, B.F. 1957. *Verbal Behavior*. New York: Appleton Century Crofts.

Skinner, B.F. 1960. 'Pigeons in a pelican', *American Psychologist*, 15: 28–37.

Skolnikoff, E.B. 2011. Video interview for MIT 150 Infinite History Project, available at: www.youtube.com/watch?v=J0cvFxzvz_c#t=22 (accessed 9 April 2016).

Small, M.F. 1993. *Female Choices: Sexual behavior of female primates*. Ithaca, NY: Cornell University Press.

Smith, A.D.M. and S. Höfler. 2014. 'The pivotal role of metaphor in the evolution of human language', in J.E. Díaz Vera (ed.), *Metaphor and Metonymy through Time and Culture: Perspectives on the sociohistorical linguistics of figurative language*. Amsterdam: Mouton.

Smith, A. and S. Höfler (in press, 2016). 'From metaphor to symbols and grammar: The cumulative cultural evolution of language', in C. Power, M. Finnegan and H. Callan (eds), *Human Origins: Contributions from social anthropology*. New York: Berghahn.

Smith, N. 1999. *Chomsky: Ideas and ideals*. Cambridge: Cambridge University Press.

Smuts, B. 1985. *Sex and Friendship in Baboons*. New York: Aldine.

Snead, D.L. 1999. *The Gaither Committee, Eisenhower, and the Cold War*. Columbus, OH: Ohio State University Press.

Solway, J. (ed.). 2006. *The Politics of Egalitarianism: Theory and practice*. New York and Oxford: Berghahn Books.

Sperber, D. 1985. *On Anthropological Knowledge: Three essays*. Cambridge: Cambridge University Press.

Sperber, D. and D. Wilson. 1986. *Relevance: Communication and cognition*. Oxford: Blackwell.

Sperlich, W. 2010. *Noam Chomsky*. Bath: Bath Press.

Steels, L. 1998. 'Synthesizing the origins of language and meaning using coevolution, self-organization and level formation', in J. Hurford, M. Studdert-Kennedy and C. Knight (eds), *Approaches to the Evolution of Language*. Oxford: Oxford University Press.

Steels, L. 2007. 'The recruitment theory of language origins', in C. Lyon, C.L. Nehaniv and A. Cangelosi (eds), *Emergence of Communication and Language*. London: Springer-Verlag.

Steels, L. 2009. 'Is sociality a crucial prerequisite for the emergence of language?', in R. Botha and C. Knight (eds), *The Prehistory of Language*. Oxford: Oxford University Press.

Steels, L. 2012. Self-organization and selection in cultural language evolution. In L. Steels (ed.), *Experiments in cultural language evolution*. Advances in Interaction Studies 3. Amsterdam/Philadelphia, PA: John Benjamins, pp. 1-37.

Steels, L. 2014. 'Breaking down false barriers to understanding', in D. Dor, C. Knight and J. Lewis (eds), *The Social Origins of Language*. Oxford: Oxford University Press.

Steels, L. 2015. *The Talking Heads Experiment: Origins of words and meanings*. Berlin: Language Science Press.

Steels, L. and F. Kaplan. 2001. 'AIBO's first words: The social learning of language and meaning', *Evolution of Communication*, 4(1): 3–32.

Strazny, P. (ed.). 2013. *Encyclopedia of Linguistics*. London: Routledge.

Stroud, J. 1949. 'The psychological moment in perception', *Cybernetics*, 6.

Strum, S.C. 1987. *Almost Human: A journey into the world of baboons*. London: Elm Tree Books.

Taube, M. 1961. *Computers and Common Sense: The myth of thinking machines*. New York: Columbia University Press.

Taylor, C. 1975. *Hegel*. Cambridge: Cambridge University Press.

Thompson, W.I. 1971. *At the Edge of History*. New York: Harper and Row.

Tinbergen, N. 1951. *The Study of Instinct*. New York: Oxford University Press.

Tomasello, M. 1999. *The Cultural Origins of Human Cognition*. Cambridge, MA: Harvard University Press.

Tomasello, M. 2000. 'Culture and cognitive development', *Current Directions in Psychological Science*, 9(2): 37–40.

Tomasello, M. 2006. 'Why don't apes point?', in N.J. Enfield and S.C. Levinson (eds), *Roots of Human Sociality: Culture, cognition and interaction*. Oxford and New York: Berg.

Tomasello, M., M. Carpenter, J. Call, T. Beyne and H. Moll. 2005. 'Understanding and sharing intentions: The origins of cultural cognition', *Behavioral and Brain Sciences*, 28: 675–691.

Tomasello, M. and M.J. Farrar. 1986. 'Joint attention and early language', *Child Development*, 57(6): 1454–1463.

Tomasello, M., B. Hare, H. Lehmann and J. Call. 2007. 'Reliance on head versus eyes in the gaze following of great apes and human infants: The cooperative eye hypothesis', *Journal of Human Evolution*, 52: 314–320.

Tomasello, M. and H. Rakoczy. 2003. 'What makes human cognition unique: From individual to shared to collective intentionality', *Mind and Language*, 18(2): 121–147.

Tonkin, B. 1989. 'Making a difference', *City Limits* (London), 26 January – 2 February.

Tooby, J. and L. Cosmides (eds). 1992. *The Adapted Mind: Evolutionary psychology and the generation of culture*. Oxford: Oxford University Press.

Tooby, J. and L. Cosmides. 1995. 'Foreword', in S. Baron-Cohen, *Mindblindness: An essay on autism and theory of mind*. Cambridge, MA: MIT Press.

Trask, R.L. 1999. *Key Concepts in Language and Linguistics*. London and New York: Routledge.

Trivers, R. 1985. *Social Evolution*. Menlo Park, CA: Benjamin/Cummings.

Trivers, R. 2015. *Wild Life: Adventures of an evolutionary biologist*. New Brunswick, NJ: Biosocial Research.

Trotsky, L. 1940 [1925]. 'Dialectical materialism and science: a speech on D.I. Mendeleyev', *New International*, 6(1).

Turing, A. 1950. 'Computing machinery and intelligence', *Mind*, 49: 433–460.

UPI Archives. 1989. 'MIT students allege defense conflict', available at: www.upi.com/Archives/1989/06/02/MIT-students-allege-defense-conflict/2508612763200/ (accessed February 2017).

US Department of Defense. 2017. 'Department of Defense announces successful micro-drone demonstration', available at: www.defense.gov (accessed January 2017).

Uttal, W.R. 2001. *The New Phrenology: The limits of localizing cognitive processes in the brain*. Cambridge, MA: MIT Press.

Van Schaik, C.P. and C.H. Janson (eds). 2000. *Infanticide by Males and Its Implications*. Stony Brook, NY: State University of New York.

Verdu, P., N. Becker, A. Froment, M. Georges, V. Grugni, L. Quintana-Murci, J.-M. Hombert, L. Van der Veen, S. Le Bomin, S. Bahuchet, E. Heyer and F. Austerlitz. 2013. 'Sociocultural behavior, sex-biased admixture and effective population sizes in Central African Pygmies and non-Pygmies', *Molecular Biology and Evolution*, 30(4): 918–937.

Von Neumann, J. 1958. *The Computer and the Brain*. New Haven, CT: Yale University Press.

Vroon, R. 1997. 'The poet and his voices', in R. Vroon (ed.), *Collected Works of Velimir Khlebnikov*, Vol. III, *Selected Poems*, trans. P. Schmidt. Cambridge, MA: Harvard University Press.

Vygotsky, L.S. 1986 [1934]. *Thought and Language*. Cambridge, MA: MIT Press.

Wallerstein, I. and P. Starr (eds). 1972. 'Confrontation and counterattack', in I. Wallerstein and P. Starr (eds), *The University Crisis Reader*. New York: Random House.

Watson, J.D. and F.H.C. Crick. 1953. 'A structure for deoxyribose nucleic acid', *Nature*, 3(171): 737–738.

Weaver, W. 1955a. 'Foreword: The new tower', in W.N. Locke and A.D. Booth (eds), *Machine Translation of Languages*. Cambridge, MA: MIT Press.

Weaver, W. 1955b. 'Translation', in W.N. Locke and A.D. Booth (eds), *Machine Translation of Languages*. Cambridge, MA: MIT Press.

White, G.D. 2000. *Campus Inc.: Corporate power in the ivory tower*. New York: Prometheus Books.

Whiten, A. 1999. 'The evolution of deep social mind in humans', in M. Corballis and S.E.G. Lea (eds), *The Descent of Mind: Psychological perspectives on hominid evolution*. Oxford: Oxford University Press.

Wiener, N. 1948. *Cybernetics, or, Control and Communication in the Animal and the Machine*. Cambridge, MA: MIT Press.

Wiesner, J. 1958. 'Electronics and the missile', *Astronautics*, 3: 20–21, 114.

Wiesner, J. 1961. *Report to the President-Elect of the Ad Hoc Committee on Space*, available at: www.hq.nasa.gov/office/pao/History/report61.html (accessed September 2016).

Wiesner, J. 1986. 'War and peace in the nuclear age; bigger bang for the buck, a; interview with Jerome Wiesner, [1]', WGBH Media Library & Archives, available at: http://openvault.wgbh.org/catalog/V_DD3A084107E94632B6AD7D428A966304 (accessed March 2017).

Wiesner, J. 2003. *Jerry Wiesner: Scientist, statesman, humanist: memories and memoirs*. Cambridge, MA: MIT Press.

Williamson, J. 2004. 'Chomsky, language, World War II, and me', in P. Collier and D. Horowitz (eds), *The Anti-Chomsky Reader*. San Francisco, CA: Encounter.

Wilson, E.O. 1975. 'Human decency is animal', *New York Times Magazine*, 12 October.

Wilson, E.O. 1995. 'Science and ideology', *Academic Questions*, 8.

Winkelman, M. 2002. 'Shamanism and cognitive evolution', *Cambridge Archaeological Journal*, 12(1): 71–101.

Wolpert, L. 2006. *Six Impossible Things before Breakfast: The evolutionary origins of belief*. London: Faber & Faber.

Wood, E.T., D.A. Stover, C. Ehret, G. Destro-Bisol, G. Spedini, H. McLeod, L. Louie, M. Bamshad, B. Strassman, H. Soodyall and M.F. Hammer. 2005. 'Contrasting patterns of Y chromosome and mtDNA variation in Africa: Evidence for sex-biased demographic processes', *European Journal of Human Genetics*, 13(7): 867–876.

Woodburn, J. 1982. 'Egalitarian societies', *Man: The Journal of the Royal Anthropological Institute*, 17(3): 431–451.

Yngve, V. 1956. 'Mechanical translation research at MIT', *Mechanical Translation*, 3: 44–45.

Yngve, V. 1964. 'Implications of mechanical translation research', *Proceedings of the American Philosophical Society*, 104(4): 275–281.

Zimmermann, H.J. 1991. 'Director's profile interview', available at: www.rle.mit.edu/henry-j-zimmermann-directors-profile-interview/ (accessed May 2017).

INDEX